Advances in Experimental Medicine and Biology

Volume 1252

AF832277

Advances in Experimental Medicine and Biology provides a platform for scientific contributions in the main disciplines of the biomedicine and the life sciences. This series publishes thematic volumes on contemporary research in the areas of microbiology, immunology, neurosciences, biochemistry, biomedical engineering, genetics, physiology, and cancer research. Covering emerging topics and techniques in basic and clinical science, it brings together clinicians and researchers from various fields.

Advances in Experimental Medicine and Biology has been publishing exceptional works in the field for over 40 years, and is indexed in SCOPUS, Medline (PubMed), Journal Citation Reports/Science Edition, Science Citation Index Expanded (SciSearch, Web of Science), EMBASE, BIOSIS, Reaxys, EMBiology, the Chemical Abstracts Service (CAS), and Pathway Studio.

2018 Impact Factor: 2.126.

More information about this series at http://www.springer.com/series/5584

Sadaf Alipour • Ramesh Omranipour
Editors

Diseases of the Breast during Pregnancy and Lactation

Foreword by Professor J. Michael Dixon, OBE

Professor of Surgery and Consultant Surgeon,
Edinburgh Breast Unit, University of Edinburgh,
NHS Lothian, UK

Springer

Editors
Sadaf Alipour
Breast Disease Research Center (BDRC)
Tehran University of Medical Sciences
Tehran, Iran

Department of Surgery, Arash
Women's Hospital
Tehran University of Medical Sciences
Tehran, Iran

Ramesh Omranipour
Breast Disease Research Center (BDRC)
Tehran University of Medical Sciences
Tehran, Iran

Department of Surgical Oncology,
Cancer Institute
Tehran University of Medical Sciences
Tehran, Iran

ISSN 0065-2598 ISSN 2214-8019 (electronic)
Advances in Experimental Medicine and Biology
ISBN 978-3-030-41595-2 ISBN 978-3-030-41596-9 (eBook)
https://doi.org/10.1007/978-3-030-41596-9

This Springer imprint is published by the registered company Springer Nature Switzerland AG.
The registered company address is: Gewerbestrasse 11, 6330 Cham, Switzerland

To:
My dear husband Farivar,
And my lovely daughter Newsha.

–Ramesh

To:
Baba, Maman, and Marjan
Amirhossein, Yaas, and Noora
For the warm support they offer,
And the endless love they share.

–Sadaf

Foreword

Pregnancy should be a time of rejoicing for a woman and her partner with the thrill of a new birth to look forward to. After delivery, breastfeeding allows bonding and provides the best source of nutrients for the baby. In a few patients, however, pregnancy does not come alone, and a woman's biggest fear, breast cancer, develops in a few women at a time in their life when they should be optimistically looking toward the future. Also for some women, breastfeeding can be a challenge. Although breastfeeding seems the most natural thing in the world, I am not sure we are really honest with women about how challenging breastfeeding can be.

Current books available to doctors dealing with breast problems during pregnancy and lactation are limited. They are written from one perspective and rarely give a balanced or comprehensive view of the problems. This book provides a range of views and perspectives of the problems women experience during pregnancy and lactation. Rather than me extolling the virtues of this book which are clear just from looking at the chapter listing, I thought it might be worthwhile sharing two challenging patients as a background as to why a new book dealing with pregnancy and lactation problems is so timely and necessary – some of the patient details have been modified to protect their identity. When you read their stories, you will understand why there is a need for such a book and why the authors and editors are to be congratulated for producing such a valuable text.

- RP was a 30-year-old nurse. She was pregnant and was having problems with morning sickness. She went to her GP to get advice, and just as she was on the way out of the consultation, she mentioned rather as an afterthought that she had a breast lump. GP appointments in the UK are only 10 min, and the GP indicated that normally as it was another problem, she should make another appointment to have her breast lump assessed, but the GP at that visit did perform a breast examination and confirmed a lump, reassuring that the lump felt benign. A routine referral to see a breast surgeon was made. Two weeks later, RP attended a breast clinic where a breast lump measuring 30 mm was confirmed, and while a fibroadenoma was considered the most likely, a core biopsy was performed. The pathology showed the lump was a triple-negative breast cancer and as RP was only 20 weeks pregnant. Chemotherapy was then started, and she was referred to genetics. The response to chemotherapy was good, but the gene test showed an abnormality in BRCA1, and this was thought likely pathogenic. Chemotherapy was completed, the baby was delivered, and then surgery followed. A bilateral mastectomy with reconstruction was performed after discussion of the implications of the BRCA1 abnormality. Within the breasts, there was no residual disease at the site of the initial cancer after chemotherapy. RP had a good cosmetic outcome from her surgery, and all wounds healed. She coped amazingly well during what was a hugely stressful time. Her baby was bottle-fed and thrived. Given the wish for more children, a decision was taken by RP and her husband to have IVF with preimplantation genetic testing to ensure any subsequent children did not carry the BRCA1 gene abnormality. It was during this time that the genetics team contacted RP with updated information, which had now concluded the genetic abnormality in BRCA1 present in RP had been downgraded to a variant of unknown significance. RP found this news devastating and confusing. Explaining what the downgrading meant as a risk to RP and her daughter took time and numerous visits, and slowly, RP understood this was good news in that her daughter was no longer at increased risk and the downgrading meant that she could keep her ovaries. For patients like this, a huge team of professionals is involved in treating and supporting her through very difficult times. At a time when she was just getting her life on track, RP found the news about the downgrade to be a huge psychological blow.

- HS was a 34-year-old first-time mother. She started to breastfeed with much hope and great enthusiasm. Within a week, her nipples became cracked. She was spending more than 8 hours a day breastfeeding, and her baby was struggling to get enough milk. HS received advice from breastfeeding counsellors and health visitors, but the advice was not always consistent. She developed pain, erythema and swelling but delayed seeing her GP by 48 hours. The infection did not settle, and she was given a second course of antibiotics. HP in retrospect had a breast abscess by the time of the second course of antibiotics, and when she was eventually referred to hospital, her whole breast was erythematous and oedematous with enlarged tender axillary nodes and systemic signs of sepsis. Both nipples were excoriated, and HP was exhausted and in pain. Two abscesses were drained by aspiration and irrigation with local anesthetics. After an in-depth discussion, it was decided to stop breastfeeding and suppress her milk flow with cabergoline. This is something we rarely do, but the whole breast was inflamed and very tender. Her baby was transferred to bottle feeding and coped well. HP was discharged home the next day on oral antibiotics and needed two more aspirations in the following week. After 10 days, HP was reviewed, and I could hardly believe it was the same woman. She was happy, enjoying her baby, and in no pain. HP felt very guilty that she had not managed to breastfeed for longer, but she was happy with life. Her abscesses had settled, and her breast slowly returned to normal with no obvious change in shape.
- A journalist writing in the UK newspaper *The Guardian* recently wrote: "As I type this, the clock hands are creeping towards 2.30 a.m., and my baby daughter is slumbering beside me having had her first feed of the night. There will, unquestionably, be more. I am on hand 24/7 to meet her

dietary needs, an always-open milk bar with only one thing on the menu and only one employee serving the drinks. There is very little customer feedback, though occasionally I am rewarded with a loud and strangely satisfying burp." She goes on "we must be honest about how tough breast-feeding can be. A procession of midwives – all of them kind, all of them determined to show me just how straightforward nature can be – grapple with my baby's head and my nipple and try to bring them together in perfect harmony. It does not work. I spent my daughter's first evening on earth painstakingly squeezing out tiny droplets of colostrum into a minuscule syringe. It is only the next day that, infuriated perhaps by my increasingly cack-handed attempts, she eventually complies." She finishes "nature can be marvelous and can also be capricious, but it is also supremely painful."

The key to management of any patient experiencing problems is an excellent knowledge base, communication, and teamwork. While the two patient stories and the perspective of the journalist might not be common, they are important reminders that problems do occur during pregnancy and lactation. This book provides the depth of information to allow the reader to provide care for such patients. Importantly, this book deals with both physical and psychological aspects of diseases and conditions affecting women in pregnancy and lactation. It is not just getting the patient better physically that matters; equally important is dealing with patient's fears and anxieties. Ensuring they have a good quality of life is also paramount. I hope that you find this book useful and enjoy reading it as much as I have. I know that if you do read this book and ever come across patients with such extreme problems, you will have all the knowledge, information, and skills you need to manage such patients.

Edinburgh, UK Michael Dixon

Preface

Pregnancy naturally goes with hope and happiness in the family. However, the experience might be linked up with a sense of anxiety, because of the thrilling anticipation of a dearest newcomer whose needs might be hard to fulfill by the parents. When, in these circumstances, health of the expectant mother seems endangered, the whole family gets distressed. This situation also alarms the physicians, who know that diseases are likely to be overlooked during gestation and difficult to approach because of safety issues for the mother and child.

The notion to write this book originated from the numerous women who came to us for various complaints in their breasts, from a distorted look to intractable mastitis or alarming lumps during these early phases of maternity.

Unfortunately in recent years, we are faced with increasing numbers of pregnancy-associated breast cancers; this is due both to the decreasing age of breast cancer and to the globally increasing age of childbearing. Whereas diagnosis and management of malignancy in the prenatal and postnatal settings is associated with several challenges, approaching benign conditions are sometimes no less defying in these periods.

In this book, we have tried to discuss every condition of the breast that may bother women in their prenatal and breastfeeding days. Physiologic alterations, benign disorders, and malignancies of the breasts are approached in view of concerns in pregnancy and lactation. In addition, care of the mother and fetus during and after treatment of pregnancy-associated breast cancer, fertility and gestation after treatment of the disease, as well as related psychological aspects are conversed. The authors of all chapters are academics from valued dedicated centers around the world and have written the texts based on effective medical literature, international clinical guidelines, and systematic practice.

We hope that this book will meet the needs of general physicians, gynecologists, surgeons, internists, oncologists, and every other practitioner who deals with problems of the breasts in pregnant and nursing women.

Tehran University of Medical Sciences, Tehran, Iran — Ramesh Omranipour

Tehran University of Medical Sciences, Tehran, Iran — Sadaf Alipour

Acknowledgments

The editors would like to thank *Dr. Nima Rezaei* for his great professional support from the first to last steps of preparing this book,

Dr. Bita Eslami for her technical contribution to the project, and ***Dr. Reza Saidi*** and ***Dr. Adel Yazdankhah*** for the help they provided through the work.

Contents

Contributors

Ashley Alex Department of Surgery, Division of Surgical Oncology, Virginia Commonwealth University/Massey Cancer Center, Richmond, VA, USA

Hattan A. Alghamdi Department of Surgery, Division of Surgical Oncology, University of Miami, Miller School of Medicine, Miami, FL, USA

Sadaf Alipour Breast Disease Research Center (BDRC), Tehran University of Medical Sciences, Tehran, Iran
Department of Surgery, Arash Women's Hospital, Tehran University of Medical Sciences, Tehran, Iran

Heba Alkhashnam Department of Plastic and Breast Surgery, Gustave Roussy Cancer Campus, Grand Paris, France

Tamrah Alrammah Department of Surgery, Division of Surgical Oncology, University of Miami, Miller School of Medicine, Miami, FL, USA

Jennifer Axilbund Department of Surgery, Johns Hopkins University, School of Medicine, Baltimore, MD, USA

Giulia Bellettini Pediatrician, Milan, Italy

Eva Bhandary Department of Surgery, Division of Surgical Oncology, Virginia Commonwealth University/Massey Cancer Center, Richmond, VA, USA

Vesna Bjelic-Radisic Breast Unit, Helios University Hospital, University Witten Herdecke, Wuppertal, Germany

Khadija Bouhna Cliniques Universitaires Saint-Luc, Obstétrique, Brussels, Belgium

Barbara Buonomo Gynecologic Oncology Program, European Institute of Oncology IRCCS, Milan, Italy
Gynecology and Obstetrics Unit, Department of Neuroscience, Reproductive Sciences and Dentistry, School of Medicine, University of Naples Federico II, Naples, Italy

Jennifer Chen Department of Surgery, Johns Hopkins University, School of Medicine, Baltimore, MD, USA

Giovanni Codacci-Pisanelli Department of Medical and Surgical Sciences and Biotechnology, Sapienza University of Rome, Rome, Italy

Angelica Conversano Department of Plastic and Breast Surgery, Gustave Roussy Cancer Campus, Grand Paris, France

Konstantinos D. Dinas Second Department of Obstetrics and Gynaecology, Medical School, Aristoteles University of Thessaloniki, Hippokrateion Hospital, Thessaloniki, Greece

Fereshteh Ensani Department of Pathology, Tehran University of Medical Sciences, Tehran, Iran

Mohsen Esfandbod Department of Oncology, Sina Hospital, Tehran University of Medical Sciences, Tehran, Iran

Amirhossein Eskandari Deputy of Education, Ministry of Health, Tehran, Iran

Flavia Faccio Applied Research Division for Cognitive and Psychological Science, European Institute of Oncology IRCCS, Milan, Italy

Richard Gilmore Department of Surgery, Johns Hopkins University, School of Medicine, Baltimore, MD, USA

Mehran Habibi Department of Surgery, Johns Hopkins University, School of Medicine, Baltimore, MD, USA

Farnaz Amouzegar Hashemi Radiation Oncology Research Center, Cancer Institute, Tehran University of Medical Sciences, Tehran, Iran

Jean-Francois Honart Department of Plastic and Breast Surgery, Gustave Roussy Cancer Campus, Grand Paris, France

Behnaz Jahanbin Department of Pathology, Cancer Institute, Imam Khomeini Hospital Complex, Tehran University of Medical Sciences, Tehran, Iran

Anna L. V. Johansson Department of Medical Epidemiology and Biostatistics, Karolinska Institute, Stockholm, Sweden

Frédéric Kolb Department of Plastic and Breast Surgery, Gustave Roussy Cancer Campus, Grand Paris, France

Dhananjay Kulkarni Western General Hospital, Edinburgh, Scotland

Adriana K. Langer Radiology Department, Institut Curie, Saint Cloud, France

Marie Catherine Lee Comprehensive Breast Program, Moffitt Cancer Center, Tampa, FL, USA
Department of Surgery, University of South Florida, Tampa, FL, USA

Nicolas Leymarie Department of Plastic and Breast Surgery, Gustave Roussy Cancer Campus, Grand Paris, France

Samantha Linhares University of Miami, Miller School of Medicine, Miami, FL, USA

Chafika Mazouni Department of Plastic and Breast Surgery, Gustave Roussy Cancer Campus, Grand Paris, France

Kandace P. McGuire Department of Surgery, Division of Surgical Oncology, Virginia Commonwealth University/Massey Cancer Center, Richmond, VA, USA

Mina Mhallem Gziri Cliniques Universitaires Saint-Luc, Obstétrique, Brussels, Belgium

Mecker G. Möller Department of Surgery, Division of Surgical Oncology, University of Miami, Miller School of Medicine, Miami, FL, USA

Faina Nakhlis Brigham and Women's Hospital, Department of Surgery, Harvard Medical School, Boston, MA, USA

Ronak Nalini Department of Internal Medicine, Division of Hematology-Oncology, Kermanshah University of Medical Sciences, Kermanshah, Iran

Ali-Akbar Nejatisafa Department of Psychiatry, Division of Psychosomatic Medicine, Psychosomatic Research Center, Tehran University of Medical Sciences, Tehran, Iran

Micaela Notarangelo Private Lactation Consultant Practice, Lerici, La Spezia, Italy

Ramesh Omranipour Breast Disease Research Center (BDRC), Tehran University of Medical Sciences, Tehran, Iran
Department of Surgical Oncology, Cancer Institute, Tehran University of Medical Sciences, Tehran, Iran

Fedro A. Peccatori Gynecologic Oncology Program, European Institute of Oncology IRCCS, Milan, Italy

Vishnu Prasath Department of Surgery, Johns Hopkins University, School of Medicine, Baltimore, MD, USA

Kartini Rahmat Department of Biomedical Imaging, Faculty of Medicine, University of Malaya, Kuala Lumpur, Malaysia

Francoise Rimareix Department of Plastic and Breast Surgery, Gustave Roussy Cancer Campus, Grand Paris, France

Benjamin Sarfati Department of Plastic and Breast Surgery, Gustave Roussy Cancer Campus, Grand Paris, France

Vahid Soleimani Department of Pathology, Cancer Institute, Imam Khomeini Hospital Complex, Tehran University of Medical Sciences, Tehran, Iran

Hanne Stensheim Cancer Registry of Norway, Oslo, Norway

James Sun Comprehensive Breast Program, Moffitt Cancer Center, Tampa, FL, USA

Nur Aishah Taib University of Malaya, Cancer Institute, Kuala Lumpur, Malaysia
Department of Surgery, Faculty of Medicine, University of Malaya, Kuala Lumpur, Malaysia

Omid S. Tehrani Department of Hematology and Medical Oncology, Stanford University, Stanford, CA, USA

Mahtab Vasigh Breast Disease Research Center (BDRC), Tehran University of Medical Sciences, Tehran, Iran

Bruna Migliavacca Zucchetti European Institute of Oncology IRCCS, Milan, Italy
Medical Oncology Department, Hospital Sirio-Libanes, Sao Paulo, Brazil

Abbreviations

ADH	Atypical ductal hyperplasia
AH	Atypical hyperplasia
ALH	Atypical lobular hyperplasia
ALTTO	Adjuvant lapatinib and/or trastuzumab treatment optimization
AMH	Anti-Müllerian hormone
APBI	Accelerated partial breast irradiation
ASRM	American Society for Reproductive Medicine
ASCO	American Society of Clinical Oncology
BCS	Breast conservative/conserving surgery
BCSS	Breast cancer-specific survival
BCY	Breast cancer in young women
BE	Breast examination
BI	Breast inspection
BP	Breast palpation
CC	Craniocaudal
CDC	Center for Disease Control
CMF	Cyclophosphamide, methotrexate and 5 fluorouracil
CNB	Core needle biopsy
CNS	Central nervous system
CNGOF	Collège National des Gynécologues et Obstétriciens Français
CT	Computer tomography
DBR	Delayed breast reconstruction
DCIS	Ductal carcinoma in situ
DIEP	Deep inferior epigastric perforator
DIN	Ductal intraepithelial neoplasia
DFS	Disease-free survival
EGFR-2	Epidermal growth factor receptor 2
ELIOT	Electron beam intraoperative radiotherapy
ER	Estrogen receptor
ET	Endocrine therapy
ESMO	European Society for Medical Oncology
FA	Fibroadenoma
FAC	5-FU, doxorubicin, cyclophosphamide
FCC	Fibrocystic changes
FDA	Food and Drug Administration
FHR	Fetal heart rate

fLCIS	Florid lobular carcinoma in situ
FNA	Fine needle aspiration
FNAB	Fine needle aspiration biopsy
FSH	Follicle-stimulating hormone
5-FU	5-Fluorouracil
GBG	German Breast Group
GnRH	Gonadotropin-releasing hormone
hCG	Human chorionic gonadotropin
HR	Hazard ratio
HRT	Hormone replacement therapy
IBR	Immediate breast reconstruction
ICSI	Intracytoplasmic sperm injection
I&D	Incision and drainage
IBC	Inflammatory breast cancer
IDC	Invasive ductal carcinoma
IGM	Idiopathic granulomatous mastitis
IGRT	Image-guided radiotherapy
IMRT	Intensity-modulated radiotherapy
INCIP	International Network on Cancer, Infertility and Pregnancy
IVF	In vitro fertilization
IVM	In vitro maturation
LA	Local anesthesia
LCIS	Lobular carcinoma in situ
LD	Latissimus dorsi
LIN	Lobular intraepithelial neoplasia
LN	Lobular neoplasia
LVI	Lympho-vascular invasion
MLO	Mediolateral oblique
MRI	Magnetic resonance imaging
MRSA	Methicillin-resistant *Staphylococcus aureus*
NAC	Nipple-areolar complex
NCCN	National Comprehensive Cancer Network
NeoALTTO	Neoadjuvant Lapatinib and/or trastuzumab treatment optimization
NPI	Nottingham Prognostic Index
NSAID	Nonsteroidal anti-inflammatory drug
OS	Overall survival
PABC	Pregnancy-associated breast cancer
PARP	Poly (ADP-ribose) polymerase
PDB	Paget's disease of the breast
PET	Positron emission tomography
PGD	Pre-implantation genetic diagnosis
pLCIS	Pleomorphic LCIS
PPBC	Postpartum breast cancer
RA	Regional anesthesia
RT	Radiation therapy
SA	Sclerosing adenosis
SART	Society for Assisted Reproductive Technology

SEER	Surveillance, epidemiology and end results
SLNB	Sentinel lymph node biopsy
SLN	Sentinel lymph node
TDAP	Thoracodorsal artery perforator
TDLU	Terminal ductal lobular unit
TIL	Tumor-infiltrating lymphocytes
TRAM	Transverse rectus abdominis myocutaneous
UDH	Usual ductal hyperplasia
US	Ultrasonographic scan/ultrasound
VAB	Vacuum-assisted biopsy
VUS	Variants of uncertain significance
WHEL	Women's Healthy Eating and Living
WHO	World Health Organization

Part I

Basic Considerations

Anatomy and Physiology of the Breast during Pregnancy and Lactation

1

Ashley Alex, Eva Bhandary, and Kandace P. McGuire

Abstract

The mature breast is located within the anterior thoracic wall, lying atop the pectoralis major muscle. Pubertal changes lead to incomplete development of the breast, a process which is only completed during pregnancy. The incomplete breast consists mostly of adipose tissue but also lactiferous units called lobes. These eventually drain into the lactiferous ducts and then into the lactiferous sinus and then to the nipple-areolar complex. During pregnancy, the breast undergoes both anatomic and physiologic changes to prepare for lactation. During the first trimester, the ductal system expands and branches out into the adipose tissue in response to the increase of estrogen. Elevated levels of estrogen also cause a decrease in adipose tissue and ductal proliferation and elongation. Estrogen also stimulates the pituitary gland which leads to elevated levels of prolactin. By the twentieth week of gestation, mammary glands are sufficiently developed to produce components of milk due to prolactin stimulation. Milk production is inhibited by high estrogen and progesterone levels and colostrum is produced during this time. In the third trimester and then rapidly after birth, these levels decrease, allowing for milk production and eventual letdown to allow for breastfeeding. Most pregnancies cause the areola to darken, the breast to increase in size, and the Montgomery glands to become more prominent. Postlactational involution occurs at the cessation of milk production caused by a decline in prolactin.

Keywords

Anatomy · Physiology · Pregnancy · Colostrum · Breastfeeding · Involution

1.1 Overview

The mature breast is located within the anterior thoracic wall, between the second and sixth intercostal cartilage. More specifically, the innermost portion of the breast lies atop the pectoralis fascia of the pectoralis major, serratus anterior, external oblique abdominal muscles, and the upper extent of the rectus sheath; it measures from 10 to 12 cm in diameter. Breasts are cone shaped structures

A. Alex, B.S. · E. Bhandary, B.S. · K. P. McGuire, M.D. (✉)
Department of Surgery, Division of Surgical Oncology, Virginia Commonwealth University/ Massey Cancer Center, Richmond, VA, USA
e-mail: bhandarye@mymail.vcu.edu; kandace.mcguire@vcuhealth.org

S. Alipour, R. Omranipour (eds.), *Diseases of the Breast during Pregnancy and Lactation*, Advances in Experimental Medicine and Biology 1252,
https://doi.org/10.1007/978-3-030-41596-9_1

that extend from each lateral border of the sternum to the anterior axillary line. During puberty, the onset of the menstrual cycle and changes in hormones such as estrogen and progesterone in the body leads to the incomplete development of the breast. Only during pregnancy will the female breast mature to its full capacity.

Most of the incompletely developed breast consists of adipose tissue, but also comprises fibroglandular parenchyma and connective tissue. The breast parenchyma contains 15–20 units called lobes. These lobes are made up of 20–40 lobules and each lobule consists of 10–100 hollow cavities known as alveoli that are a few millimeters in size [1]. Cuboidal epithelium capable of synthesizing the protein and lipid components of breast milk, and myoepithelial cells capable of contracting epithelial cells compose each alveolus. The lobes of the breast drain into lactiferous ducts which broaden to form a sinus prior to converging with the nipple. Major ducts are made up of double layers of cuboidal epithelial cells and minor ducts are made up of a single layer of cuboidal cells while the lactiferous sinus is lined by stratified squamous epithelial cells. (Fig. 1.1).

The lactiferous sinus drains to the nipple-areola complex- the more pigmented circular area on the vertex of the breast. Underneath the areola, smooth muscle fibers lie in a circular pattern in the dense connective tissue and parallel to lactiferous ducts in order to erect nipples in response to appropriate stimuli. The areola also contains sweat, sebaceous and accessory glands called the Montgomery tubercles, which secrete oils [2] (Fig. 1.2).

1.2 Breast Changes during Pregnancy and Lactation

Anatomical and physiological changes occur in the mature breast as a result of elevated hormone levels during pregnancy. The alveolar epithelium increases in size and begins the secretion of components of the milk in response to elevated estrogen levels as early as ovulation. During the second week of pregnancy, the corpus luteum secretes estrogen and progesterone; while the placenta takes on this role during later stages of pregnancy. Prior to pregnancy, the ratio of adipose tissue to glandular and ductal tissue is large in mammary glands. During the first trimester, the ductal system expands and branches out into the adipose tissue in response to the increase of estrogen. Elevated levels of estrogen also cause a decrease in adipose tissue. Approximately 8 weeks after fertilization, trophoblasts, the cells that eventually become the placenta, produce the human chorionic gonadotropin (hCG) hormone. HCG works to prevent degradation of the corpus luteum and stimulate the corpus luteum to continue the production of progesterone and estro-

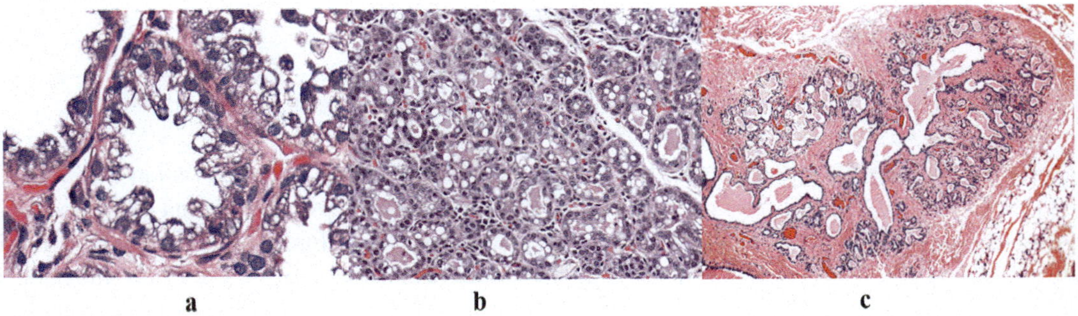

Fig. 1.1 Physiological changes in the breast during pregnancy, lactation and involution. (**a**) Breast in mid-pregnancy. Acinar cells have a hobnail appearance with vacuolated cytoplasm. Note absence of luminal secretions. Hematoxylin and eosin stain. (**b**) Lactating breast. The acini are expanded with accumulation of secretions. Acinar epithelium appears finely vacuolated. Hematoxylin and eosin stain. (**c**) Involution of breast 1 month after cessation of lactation. Acini are dilated with accumulation of secretions. Hematoxylin and eosin stain. (Reprinted from "Breast Pathology", Second Edition, Syed A. Hoda, Normal Breast and Developmental Disorders, Pages 1–23, Copyright 2017, with permission from Elsevier)

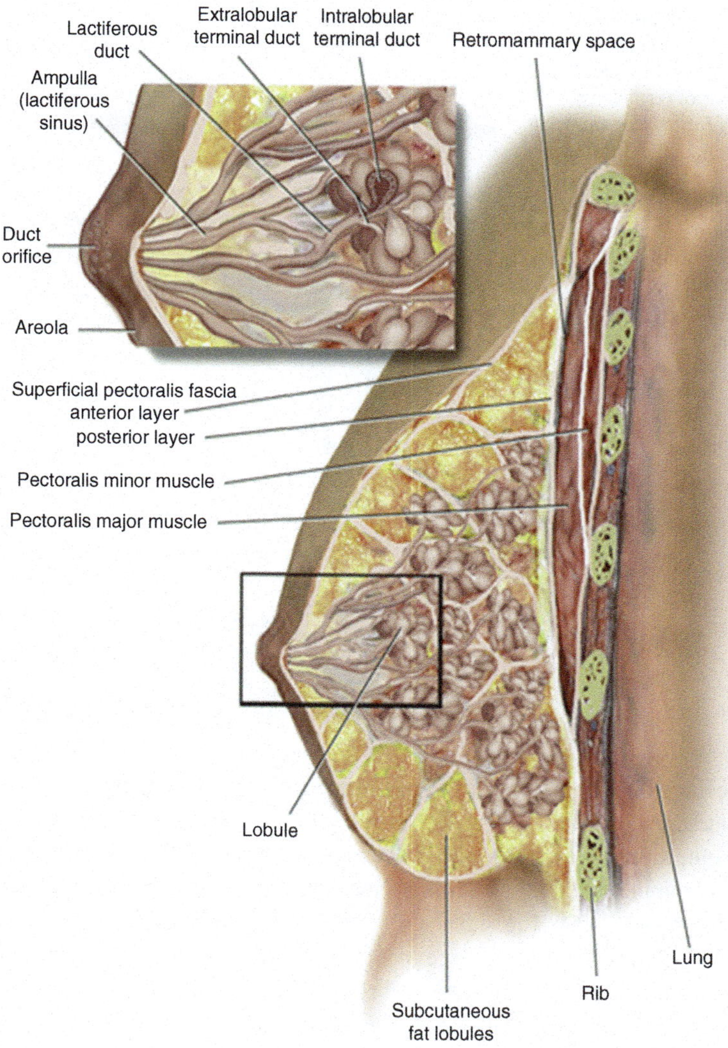

Fig. 1.2 The anatomy of the mature breast prior to pregnancy. (Reprinted from "Atlas of Pelvic Anatomy and Gynecologic Surgery", Fourth Edition, Donna L. Stahl, Karen S. Columbus, Michael S. Baggish, Anatomy of the Female Breast, Pages 1169–1180, Copyright 2016, with permission from Elsevier)

gen. HCG peaks at around the ninth week of pregnancy and declines afterward [3].

During early pregnancy, serum progesterone levels increase from 27 ng/ml (range, 20.0–42.2 ng/ml) to 138 ng/ml (range 105–215, ng/ml) during late pregnancy. Elevated concentration of progesterone induces lobular branching and enlargement in the breast. On the other hand, serum estrogen levels rise from 2 ng/ml (range, 1.19–4.00 ng/ml) during early pregnancy to 22 ng/ml (range, 13.6–35.6 ng/ml) during late pregnancy [4]. Increased concentration of estrogen leads to some increase in the proliferation of adipose tissue but mainly leads to ductal proliferation and elongation. These lobular-ductal units replace a fair amount of adipose tissue during mammary gland development in pregnancy [5]. As these hormone concentrations rise, ductal-lobular proliferation continues to occur.

Estrogen levels also impact the size and activity of the anterior pituitary gland in the brain. The pituitary gland can increase up to 36% in size when estrogen stimulation leads to an increase in the number and size of lactotroph cells within the gland [6]. This stimulation results in the synthesis and secretion of prolactin by lactotroph cells. Prolactin is a hormone that induces lactation in alveolar cells of lobules in mammary glands [7]. By the twentieth week of gestation, mammary glands are sufficiently developed to produce components of milk due to prolactin stimulation. Some ejection of milk occurs when myoepithe-

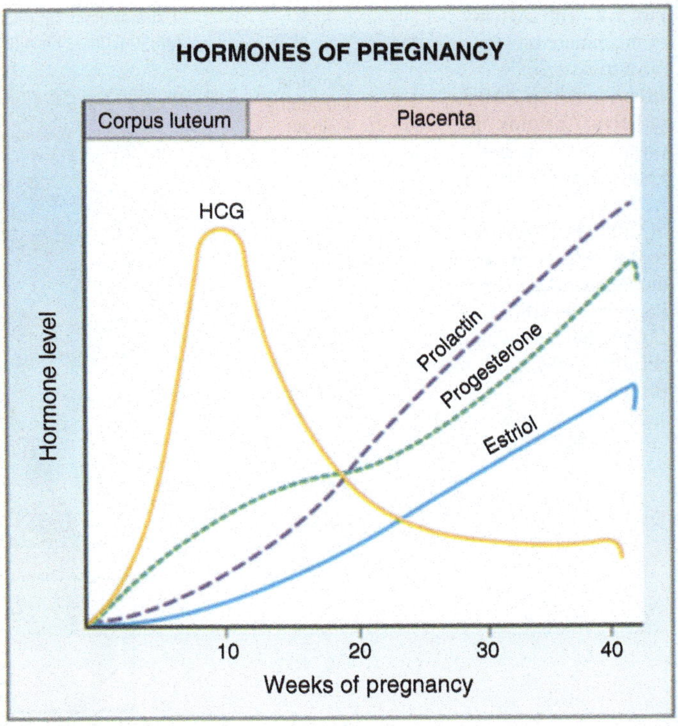

Fig. 1.3 The hormones of pregnancy. HCG, prolactin, progesterone, and estriol (primary form of estrogen) shown throughout the different stages of pregnancy. (Reprinted from "Physiology", Sixth Edition, Linda S. Costanzo, Reproductive Physiology, Pages 461–482, Copyright 2018, with permission from Elsevier)

lial cells respond to oxytocin and contract milk-producing alveolar cells. However, higher concentrations of estrogen and progesterone circulating in the blood inhibit milk production during pregnancy [8, 9] (Fig. 1.3).

The second trimester of pregnancy involves the accumulation of colostrum through milk acini, which include milk-producing cuboidal epithelial cells and myoepithelial cells that contract them [10]. Colostrum is any milk that is released during the first couple of days after parturition. A key characteristic of colostrum is the presence of an abundance of antibodies produced by lymphocytes compared to the presence of low quantities of lipids produced by epithelial cells. When the breast increases in size as a response to increased hormone levels, lymphocytes, eosinophils, and plasma cells aggregate within the connective tissues contributing to the release of antibacterial compounds into the alveoli. As immune cells and plasma cease to accumulate in the breast, the production of colostrum decreases and lipid-rich breast milk increases.

During the third trimester, the ductal system continues to expand, dilate and fill with colostrum. After birth, there is a rapid decrease in progesterone while there is an increase in prolactin and oxytocin. Prolactin pushes forward milk production, while oxytocin triggers the let-down reflex that allows the infant to withdraw breast milk from the milk ducts. The let-down reflex is a neuroendocrine reflex that results in the release of milk when the nipple-areola complex is stimulated. When an infant sucks on the nipple, it stimulates the fourth intercostal nerve present in the breast, causing the hypothalamus to release oxytocin [11]. Myoepithelial cells around the alveoli contract and squeeze the milk out, pushing it down the ducts and out of the nipple in response to oxytocin [12].

Most pregnancies cause the areola to darken, the breast to increase in size, and the Montgomery glands to become more prominent; [3] indicating that the body is ready for lactation. Failure in these changes can create problems with breastfeeding, typically associated with producing inadequate milk [13]. One point to notice here is that all of these changes lead to the enlargement of the breasts, but the size of the breast does not equate to its functional capacity. The maximum capacity of storing milk varies from 80 ml to 600 ml in volume. During breastfeeding, breasts capable of storing a lower volume of breast milk empty quickly, but alveolar cells within the mammary glands synthesize breast milk rapidly in contrast to alveolar cells in breasts with larger capacity [11].

Post-lactational involution occurs at the cessation of milk production caused by a decline in prolactin. Massive apoptosis and cell death occur in the mammary gland and the tissue in the breast is remodeled. The connective tissue of the lobules goes from a loose to a dense structure. Acini lose lining cells and the basement membrane of the acini becomes thicker [10].

References

1. Stahl DL, Columbus KS, Baggish MS (2016) The breast. In: Atlas of pelvic Anatomy and gynecologic surgery, 4th edn. Elsevier Inc., Philadelphia, pp 1169–1180
2. Farhadieh R, Bulstrode N, Cugno S (2015) Anatomy and physiology of the breast. In: Plastic and reconstructive surgery: approaches and techniques, 1st edn. Wiley, New York, p 480
3. Costanzo LS (2018) Reproductive physiology. In: Physiology, 6th edn. Elsevier Inc, Philadelphia, pp 461–482
4. Schock H, Zeleniuch-Jacquotte A, Lundin E, Grankvist K, Lakso HÅ, Idahl A et al (2016) Hormone concentrations throughout uncomplicated pregnancies: a longitudinal study. BMC Pregnancy Childbirth 16(1):146
5. Truchet S, Honvo-Houéto E (2017) Physiology of milk secretion. Best Pract Res Clin Endocrinol Metab 31(4):367–384
6. Foyouzi N, Frisbaek Y, Norwitz ER (2004) Pituitary gland and pregnancy. Obstet Gynecol Clin N Am 31:873–892
7. Shiu RP (1980) The prolactin target cell and receptor. Prog Reprod Biol 6:97
8. Neville MC, Morton J, Umemura S (2001) Lactogenesis. The transition from pregnancy to lactation. Pediatr Clin N Am 48(1):35–52
9. Adriance MC, Inman JL, Petersen OW, Bissell MJ (2005) Myoepithelial cells: good fences make good neighbors. Breast Cancer Res 7(5):190–197
10. Jones JL (2019) Breast. In: Underwood's Pathology, 7th edn. Elsevier Ldt., Edinburgh, pp 416–437
11. Sriraman NK (2017) The nuts and bolts of breastfeeding: anatomy and physiology of lactation. Curr Probl Pediatr Adolesc Health Care 47(12):305–310
12. Neville MC (2001) Anatomy and physiology of lactation. In: Pediatric clinics of North America, vol 48. Elsevier Inc, New York, pp 13–34
13. Balest AL, Riley MM, Bogen DL (2018) Neonatology. In: Zitelli and Davis'Atlas of pediatric physical diagnosis, 7th edn. Elsevier Inc., Philadelphia, pp 44–70

2 Physical Breast Examination in Pregnancy and Lactation

Sadaf Alipour

Abstract

Physical exam of the breast is a very important part of breast assessment both for breast cancer screening, and when approaching breast lesions. Examination during pregnancy and breastfeeding follows exactly the same method as non-pregnancy periods. However, physical changes that occur in the breast during these times due to hormonal effects cause alterations that can on one hand conceal some pathologic disorders, and may on the other hand appear as pathologic findings while being purely physiologic. This chapter focuses first on some key points for an accurate breast examination, and then reviews some challenging controversial findings that may be noticed during breast exam in a pregnant or lactating woman.

Keywords

Breast exam · Breast palpation · Breast inspection · Breastfeeding · Pregnancy

S. Alipour (✉)
Breast Disease Research Center (BDRC), Tehran University of Medical Sciences, Tehran, Iran

Department of Surgery, Arash Women's Hospital, Tehran University of Medical Sciences, Tehran, Iran
e-mail: salipour@tums.ac.ir

2.1 Overview

Breast examination (BE) is a skill that needs both experience and attention. One should not only put adequate time but also care enough to examine all parts that are involved in breast disease, that is whole breasts, axillary regions, and supra- and infraclavicular areas. BE includes taking related history from the patient, breast inspection (BI), and breast palpation (BP) [1, 2].

During the interview, in addition to general items such as age and queries about systemic and underlying diseases or medications, questions should be asked about menstrual and reproductive history including consumption of exogenous hormones; past history of irradiation to the chest or recent trauma, any type of breast surgery, previous breast imaging, and breast symptoms including pain, swelling, skin discoloration, recent asymmetry, or enlargement; as well as nipple eczema, itching, desquamation, retraction, or secretions. Furthermore, previous history of benign breast disease and breast, ovarian or other cancer, and a family history of any of these malignancies should be enquired [3, 4].

During BI, the examiner should inspect breasts and lymph node–bearing areas in different positions for various pathologic changes such as breast edema or asymmetry; skin edema, dimpling, ulcer, retraction, or color change; nipple

S. Alipour, R. Omranipour (eds.), *Diseases of the Breast during Pregnancy and Lactation*, Advances in Experimental Medicine and Biology 1252,
https://doi.org/10.1007/978-3-030-41596-9_2

desquamation or disfiguration; as well as any ulcer or swelling [3–5].

While performing BP, the upper trunk should be disrobed so that relevant areas are completely exposed. The examiner should be aware that some breasts contain much glandular tissue and feel very nodular on palpation. It is the skill of the physician to be attentive about the symmetry of thick or nodular areas of the breast, as well as any abnormality that might be palpated as hard, or firmer, among those areas. Thus, the presence of any lump, thickening, or nodularity in the breast should be assessed. Thereafter, palpation of axillary, supraclavicular and infraclavicular areas, as well as slight nipple squeezing for any significant discharge should be performed [3–5].

The following four aspects should be definitely considered in a perfect BE:

First, in view of the appropriate timing, BE is better performed when the breast is not edematous, engorged, and tender because of cyclical hormonal effects. Normally, women feel uncomfortable with their breasts from 1 to 2 days before their menstrual period to the first or second days of the menses. Some women experience a sense of heaviness for a longer duration in their breasts, beginning even 1 week before, and going on throughout menstruation. BE should, therefore, preferably be performed at a time where breasts are at ease, most ideally the week after the menses. In extreme cases, breasts tenderness occurs during the major part of the hormonal cycles, which is considered pathologic, and the BE is inevitably performed on sensitive breasts.

Second, from the aspect of the anatomic extent of the breasts (see also Chap. 1), it should be kept in mind that despite its more limited average anatomic definition, breast tissue might extend from below the clavicle to a few centimeters under the inframammary line from top to bottom, and from midsternum to midaxillary lines from side to side (Fig. 2.1) [3, 6–8]. All these parts should be carefully examined during a BE. For this purpose and to avoid leaving out any part, it has been proposed to always follow a definite pattern for palpating the entire breast; suggested patterns are depicted in Fig. 2.2a. Practically, one can use a combination of patterns (Fig. 2.2b), but it seems wise to follow approximately the same combined pattern for all women to perform BE accurately.

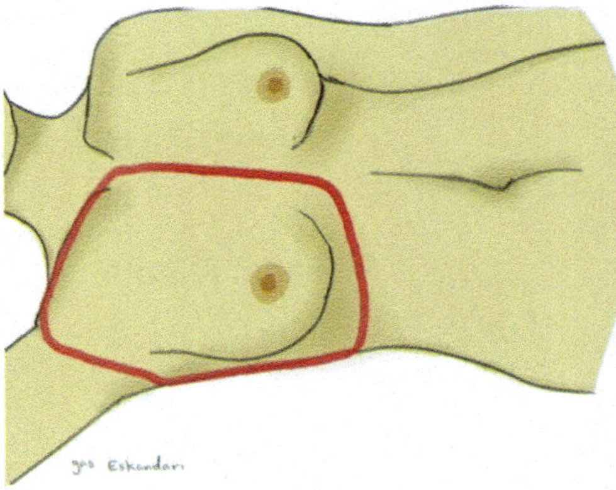

Fig. 2.1 Boundaries of the expected extent of breast examination

Third, in terms of how to palpate the breast, it is important not to pinch breast tissue during BE because fibroglandular components will be felt similar to pseudomasses through fatty tissue (Fig. 2.3). Breast tissue must be palpated by the pads of the 3 middle fingers (Fig. 2.4), pressing the breast to the chest wall in different levels of pressure, with small circular motions of the fingers in each point.

All findings of BI and BP should be documented. Picturing the roundness of the breast as a clock, location of each pathologic or even challenging physiologic finding should be noted down. It is preferable to show the size of the lesions, distance from the areola, and placement as hour on the clock in a sketch (Fig. 2.5). In addition, shape, consistency, mobility, and circumference of any lump must be noted.

2.2 Concerns in Pregnancy and Lactation

Physical examination of the breasts during pregnancy or at the time of breastfeeding follows basically the same rules and methods as a usual BE. However, physiologic changes that occur during these periods affect the size, shape, and consistency of the breasts (see also Chap. 1),

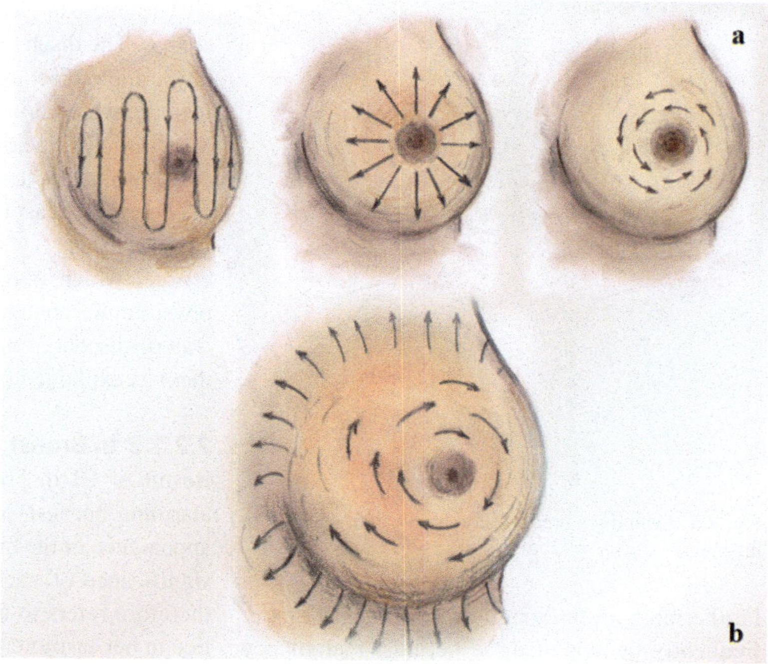

Fig. 2.2 (**a**) Suggested patterns of breast palpation. (**b**) Practical combined pattern of breast palpation

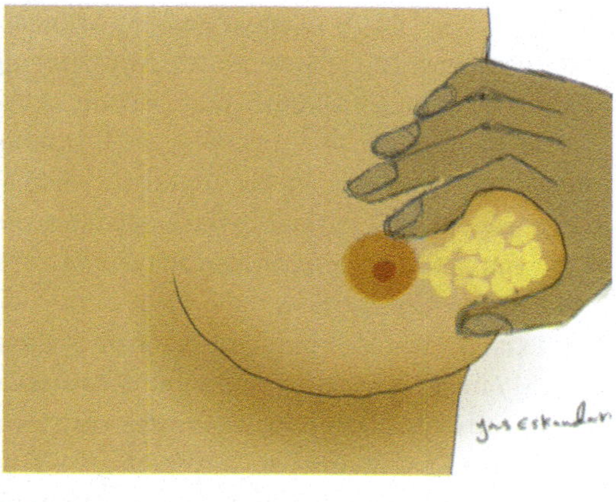

Fig. 2.3 Pseudomass detected when pinching the breast during breast palpation

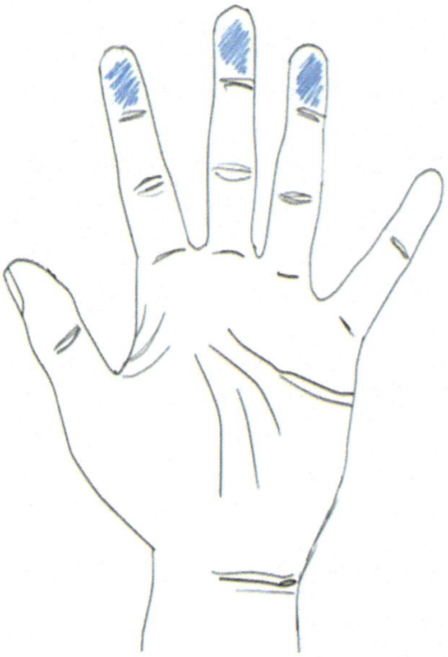

Fig. 2.4 Pads of the middle 3 fingers for appropriate breast exam

making BE much more difficult because of increased nodularity and diffuse firmness [1, 9–11]. Therefore, it is strictly recommended to perform BE in women who are planning for pregnancy, and at the first prenatal visit [12, 13]. A record should always be kept of results of these examinations.

Most of the apparently abnormal findings in BE of gravid or nursing women are physiologic alterations (Table 2.1). However, breast masses do occur in this period. Although lumps that occur in all women might occur and even enlarge in pregnant and nursing mothers, some masses are unique to this time, including galactoceles, lactating adenomas, or infarctions [1, 14, 15].

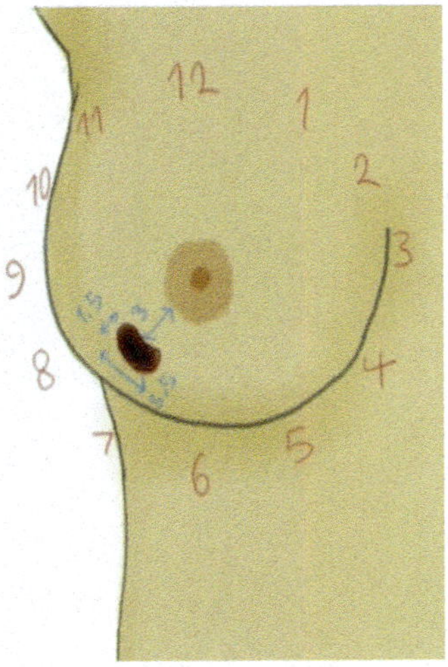

Fig. 2.5 Defining a breast lesion by location and size in the breast

Furthermore, even though breast lumps are more frequently benign in these periods, malignancy does occur and might impose a diagnostic challenge [9, 16, 17]. It should be emphasized that pregnant women are generally younger than 40 years, and thus any cancer would probably first be suspected based on physical findings and not on mammography [5]. For every indeterminate finding, the physician should not delay further evaluation by mammography or biopsy in case of persistent doubt [9, 12, 18] (see also Chaps. 3 and 4).

2.2.1 During Pregnancy

Physiologic changes of pregnancy influence both BI and BP (see also Chap. 1). These are summarized in Table 2.1.

2.2.1.1 In History Taking

Complaints of pain might aggravate, which warrants no more investigation if bilateral and associated with breast engorgement, except for a complete BE. Unilateral pain and tenderness needs to be assessed to rule out mastitis and breast abscess (see also Chaps. 5 and 7).

The woman may complain of nipple discharge. The discharge may be colored, similar to physiologic discharge observed in non-pregnant women, which does not need further evaluation. Milk discharge beginning in pregnancy is considered physiologic and is more commonly expected to occur in the last trimester [19, 20]. However, a most alarming symptom is bloody nipple discharge, which may occur as a consequence of physiologic changes of pregnancy. This and watery discharge need a more thorough assessment as explained below (see also Chap. 5).

2.2.1.2 In Breast Inspection

Result of BI in pregnant women is sometimes alarming because of pronounced changes in the appearance of the breasts, and evidence about the significance of each sign is scarce. The author therefore refers to the experience based on practice in her institute. Unexpectedly, findings of BI are unilateral in some circumstances, which cause much distress. This phenomenon may perhaps be secondary to unequal amount of breast tissue components on the two sides. Any previous asymmetry in breast volumes may become exaggerated, causing one to notice it for the first time. Engorged breasts may appear slightly erythematous; this redness is ordinarily intensified in more dependent areas, such as around the areola or in lower halves (Fig. 2.6).

Edema occurs both in breast parenchyma and skin. Skin edema may cause dimpling, which is occasionally localized to a small area such as a few centimeters, in any part of the breasts. This might be a normal variant if not accompanied by other pathologic findings in BE and ultrasound (US). Edema of the nipples and areolas produce disturbing pictures not unlike malignant infiltration in some circumstances, but these must not alarm the examiner if seen bilaterally while other parts of BE and US are within normal limits. Another expected finding in pregnant women is axillary swelling, which is secondary to physiologic changes that occur correspondingly in preexisting axillary breasts and is sometimes

Table 2.1 Positive findings in breast exam caused by physiologic changes of pregnancy and lactation

Findings in breast exam	Physiologic correlate	Hints	Should be ruled-out	Diagnostic approach	Follow-up
During Pregnancy					
In Breast Inspection					
Pain and tenderness	Breast engorgement and edema	May be unilateral	Mastitis and breast abscess	US if severe and unilateral	Observe for other presentations of infection
Bloody nipple discharge	Proliferation of duct epithelium	May be unilateral	Papilloma, breast cancer	US, cytology recommended, rarely mammo IDP	Observe with serial exam
Breast asymmetry	Breast engorgement and edema	Is exaggeration of previous asymmetry	Breast lump causing enlargement	US if clinically suspicious	No
Erythema	Breast engorgement	More severe in dependent parts	Mastitis and breast abscess, inflammatory breast cancer	US only if localized and suspicious, rarely mammo IDP	No
Dimpling	Skin edema	Occasionally localized to a small area	Mastitis and breast cancer	US, and mammo IDP	Observe with serial exam +/- US
Swollen nipple-areola	Edema of the nipple and areola	May be asymmetric	Mastitis and breast cancer	US if clinically suspicious, rarely mammo IDP	No
Axillary swelling	Hypertrophy of ectopic axillary breast	May be asymmetric or unilateral	Pathologic lymphadenopathy, axillary mass	US if clinically suspicious	No
In Breast Palpation					
Asymmetric focal nodularity	Development of alveoli	May feel like a mass	Factual breast lumps	US, rarely mammo IDP	No
Watery discharge	Milk secretion	Frequently precedes milk secretion	Factual watery discharge	If watery, unilateral, and spontaneous: US, and mammo IDP	No
Bloody discharge	Proliferation of duct epithelium	in up to 20% of gravid women, especially in 2nd and 3rd trimesters	Papilloma, breast cancer	US, and mammo IDP	Observe with serial exam
During Lactation [a]					
Nipple erosion	Suckling of the infant	May lead to nipple ulcer	Paget's disease	Biopsy only in extreme suspicious cases	Local care of the nipple
Bloody nipple discharge, mixed with milk	Epithelial proliferation of the ducts	In up to 15% of nursing mothers	Papilloma/ breast cancer	If it continues for more than 2 months after delivery, US ± mammo ± biopsy	can go on with breastfeeding

US = ultrasound, mammo = mammography; a. All findings of breast exam during pregnancy apply in breastfeeding also. The mother should empty her milk as much as possible before breast exam. IDP = if doubt persists.

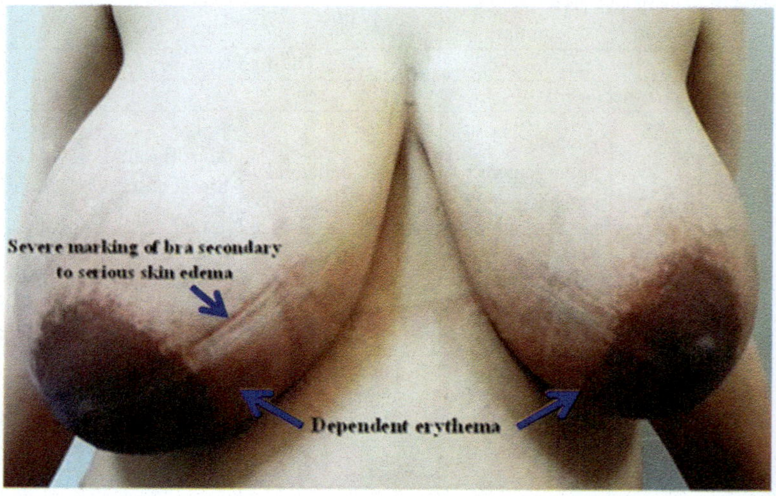

Fig. 2.6 Enlargement, edema, erythema, and areolar darkening in the breasts of a 27 weeks pregnant woman

asymmetric or even unilateral because of different amounts of breast tissue in axillae. These can happen actually in any other ectopic breast tissue as well, although seen rarely (see also Chap. 5).

2.2.1.3 In Breast Palpation

Development of alveoli during pregnancy might not occur symmetrically, and these can be felt as focal nodularities, or even feel similar to masses. Whereas extra work-up should not be done for physiologic nodularities, any suspicion of a mass should be assessed with US. If US illustrates glandular tissue and this correlates with the examiner presumption, no further work-up needs to be done. However, if presence of a mass is highly assumed, further follow-up by mammography, and biopsy when necessary, must be performed [11, 15, 21, 22].

Nipple discharge should be checked by squeezing of the nipple. According to the author's experience, watery discharge frequently precedes milk secretion, and squeezing should be repeated several times to find this out. Even when milk is not seen coming out of the breast after multiple attempts, if the discharge is from both breasts, it can be followed up for few weeks to get sure that it is not milk. If not, the same approach as that followed for non-pregnant women with discharge is followed. When the discharge is bloody, complete BP and US should be done. If no pathology is found, physiologic discharge of pregnancy should be considered because this occurs in up to 20% of gravid women, especially in the second and third trimesters [9, 11, 13, 17, 23, 24] (see also Chaps. 4, 5, 6 and 8).

2.2.2 During Lactation

Changes are about the same in lactation, except that milk fills the spaces, BE becomes more challenging, and lactating adenoma, galactoceles, and mastitis are much more common compared with pregnant women (see also Chaps. 6 and 7) [1, 14, 15]. In addition, axillary breasts might get filled with milk, and get more enlarged and more asymmetric; breast asymmetry may also get aggravated in breastfeeding mothers (Fig. 2.7).

However, same criteria as above imply for BE. The specific point is that the breasts should be emptied, whether by nursing or by draining milk out, and palpation be done afterward. Bloody nipple discharge, mixed with milk, may occur in 15% of nursing mothers, who can go on with breastfeeding. However, if it continues for more than 2 months after delivery, further work-up should be carried out [11, 14, 17, 18, 20, 24]. Breastfeeding may cause nipple erosion or ulcer. These are not considered dangerous and need no

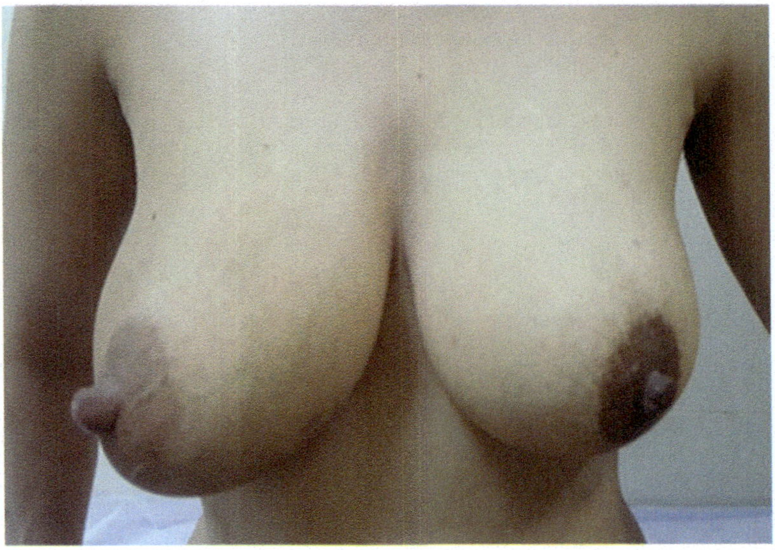

Fig. 2.7 Breast asymmetry in a mother nursing her fifth child; she only used her right breast for breastfeeding

further follow-up except for an intractable expanding case which would need tissue sampling; actually when nipple eczema and desquamation persist despite appropriate local care and treatment, Paget's disease should be considered (see also Chap. 18).

References

1. Faguy K (2015) Breast disorders in pregnant and lactating women. Radiol Technol 86(4):419M–438M
2. Veitch D, Goossens R, Owen H, Veitch J, Molenbroek J, Bochner M (2019) Evaluation of conventional training in clinical breast examination (CBE). Work 62(4):647–656
3. Henderson JA, Ferguson T (2019) Breast examination techniques. StatPearls Publishing LLC, Treasure Island
4. Morrow M (2014) Physical examination of the breast. In: Harris JR, Morrow M, Lippman ME, Osborne CK (eds) Diseases of the breast, 5th edn. Wolters Kluwer, Philadelphia, pp 25–28
5. Goodson WH 3rd. (1996) Clinical breast examination. West J Med 164(4):355–358
6. Cooper AP (1840) On the anatomy of the breast. Longman
7. de la Pared TA, y Mama A (2006) Anatomy of the thoracic wall, axilla and breast. Int J Morphol 24(4):691–704
8. McGuire KP (2016) Breast anatomy and physiology. In: Aydiner A, İğci A, Soran A (eds) Breast disease-diagnosis and pathology. Springer, Cham, pp 1–14
9. Lee SS, Hartman HJ, Kuzmiak CM, Crosby KL (2013) The management of breast symptoms in the pregnant and lactating patient. Curr Obstet Gynecol Rep 2(1):53–58
10. Damrich D, Glasser G, Dolan M (1998) The characteristics and evaluation of women presenting with a breast mass during pregnancy. Prim Care Update for Ob/Gyns 5(1):21–23
11. Nahklis F, Iglehart JD (2010) Evaluation and management of women presenting with breast symptoms during pregnancy. In: Breast surgical techniques and interdisciplinary management. Springer, New York, pp 521–532
12. Hogge JP, De Paredes ES, Magnant CM, Lage J (1999) Imaging and management of breast masses during pregnancy and lactation. Breast J 5(4):272–283
13. Vashi R, Hooley R, Butler R, Geisel J, Philpotts L (2013) Breast imaging of the pregnant and lactating patient: physiologic changes and common benign entities. Am J Roentgenol 200(2):329–336
14. Sorosky JI, Scott-Conner CE (1998) Breast disease complicating pregnancy. Obstet Gynecol Clin North Am 25(2):353–363
15. Langer A, Mohallem M, Berment H, Ferreira F, Gog A, Khalifa D et al (2015) Breast lumps in pregnant women. Diagn Interv Imaging 96(10):1077–1087
16. Sangri AM, Shaikh AG, Unar F (2017) Benign breast diseases in pregnancy. J Surg Pak 22:4
17. Scott-Conner CE, Schorr SJ (1995) The diagnosis and management of breast problems during pregnancy and lactation. Am J Surg 170(4):401–405
18. Yu JH, Kim MJ, Cho H, Liu HJ, Han SJ, Ahn TG (2013) Breast diseases during pregnancy and lactation. Obstet Gynecol Sci 56(3):143–159
19. Sakorafas G (2001) Nipple discharge: current diagnostic and therapeutic approaches. Cancer Treat Rev 27(5):275–282
20. Zervoudis S, Latrakis G, Economides P, Polyzos D, Navrozoglou I (2010) Nipple discharge screening. Womens Health 6(1):135–151

21. Parker S, Saettele M, Morgan M, Stein M, Winkler N (2017) Spectrum of pregnancy-and lactation-related benign breast findings. Curr Probl Diagn Radiol 46(6):432–440
22. Joshi S, Dialani V, Marotti J, Mehta TS, Slanetz PJ (2013) Breast disease in the pregnant and lactating patient: radiological-pathological correlation. Insights Imaging 4(5):527–538
23. Sauter ER, Schlatter L, Lininger J, Hewett JE (2004) The association of bloody nipple discharge with breast pathology. Surgery 136(4):780–785
24. Lafreniere R (1990) Bloody nipple discharge during pregnancy: a rationale for conservative treatment. J Surg Oncol 43(4):228–230

Breast Imaging in Pregnancy and Lactation

3

Adriana K. Langer

Abstract

All breast disorders found during pregnancy and lactation should be carefully evaluated. Most of them are benign, but it is essential to exclude pregnancy-associated breast cancer (PABC), which is too often diagnosed late. The first-line imaging technique is ultrasound (US), which must be completed by mammography if there is any clinical or US suspicious sign. In lactating patients with PABC, breast magnetic resonance imaging (MRI) can be useful for local assessment.

Management depends on the precise analysis and BI-RADS[1] classification of the lesion. During pregnancy and lactation, there is an overlap in imaging: many benign lesions can grow, infarct, become heterogeneous and thus suspicious, and on the other hand, PABC does not always present with typical malignant features. That is why biopsy must be performed if after the clinical and radiological evaluation the doubt persists, i.e. for all BI-RADS 4 and 5 lesions, and for some BI-RADS 3 lesions.

Keywords

Adenoma · Breast · Galactocele · Lactation · Fibroadenoma · Mammography · MRI · PABC · Pregnancy · Ultrasound

3.1 Ultrasound

US is the first-line imaging technique for young women, pregnant or lactating, because of its safety, availability, and the essential information it gives [1–4]:

- When there is doubt about a lump, US can confirm the clinician's feeling that there is

[1] BI-RADS (Breast Imaging-Reporting and Data System) is a standardized breast imaging terminology, and a classification system for mammography, ultrasound, and MRI of the breast.

BI-RADS 1: normal finding.

BI-RADS 2: benign finding.

BI-RADS 3: probably benign (likelihood of cancer ≤2%).

BI-RADS 4: indeterminate or suspicious of malignancy (>2 to <95%).

BI-RADS 5: highly suggestive of malignancy (≥95%).

A. K. Langer (✉)
Radiology Department, Institut Curie, Saint Cloud, France
e-mail: adriana.langer@curie.fr

S. Alipour, R. Omranipour (eds.), *Diseases of the Breast during Pregnancy and Lactation*, Advances in Experimental Medicine and Biology 1252, https://doi.org/10.1007/978-3-030-41596-9_3

actually no lump but just normal fibroglandular tissue.

- US accurately diagnoses simple cystic lesions (cyst, galactocele).
- US investigates all other (solid and atypical cystic) lesions with precise description and BI-RADS classification.
- During pregnancy, duct ectasia is frequent, and breasts are more hypoechoic in US (because of lobular hyperplasia and duct dilation).
- During lactation, there is diffuse hyperechogenicity, prominent ductal system, and increased vascularity in US.
- Benign lesions, particularly fibroadenomas and hamartomas may increase in size during pregnancy, become heterogeneous in US, and undergo infarction. They may therefore look suspicious [1–6].

3.2 Mammography

It is important not to perform mammography in young patients when unnecessary, for example for breast pain, if clinical examination and US are diagnostic and reassuring. However, it should be performed whenever there is a persistent doubt after clinical and US examination, as it is often helpful and is not dangerous, whereas the delay in diagnosis of PABC is very deleterious for both the mother and the infant.

If biopsy of a lesion shows breast cancer, and mammography has not been performed, it should be done for assessment of overall extension and therapeutic decision, whatever the age of the patient, during pregnancy and lactation. In this case, mammography must be bilateral and include craniocaudal (CC) and mediolateral oblique (MLO) views of the affected breast and at least MLO view of the contralateral breast [1–8].

3.2.1 Concerns in Pregnancy

For a 4-view mammogram, the mother receives a dose of radiation to the breast of 3 mGy, and the uterus less than 0.03 μGy [7]. The fetus is thus exposed to a negligible amount of radiation of about 0.001–0.01 mGy, depending on fetal weight and gestational age [1, 2, 4], as compared with natural weekly radiation of 0.02 mGy received by the fetus [5]. There is a consensus that when fetal dose is less than 50 mGy, the risk is negligible. During pregnancy, mammography can eventually be performed with a lead screen or apron (mostly for patient reassurance).

3.2.2 Concerns in Lactation

It is recommended to perform mammography immediately after breastfeeding, as breast density will be lower.

Highlights

- During pregnancy and lactation, there is diffuse and often marked increase in breast density and decrease in adipose tissue.
- Benign microcalcifications can be found, secondary to gestational or secretory hyperplasia. They are generally round, regularly shaped, diffuse or focal.
- If a lesion shows fatty density (galactocele, lactating adenoma, hamartoma), benignity can be affirmed and biopsy avoided.
- In PABC, mammography is suggestive in 75%–80% of cases, showing suspicious microcalcifications or architectural distortion, often undetected with US [6, 8].

3.3 Breast Magnetic Resonance Imaging

3.3.1 Concerns in Pregnancy

On the basis of current knowledge, the injection of Gadolinium must be avoided during pregnancy, as it enters the fetal circulation and, although no cases of adverse fetal effects have been reported, animal studies have shown adverse effects [9]. Other concerns with MRI are heating, which may affect cell migration during the first trimester, and acoustic noise, which may damage fetal hearing (by 24 weeks, once the fetal ear develops).

MRI (without injection) is not forbidden during pregnancy if necessary. However, as breast MRI needs Gadolinium injection to be contributive, it is not performed. In cases of PABC, the adequate assessment is done with bilateral mammography and US, which are complementary. Nevertheless, some researchers [10] have recently studied special sequences that do not require Gadolinium injection and may thus be interesting for PABC during pregnancy.

3.3.2 Concerns in Lactation

An MRI can be performed during lactation. Its main indication is diagnosed PABC, as its extension can be underestimated in these young patients.

A small percentage of Gadolinium is excreted in breast milk and absorbed by the infant, with no reported cases of direct toxicity [11]. Cessation of breastfeeding seems nevertheless preferable, with a recommendation to the mother to express and discard the milk of both breasts for 12–24 h. Contrast agent is undetectable in the mothers' circulation after 24 h [9].

Highlights

- Increased background parenchymal enhancement during lactation, which may be asymmetric, and diffuse high signal intensity on T2-weighted images can decrease efficacy.
- In Myers' study [12], MRI's sensitivity in PABC was found to be excellent (98%), the most common findings being solitary mass (55%), multiple masses (15%), and non-mass enhancement (23%).
- Slightly less than one-fourth (23%) of patients had larger tumor extension showed by MRI (and histologically proved), and it changed surgical management in 28% of cases.

3.4 Radiological Appearances of Breast Diseases during Pregnancy and Lactation

3.4.1 Lactating Adenoma

The US shows a solid mass of benign appearance (BI-RADS3) resembling fibroadenoma (well delimited, homogeneous hypoechoic, with main axis parallel to the skin). Sometimes its appearance can be misleading, and biopsy may be needed: microlobulated (Fig. 3.1) or

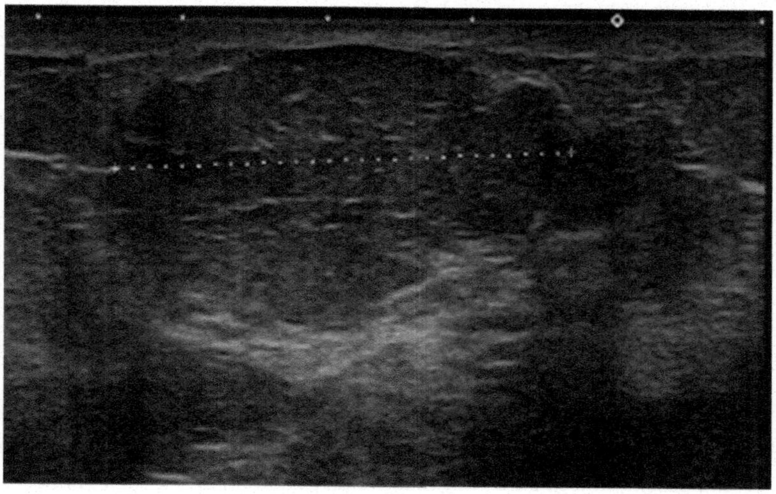

Fig. 3.1 A 27 years-old, 6 month pregnant woman with palpable mass. US shows oval mass with microlobulated borders. Biopsy: lactating adenoma

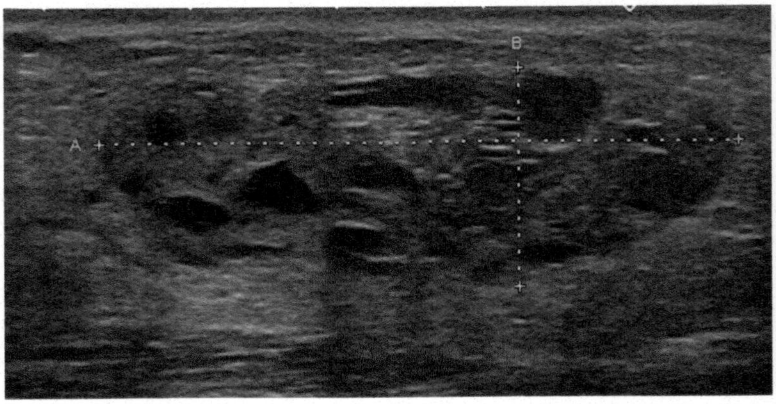

Fig. 3.2 A 31 years-old lactating woman with painless palpable mass. US shows oval, well circumscribed 45 mm mass containing multiple small anechoic foci. Biopsy: lactating adenoma

poorly defined borders, zones of anechoic fluid (Fig. 3.2), or heterogeneous content (BI-RADS 4).

If mammography is performed, it can be visible as a well-defined mass. When it contains fat density (because of the colostrum in the secretory lobules), benignity can be affirmed.

3.4.2 Galactocele

It is the most common benign breast lesion in lactating women, generally diagnosed after delivery (see also Chap. 6). The US shows a round or oval, well-delimited structure, whose appearance depends on the proportions of fluid, fat, and protein. It can be anechoic, hypoechoic, contain a fluid-fat level, or fine echoes. If there is inflammation, it may appear as an atypical cystic lesion with a thick wall (Fig. 3.3).

If mammography is performed, galactocele can be visible as a well-defined mass. As with lactating adenoma, when it contains fat density (sometimes a fat-fluid level), benignity can be affirmed. Clinical examination and US are usually sufficient. If in doubt, aspitarion affirms the diagnosis when it brings milky fluid.

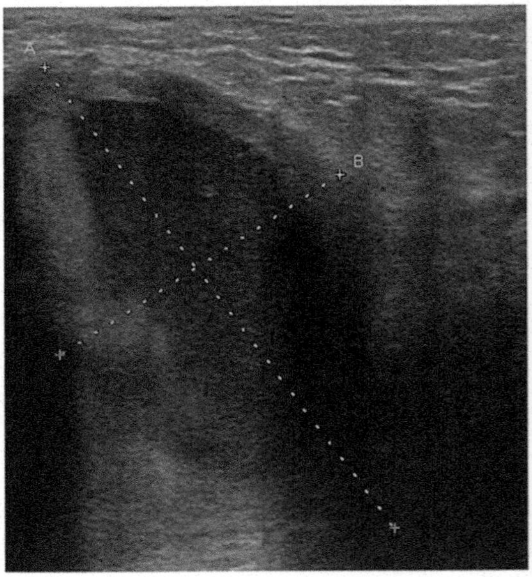

Fig. 3.3 A 28 years-old lactating woman with palpable mass. US shows heterogenous mass, with thick echoic wall and hypoechoic center, whose long axis is not horizontal. Biopsy: galactocele

3.4.3 Breast Infarction

Necrosis and bleeding can occur during pregnancy and lactation in hypertrophic breast tissue or in a preexisting mass as fibroadenoma, hamartoma, or lactating adenoma (see also Chap. 6). Patients present with an often painful mass, which appears solid and heterogeneous on US (BI-RADS 4) (Fig. 3.4). Associated lymphadenopathy may further divert the diagnosis. Biopsy is required.

3.4.4 Fibroadenoma

It is the most frequently observed tumor during pregnancy and generally has a BI-RADS 3 benign appearance on US (oval, main axis parallel to the skin, hypoechoic homogeneous, well delimited) (see also Chap. 6).

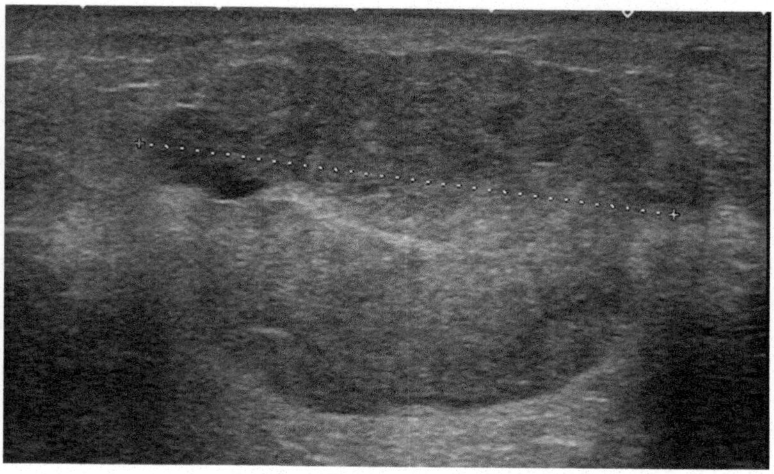

Fig. 3.4 A 26 years-old lactating woman (2 weeks postpartum) with palpable mass, local pain and slightly inflammatory skin. US shows oval, microlobulated, slightly heterogeneous 45 mm mass. Biopsy: lactating adenoma with large areas of ischemic necrosis

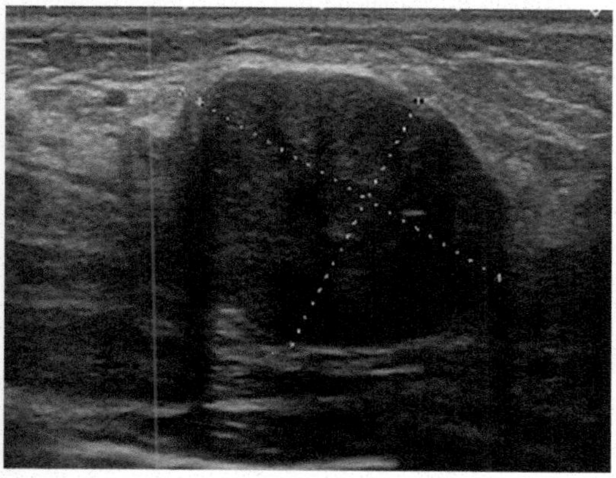

Fig. 3.5 A 26 years-old, 4 months pregnant woman. The breast mass was known for years but grew recently. US shows well defined, heterogeneous round mass. Biopsy: fibroadenoma

However, during pregnancy and lactation, it can grow, bleed, and undergo ischemic changes. In those cases, its appearance becomes ambiguous (BI-RADS 4): it may have poorly defined borders, fluid component, and heterogeneous structure, thus requiring biopsy to rule out malignancy (Fig. 3.5).

3.4.5 Fibrocystic Disease

As in all women, cysts can be simple or complicated (thick fluid contents, no wall). Puncture can be useful for diagnostic purposes (if there is doubt about a complicated cyst) or to drain pressurized and/or painful cysts.

3.4.6 Hamartoma

It may have a typical BI-RADS 2 “breast in breast” appearance on US. However, diagnosis can be more challenging if there is associated infarction, and the mass appears atypical, BI-RADS 4. If mammography is done and there is a fatty component, benignity can be affirmed; if not, biopsy should be performed to rule out malignancy.

3.4.7 Mastitis and Abscess

These conditions are rare during pregnancy; they are more frequent during lactation. *Staphylococcus aureus* often causes abscesses, whereas Streptococcus infections often manifest as diffuse mastitis. Clinical examination is usually sufficient for diagnosis (see also Chaps. 5 and 7).

US should be performed if abscess formation is suspected or in the absence of a quick response to antibiotic treatment. Abscess on US appears as an irregular hypo or anechoic mass, with a thick wall, which can contain fluid-debris levels, and has posterior acoustic enhancement (Fig. 3.6).

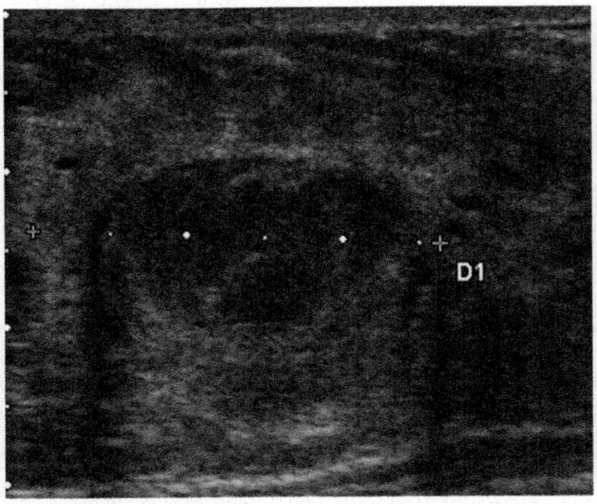

Fig. 3.6 A 39 years-old lactating woman (2 months postpartum) with painful inflammatory mass. US shows 30 mm oval well-circumscribed mass with regular, thick echogenic wall and posterior acoustic enhancement. Needle aspiration result: abscess

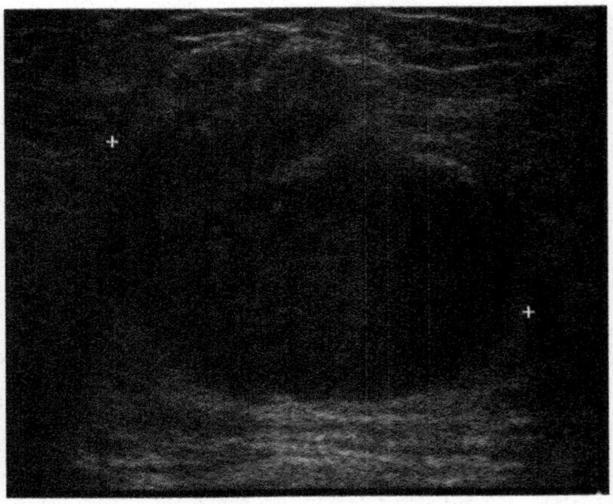

Fig. 3.7 A 38 years-old woman, 5 months pregnant, with palpable mass. US shows a mass with parallel orientation and posterior acoustic enhancement, but borders are microlobulated and lateral part of the mass is irregular. Biopsy: Grade 3 invasive ductal carcinoma

US is also very useful to guide needle aspiration and, if necessary, catheter drainage. In case of mastitis, US can show ill-defined hypoechoic areas of parenchyma.

If the patient does not respond quickly to antibiotic therapy, mammography and biopsy must be done, as inflammatory breast cancer must be ruled out in these young patients.

3.4.8 Pregnancy-Associated Breast Cancer

Delayed diagnosis of PABC is unfortunately frequent (often more than 6 months) (see also Chap. 11). This delay has many causes: lack of awareness- both by the patient and the physician- of the possibility of breast cancer in these young patients; fear of mammography, wrongly considered as dangerous and useless; and probably the wish to be, in this context, reassuring. It is essential to avoid delay in diagnosis: "Let's wait until delivery" must not be accepted. Patients generally present with a large palpable mass. All masses should be evaluated with US, and biopsy specimens from all suspicious masses should be collected and examined without postponement. Mammography is safe and useful in pregnant women, and both mammography and MRI are safe and useful in lactating women.

US is often diagnostic, showing a typical BI-RADS 5 lesion: heterogeneous solid mass with irregular borders, vertical main axis, and acoustic shadowing. However, PABC is not always that typical [1] because of the frequency of high-grade infiltrating ductal carcinomas that progress very quickly and do not always induce stromal reaction. Ayyappan et al. [13] reported a high frequency of lesions with horizontal main axis (60%) and posterior enhancement (60%). This falsely reassuring appearance can be misleading (Fig. 3.7), especially in these young patients in whom benign lesions are much more frequent.

That is why it is crucial that radiologists pay close attention to the analysis of lesion borders. Whenever microlobulated and/or irregular borders are seen, lesion must be categorized as BI-RADS 4 and a biopsy should be conducted. In high-risk patients, particularly *BRCA1* mutation carriers (Fig. 3.8), breast cancers often have a pseudo-benign appearance.

In rare cases, US can be falsely reassuring [8] either because the cancer is exclusively or predominantly intraductal (in those cases, mammography is diagnostic) or because it is benign-looking, particularly in patients with *BRCA1* mutation, whose mutational status is not always known at diagnosis (see also Chap. 17).

Mammography must thus be performed when clinical or US results are uncertain or suspicious. In our series of 117 PABC [8], despite high breast density, mammography showed, in 80% of cases, signs that immediately orientated the diagnosis: 55% of lesions demonstrated malignant microcalcifications and 15% had asymmetric density or architectural distortion, which may be overlooked in US. Mammography is also important to determine the extension of breast cancer, particularly with the microcalcifications, and helps rule out contralateral disease [14]. US and mammography are thus necessary and complementary (Fig. 3.9) in PABC during pregnancy and lactation.

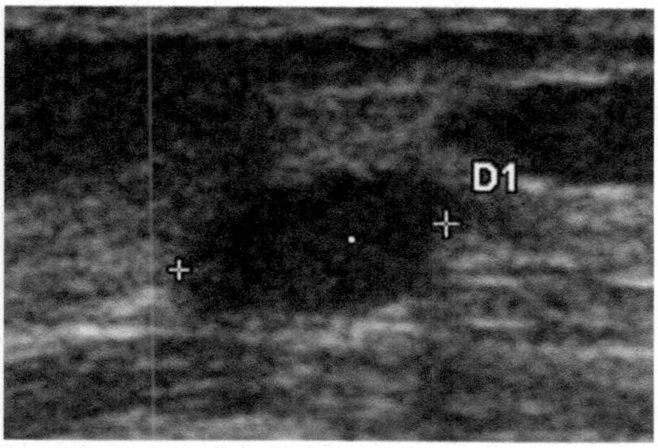

Fig. 3.8 A 39 years-old woman, 7 months postpartum, with high-risk family history. US shows small oval, benign-appearing left breast mass (BI-RADS3). Biopsy: Grade 3 invasive ductal carcinoma. Subsequent genetic evaluation revealed BRCA1 mutation

During lactation, when PABC is diagnosed in these young patients, mammography and US can be completed by breast MRI for optimal determination of lesion extension, and often in a neoadjuvant chemotherapy program. In Myers' study [12] of PABC, despite moderate or marked background parenchymal enhancement, breast MRI had a very high sensitivity (98%) for PABC, with the lesion appearing as a solid mass (55%) or as non-mass enhancement (23%). They found new contralateral disease in 4% of patients (which is similar to the rate found in breast cancer not associated with pregnancy), and surgical management was changed in 28% of their patients.

3.5 Common Practices in Imaging during Pregnancy and Lactation

When a patient presents with a complaint, in most cases a breast lump, a precise clinical and radiological assessment must be performed. Both physician and patient may want to postpone it until delivery or end of lactation, but this has to be avoided. In most cases, it is benign (80% of masses during pregnancy and lactation [2]), and the patient can be easily reassured, which is very important during those crucial moments of life.

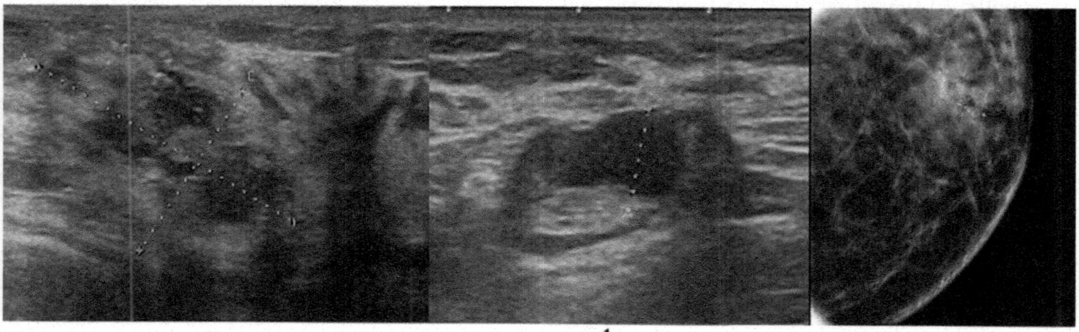

a b c

Fig. 3.9 A 31 years-old woman, 12 months postpartum, with palpable retroareolar mass. Ultrasound shows (**a**) Heterogenous 30 mm retroareolar mass and (**b**) Suspicious focal cortical thickening of axillary node. (**c**) Mammography (magnification view) shows irregular microcalcifications extending on 7.5 cm. Biopsy: Grade 3 invasive ductal carcinoma with high grade extensive DCIS and axillary node metastasis

When it is malignant, quickness in assessment is essential, as delay in management is a frequent and detrimental prognostic factor. PABC is often aggressive and needs fast multidisciplinary management. Efficient treatment is possible and can be implemented during pregnancy (see also Chaps. 12–16).

In non-pregnant patients, generally categories BI-RADS 4 and 5 require biopsy, and a 6-month follow-up is recommended for BI-RADS 3 lesions. In pregnant and lactating patients, BI-RADS 3 lesions will be either biopsied or very closely monitored (every 1–2 months).

If discovered during pregnancy or lactation, a BI-RADS 3 lesion resembling fibroadenoma can either be closely monitored (US every 1–2 months) or examined by a biopsy to avoid the stress of this close monitoring. In decision making, different factors must be evaluated: personal risk, anxiety, lesion size, as well as local customs. A 10-mm BI-RADS 3 lesion will more often be monitored, and larger lesions (>3 cm) will more easily be biopsied.

If a fibroadenoma was known before pregnancy, no follow-up is needed if the lesion is stable. It often grows, and a 20% growth, if appearance remains benign, is not worrisome: patient should be reassured, and clinical and US follow-up is enough. Biopsy is necessary to rule out malignancy if the increase in size is very important or if the lesion's aspect becomes ambiguous.

In case of multiple fibroadenomas, monitoring is usually sufficient. If there is doubt, one can first conduct a biopsy for the most suspicious nodule, followed up by a biopsy for the others.

It is important to know that during pregnancy and lactation, there is an overlap in imaging appearance of lesions. Benign lesions can grow and become heterogeneous (often because of infarction) and be classified as BI-RADS 4: only biopsy can clarify the diagnosis. On the other hand, PABC does not always present with typical malignant features, partly because of the frequency of grade 3 infiltrating ductal carcinomas, which progress too rapidly to induce the characteristic stromal reaction generally associated with breast cancer. Imaging work-up of PABC consists of bilateral mammography and US, when diagnosed during pregnancy. It can be completed by breast MRI if needed after delivery.

References

1. Vashi R, Hooley R, Butler R, Geisel J, Philpotts L (2013) Breast imaging of the pregnant and lactating patient: imaging modalities and pregnancy-associated breast cancer. AJR Am J Roentgenol 200(2):321–328
2. Vashi R, Hooley R, Butler R, Geisel J, Philpotts L (2013) Breast imaging of the pregnant and lactating patient: physiologic changes and common benign entities. AJR Am J Roentgenol 200(2):329–336
3. Sabate JM, Clotet M, Torrubia S, Gomez A, Guerrero R, de Las HP et al (2007) Radiologic evaluation of breast disorders related to pregnancy and lactation. Radiographics 27:S101–S124
4. Wang PI, Chong ST, Kielar AZ, Kelly AM, Knoepp UD, Mazza MB et al (2012) Imaging of pregnant and lactating patients: part 1, evidence-based review and recommendations. AJR Am J Roentgenol 198(4):778–784
5. Robbins J, Jeffries D, Roubidoux M, Helvie M (2011) Accuracy of diagnostic mammography and breast ultrasound during pregnancy and lactation. AJR Am J Roentgenol 196(3):716–722
6. Langer A, Mohallem M, Berment H, Ferreira F, Gog A, Khalifa D et al (2015) Breast lumps in pregnant women. Diagn Interv Imaging 96(10):1077–1087
7. Sechopoulos I, Suryanarayanan S, Vedantham S, D'Orsi C, Karellas A (2008) Radiation dose to organs and tissues from mammography: Monte Carlo and phantom study. Radiology 246(2):434–443
8. Langer A, Mohallem M, Stevens D, Rouzier R, Lerebours F, Chérel P (2014) A single-institution study of 117 pregnancy-associated breast cancers (PABC): presentation, imaging, clinicopathological data and outcome. Diag Interv Imaging 95(4):435–441
9. Tirada M, Dreizin D, Khati NJ, Akin EA, Zeman RK (2015) Imaging pregnant and lactating patients. Radiographics 35(6):1751–1765
10. Nissan N, Furman-Haran E, Allweis T, Menes T, Golan O, Kent V et al (2019) Noncontrast breast MRI during pregnancy using diffusion tensor imaging: a feasibility study. J Magn Reson Imaging 49(2):508–517

11. Beckett KR, Moriarity AK, Langer JM (2015) Safe use of contrast media: what the radiologist needs to know. Radiographics 35(6):1738–1750
12. Myers KS, Green LA, Lebron L, Morris EA (2017) Imaging appearance and clinical impact of preoperative breast MRI in pregnancy-associated breast cancer. AJR Am J Roentgenol 209(3): W177–W183
13. Ayyappan AP, Kulkarni S, Crystal P (2010) Pregnancy-associated breast cancer: spectrum of imaging appearances. Br J Radiol 83(990):529–534
14. Genin AS, De Rycke Y, Stevens D, Donnadieu A, Langer A, Rouzier R et al (2016) Association with pregnancy increases the risk of local recurrence but does not impact overall survival in breast cancer: a case-control study of 87 cases. Breast 30:222–227

Breast Cytology and Pathology in Pregnancy and Lactation

4

Vahid Soleimani and Behnaz Jahanbin

Abstract

Breast tissue reveals some physiologic changes during pregnancy and lactation due to hormonal alterations. Whole range of breast diseases including inflammatory, benign and malignant neoplasms can be seen in pregnancy but due to concurrent physiologic changes, may lead to diagnostic challenges. This chapter reviews sampling methods and histologic features of common benign breast lesions in pregnancy and lactation periods.

Keywords

Breast · Benign lesions · Biopsy · Cytology · Histology · Pathology · Pregnancy · Sampling

4.1 Overview

Female breasts confront several physiological alterations in pregnancy and lactation, such as vascular and lobular hyperplasia resulting from hormonal effects. These changes interfere with physical examination and analysis of imaging studies [1].

In the first trimester of pregnancy, terminal duct lobular units get enlarged and expanded, whereas fibrofatty stroma is reduced. In addition, stromal vascularity enhances and mononuclear cell infiltration occurs. Dilatation of superficial cutaneous venous system and noticeable areolar pigmentation accompanied by small lobular secretions may be observed by the end of the first trimester. In second and third trimesters, lobular growth advances with cellular expansion. Cytoplasmic vacuolization of lobular epithelial cells occurs, and secretions are deposited in swollen lobular glands [2] (see also Chap. 1).

4.2 Breast Sampling

Pre-surgical assessment of breast lesions improves decision making and helps clinicians in designing treatment plans. The pathologist must be informed that the patient is pregnant or in lactating state before assessing the specimen. Three most conventional procedures include fine-needle aspiration biopsy (FNAB), core needle biopsy (CNB), and vacuum-assisted biopsy (VAB).

V. Soleimani (✉) · B. Jahanbin
Department of Pathology, Cancer Institute, Imam Khomeini Hospital Complex, Tehran University of Medical, Tehran, Iran
e-mail: vsoleimani@tums.ac.ir

S. Alipour, R. Omranipour (eds.), *Diseases of the Breast during Pregnancy and Lactation*, Advances in Experimental Medicine and Biology 1252,
https://doi.org/10.1007/978-3-030-41596-9_4

4.2.1 Fine-Needle Aspiration Biopsy

FNAB was first developed in 1930, but it was not accepted throughout the next 25 years, till regularly applied for diagnostic purposes in palpable breast lesions in the 1950s. This procedure extracts cells from the lesion, and the assessment will therefore consist of cytologic evaluation of the specimen.

Pregnancy and lactation can cause false-negative and false-positive results in cytologic examination, and breast FNAB is not widely approved in these periods. Atypical nuclear features such as hyperchromasia and nuclear enlargement, as well as increased mitosis and lobular hyperplasia in glandular epithelium with secretory changes lead to challenging interpretation. If atypical findings are present, they should not be attributed to pregnancy or lactation without additional evidence, and therefore biopsy would be necessary in this setting.

FNAB has several benefits, including speed of performance, minimally invasiveness, safety, and low cost. Furthermore, FNAB needs no anesthesia because it is accompanied by minimal tenderness and pain. In addition, in patients who consume anticoagulants, there is no need for discontinuation of therapy before FNAB. Results of FNAB can be available in few hours, which is a substantial advantage of this procedure. FNAB is practical and helpful for probable abscesses in an inflamed breast or painful cysts containing thick fluid.[1, 3] Moreover, effectiveness of FNAB has been proved for cystic lesions, and for evaluating axillary lymph nodes where increased cortical thickness is detected on ultrasound (US).

Despite these benefits, FNAB has also some restrictions. Disadvantages of FNA include inability to differentiate between in situ and invasive lesions, poor sensitivity, and high costs for determining proliferation index as well as ER, PR, and HER2 status. FNAB also harbors lower sensitivity and specificity than procedures that involve tissue sampling and involves a higher percentage of extracting non-diagnostic specimens, especially in non-palpable lesions.

Nowadays, FNAB is suggested as a diagnostic tool for cystic lesions and suspected lymph node metastasis. In addition, it may be useful for evaluating lesions in the vicinity of the chest wall and superficial palpable masses to rule out regional recurrence. FNAB is not practical for solid lesions, including papillary, atypical, lobular, and fibrous lesions, such as radial scar or sclerosing adenosis.

4.2.2 Core Needle Biopsy

CNB can extract adequate tissue from the lesion, and is of a greater diagnostic value than FNAB (78% versus 55%) in pre-surgical assessment. Tissue biopsy is the best technique for evaluating breast masses under US or mammographic guidance.

CNB was first carried out in the 1990s, primarily for recognition of silent breast lesions; however, it soon replaced FNAB, not only for its validity in diagnosing benign lesions but also because of its ability in discriminating between in situ and invasive carcinoma. CNB is an invasive process that can be performed under US or mammography with local anesthesia. In some cases, the lesion is small and CNB can fragment or completely remove it; therefore, a marker should be placed in the bed of the lesion to make further localized treatment possible.

Present recommendations suggest at least 3 sample collections from focal lesions and 5 samples for lesions containing microcalcification. In one extensive review of 20 publications, sensitivity and specificity of FNAB were detected as 35–95% and 48–100%, respectively; whereas sensitivity and specificity of CNB were 85–100% and 86–100%, respectively [4]. This study showed more reproducibility for CNB specimens.

4.2.3 Vacuum-Assisted Biopsy

The use of VAB for removal of suspicious clusters of microcalcification under the guidance of mammography started in the 1990s, and it was also used for biopsy under US supervision thereafter. Many studies have shown lower false-negative results for VAB than CNB (slightly less than 10% for CNB), with higher sensitivity and specificity for diagnosing atypical ductal hyperplasia–like lesion or ductal carcinoma in situ, because of the higher amounts of tissue removal that is achieved in VAB. On average, a 11- or 12-gauge needle in VAB takes 40 mg of tissue, whereas CNB carries only 17 mg. However, VAB is 10–15 times more expensive than CNB.

4.2.4 Practical Use of Core Needle Biopsy versus Vacuum-Assisted Biopsy

BI-RADS 4 or 5 lesions should be evaluated by CNB, a technique with a very high negative predictive value (up to 99.4%). Indications of VAB for diagnostic purposes consist of determining nature of small lesions (<5 mm) and groups of microcalcification, as well as judgment of non-diagnostic results of other biopsy methods. However, VAB can also be used for some therapeutic intentions, including removal of lesions with a low risk of malignancy (BI-RADS 3 and 4a) such as fibroadenomas and intraductal papillomas, which have even been reported to be completely removed by an 8G needle in lesions less than 3 cm in size. This procedure might also be performed for lesions with distressing symptoms of pain, anxiety, or discomfort, or when the patient desires so, in some instances. VAB should not be used with therapeutic intent in malignant lesions because of fragmentation of the specimen in time of resection, which precludes assessment of margins [4].

4.3 Histologic Characteristics of Benign Breast Lesions in Pregnancy and Lactation

Breast lumps in pregnancy and lactation are mostly benign (80%) [5]. Throughout pregnancy and the lactating period, the breast can be affected by some distinct conditions such as lactating adenoma, galactocele, mastitis and lactational breast abscess that are specific to this period (see also Chaps. 6 and 7). In addition, other disorders that involve non-pregnant women such as fibrocystic changes or fibroadenoma may be seen during lactation and pregnancy.

4.3.1 Lactating Adenoma

Lactating adenoma is a benign lesion, consisting of the most common breast lump within pregnancy and puerperium. Clinically, the lesion is mobile, usually painless and slow growing, measuring less than 5 cm in size. It occurs mostly in the third trimester of pregnancy or the lactation period, and often in women in their 30s (see also Chap. 6). The etiology of lactating adenoma is ambiguous; some believe that it is a variant of fibroadenoma, tubular adenoma, or lobular hyperplasia, whereas others suggest that it is a neoplastic process that arises de novo.

Definitive diagnosis of lactating adenomas is by histologic examination of the specimen, which can be provided by CNB. Microscopic examination exhibits a well-delineated lesion composed of secretory lobules lined by epithelial cells containing granular, foamy, and vacuolated cytoplasm traversed by delicate connective tissue septa between lobules (Fig. 4.1). In some cases, CNB may not be demonstrative, and histopathological evaluation of the excisional biopsy specimen is required to rule out malignancy [6–10].

Lactating adenomas commonly subside after pregnancy and lactation. Similar to fibroadenoma, infarction may occur in these lesions.

4.3.2 Galactocele

Galactoceles are the commonest benign lesions of lactating breasts and usually appear after the cessation of breastfeeding, or while milk is not completely evacuated from the lactating breast (see also Chap. 6). Galactocele is a cystic structure filled with milk-like fluid, sometimes multiloculated. It might contain inflammatory cells including many foamy macrophages and also fat droplets and necrotic debris. Cuboidal or flat epithelial cells line the internal surface of the cyst.

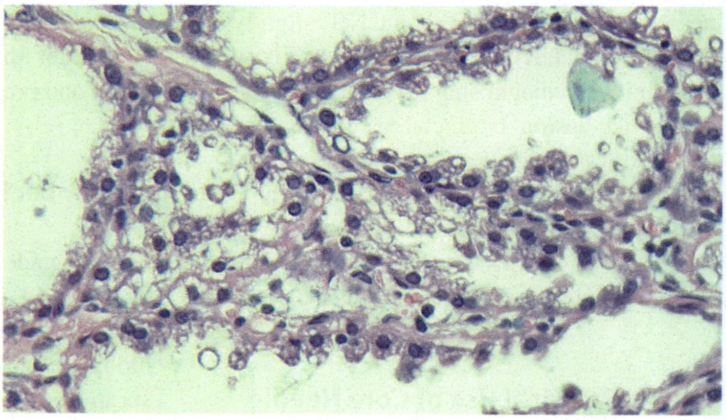

Fig. 4.1 Lactational changes in a lactating adenoma

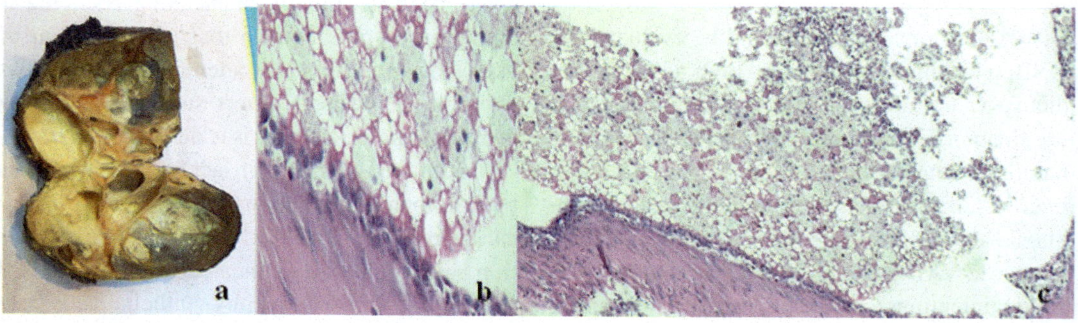

Fig. 4.2 Galactocele. (**a**) Gross picture of an opened multicystic galactocele, (**b, c**) Microscopic features show the cuboidal or flat epithelial cell-lined internal surface of the cysts containing numerous foamy macrophages, fat droplets, necrotic debris and dispersed mononuclear inflammatory cells (top right and lower part of picture)

Removing the milk (within lactation) or thick fluid (in old lesions) by aspiration is both diagnostic and therapeutic. Cytology smears show dispersed macrophages between benign acinar cell clusters and fat droplets as vacuoles in background. In fact, the cyst is composed of expanded epithelial and myoepithelial cell–lined terminal ducts or ductules containing milk. Enzymes denature milk ingredients during time and cause integration of fat globules [4, 9, 11]. The suspected etiology of galactocele is blockage of mammary duct in the lactating breast, mostly by inflammation and very rarely by a tumor [3] (Fig. 4.2).

4.3.3 Infarction

Breast infarction in pregnancy or lactation may occur as a result of hypertrophic physiologic changes within prior lesions such as fibroadenomas, lactating adenomas, or hamartomas. It is detected as a palpable painful ill-defined mass with variable consistency from soft to hard; and is sometimes misinterpreted as carcinoma (see also Chap. 6). In this situation, biopsy but not aspiration cytology is recommended [12, 13]. Microscopic examination shows a ghostly framework of underlying structures in hematoxylin and eosin–prepared slides. Structure of the tissue can be seen better by Reticulin stain or CK AE1/3 if it is not completely destroyed. Thrombosed vessels may be detected in some instances [14, 15].

4.3.4 Fibroadenoma

Overall, histologic features of fibroadenomas during pregnancy or lactation are not different from other periods. Fibroadenomas are one of the most prevalent benign breast lesions in women younger than 35 years. In pregnancy, fibroadenoma is seen commonly, sometimes produced de novo; but more frequently developing from enlarging previous lesions. Increase in size occurs secondary to stimulation of hormone receptors in the lump in response to the hormonal changes of pregnancy and breastfeeding. Infarction may occur, mostly in large ones [13] (see also Chap. 6). Grossly in the non-pregnant, non-lactating woman, fibroadenoma has a bosselated surface with firm white, gray, or tan bulging and rather homogeneous cut sections containing very fine clefts; observable by sharp eyes or under a magnifying glass.

Fibroadenomas greater than 5 cm (about 4% of the total) are named giant fibroadenoma; nonetheless, this terminology is not universally accepted. Giant fibroadenomas usually appear in pregnant or lactating women [16]. Histologic study reveals both glandular and stromal proliferation with two growth patterns including intracanalicular (when the stroma is plentiful and compresses glands) and pericanalicular (when the glands have original shape); but none of them has prognostic or clinical significance, and some fibroadenomas have both architectures simultaneously [17].

4.3.5 Fibrocystic Changes

Fibrocystic breast changes are very common and develop in 90% of women within their lives (see also Chap. 6). Despite the effect of estrogen, progesterone, and prolactin in creating fibrocystic disease, there are no dissimilarities between cystic changes in pregnant and non-pregnant women. Cysts originate from terminal duct lobular units with variable size, lined by flattened epithelial cells that are absent in some foci. Cyst content may consist of clear, cloudy yellow or bluish fluid. Under some circumstances, cysts rupture, and an inflammatory response composed of many foamy macrophages and cholesterol clefts occurs [13, 18, 19].

4.3.6 Nipple Discharge

Nipple discharge is classified as physiologic and pathologic (see also Chaps. 2, 5, and 6). The former can be observed in pregnancy and lactation when it is commonly bilateral, milky, green, yellow or white, from multiple lactiferous duct orifices and may continue up to 1 year after childbirth or ending lactation. Bloody nipple discharge in pregnancy or lactation is uncommon and probably is due to vascular engorgement and trauma, but other causes such as intraductal papilloma, duct ectasia, and malignant lesions should be ruled out especially if persisting after weaning or aggravated within lactation. Pathologic nipple discharge is usually unilateral, spontaneous, bloody or serous, and comes from one duct orifice. Any of these characteristics may be sufficient to assume a pathologic nature for the discharge, especially when an underlying mass lesion is detected. Evaluation of pathologic nipple discharge involves taking exact history, physical examination, and imaging studies including mammography and US. In non-pregnant women of all ages, the most common etiology of pathologic nipple discharge is intraductal papilloma (in 35–48% of the cases) and duct ectasia (17–36%), but underlying malignancy (especially in situ carcinoma) constitutes about 5–21% of circumstances. Cytologic examination for detecting malignancy is not suggested owing to the low sensitivity detected in several studies, but a high negative predictive value was observed in several works [20–23].

References

1. Yu JH, Kim MJ, Cho H, Liu HJ, Han SJ, Ahn TG (2013) Breast diseases during pregnancy and lactation. Obstet Gynecol Sci 56(3):143–159
2. Rosen PP (2014) Anatomic and physiologic morphology. In: Rosen PP (ed) Rosen's breast pathology, 4nd edn. Lippincott Williams & Wilkins Inc, Wolters Kluwer, pp 14–15
3. Sharma M, Gupta A, Kaul R (2017) Cytological evaluation of breast masses during pregnancy and lactation : a retrospective analysis. Glob J Reprod Med 555594
4. Łukasiewicz E, Ziemiecka A, Jakubowski W, Vojinovic J, Bogucevska M, Dobruch-Sobczak K (2017) Fine-needle versus core-needle biopsy–which one to choose in preoperative assessment of focal lesions in the breasts? Literatur Rev J USG 17(71):267
5. Vashi R, Hooley R, Butler R, Geisel J, Philpotts L (2013) Breast imaging of the pregnant and lactating patient: physiologic changes and common benign entities. Am J Roentgenol 200(2):329–336
6. Magno S, Terribile D, Franceschini G, Fabbri C, Chiesa F, Di Leone A et al (2009) Early onset lactating adenoma and the role of breast MRI: a case report. J Med Case Rep 3(1):43
7. Olfatbakhsh A, Gholizadeh Z, Beheshtiyan T, Hoseinpour P (2015) Five-year study of patients with lactating adenoma and review of the literature. Arch Breast Cancer 30:125–128
8. Hamza AA, Idris SA (2014) Lactating adenoma of the breast a diagnostic difficulty in pregnancy and rewarding natural history during lactation: a case report and review of literature. Med J 1(1):13–16
9. Buré LA, Azoulay L, Benjamin A, Abenhaim HA (2011) Pregnancy-associated breast cancer: a review for the obstetrical care provider. J Obstet Gynaecol Can 33(4):330–337
10. James K, Bridger J, Anthony PP (1988) Breast tumour of pregnancy ('lactating'adenoma). J Pathol 156(1):37–44
11. Sabate JM, Clotet M, Torrubia S, Gomez A, Guerrero R, de Las HP et al (2007) Radiologic evaluation of breast disorders related to pregnancy and lactation. Radiographics 27:S101–S124
12. Han B, Zhang H, Jiang P, Zheng C, Bi L, Lu LU et al (2015) Breast infarction during pregnancy and lactation: a case report. Exp Ther Med 10(5):1888–1892
13. Langer A, Mohallem M, Berment H, Ferreira F, Gog A, Khalifa D et al (2015) Breast lumps in pregnant women. Diagn Interv Imaging 96(10):1077–1087
14. Skenderi F, Krakonja F, Vranic S (2013) Infarcted fibroadenoma of the breast: report of two new cases with review of the literature. Diagn Pathol 8(1):38
15. Wadhwa N, Joshi R, Mangal N, Khan NP, Joshi M (2014) Cytopathologic diagnosis of spontaneous infarction of Fibroadenoma of the breast. Turkish J Pathol 30(3):237–240
16. Vijaykumar A, Ajitha MB, Shivaswamy BS, Srinivasan N (2012) A systematic study on fibroadenoma of the breast. Eur J Surg Sci 3(3):80–85
17. Goodman ZD, Taxy JB (1981) Fibroadenomas of the breast with prominent smooth muscle. The Am J Surg Pathol 5(1):99–102
18. Chen YY, Fang WH, Wang CC, Kao TW, Chang YW, Yang HF et al (2018) Examining the associations among fibrocystic breast change, Total lean mass, and percent body fat. Sci Rep 8(1):9180
19. Azzopardi JG (1979) Sarcoma of the breast. Prob Breast Pathol 2:355–359
20. Cacala SR (2010) Breast conditions during pregnancy and lactation: an understanding of unique breast conditions associated with pregnancy and lactation is essential for evaluation and management of breast problems in pregnant or lactating women. CME 28(11): 500–502
21. Lee SJ, Trikha S, Moy L, Baron P, Green ED, Heller SL et al (2017) ACR appropriateness criteria® evaluation of nipple discharge. J Am Coll Radiol 14(5):S138–S153
22. Abdalla S, Savag L, Masannat Y, Pinder SE, Fentiman IS, Hamed H (2014) Pathological nipple discharge. Open Access J Sci Technol 2:1–2
23. Mazzarello S, Arnaout A (2015) Nipple discharge. Can Med Assoc J 187(8):599

Clinical Presentations of Breast Disorders in Pregnancy and Lactation

5

Dhananjay Kulkarni

Abstract

The breast tissue undergoes significant physiological change during pregnancy and lactation. These changes can give rise to some unique disorders during pregnancy, puerperium and lactation or exaggerate pre-existing conditions. Clinical examination becomes less reliable due to textural change and density of breast tissue as a result of hormonal changes. The main symptoms during pregnancy and lactation are breast pain, mastitis, lactational abscess, breast lump, and blood- stained nipple discharge.

Lactational mastitis/ abscess must be treated without delay. Open incision and drainage of lactational abscess is rarely required, any lactational abscess should be treated with appropriate antibiotics and ultrasound guided aspiration of the pus.

Any breast lump during pregnancy and lactation should be investigated with triple assessment. Pregnancy associated breast cancer (PABC) must be ruled out. The choice of investigations and treatment needs careful consideration. While ultrasound is the investigation of choice, mammography can be performed with abdominal shielding if malignancy is suspected. Core biopsy is necessary for evaluation of any breast pathology but it comes with risk of infection, bleeding, hematoma and even milk fistula.

The treating clinical specialist must be aware of certain unusual unique clinical conditions in pregnancy and lactation including accessory axillary breast tissue, gigantomastia and Raynaud's phenomenon.

Keywords

Breast diseases · Breastfeeding · Clinical presentation · Pregnancy · Signs · Symptoms

5.1 Overview

The breast tissue undergoes significant changes during pregnancy and lactation. The ratio of glandular breast tissue to adipose and supporting connective tissue in the breast alters significantly with much more glandular proliferation. The breasts grow significantly in size and its vascularity almost doubles during this period. The vascularity of breast tissue can increase by 180%. Also, dilated veins can be seen under the skin of the breast [1–6]. The nipple and areola become more pigmented and enlarged as a result of estrogen effects. Towards the end of the third

D. Kulkarni (✉)
Western General Hospital, Edinburgh, Scotland
e-mail: drdhananjay@doctors.org.uk

S. Alipour, R. Omranipour (eds.), *Diseases of the Breast during Pregnancy and Lactation*, Advances in Experimental Medicine and Biology 1252,
https://doi.org/10.1007/978-3-030-41596-9_5

trimester, the skin becomes thin and the supporting stromal and adipose tissue in the breast is involuted, replaced by dense glandular tissue. This makes the breast very firm. Bilateral, multiduct, serous nipple discharge is common at this stage of pregnancy (see also Chap. 1).

These changes pose difficulties in clinical as well as radiological assessment of symptoms during this period. Although most of the lesions in the breast during pregnancy and lactation are benign, careful triple assessment (clinical, radiological and pathological) is essential to rule out pregnancy-associated breast cancer (PABC) [7, 8].

Some of these lesions are situated deep under the dense breast tissue virtually making them impossible to be deteced just by clinical examination. Hence it is important to thoroughly assess any symptom with triple assessment wherever appropriate.

As far as pathological assessment is concerned, core biopsy of the lesions is more reliable. The breast cells undergo considerable morphological changes during pregnancy and lactation making cytology difficult to interpret and false positive rates could be high.

5.2 Breast Pain

Physiologic breast pain is common during pregnancy and lactation because of ongoing changes and increased vascularity. The usual symptom is a bilateral, mild, dull ache that is generalized and diffuse in nature.

It is not common to see mastitis or breast infection during pregnancy, but lactational or puerperal mastitis and breast abscesses are very common. The causative organisms are usually gram-positive with *Staphylococcus aureus* and *Streptococcus* being the most common. It is thought that the bacteria from nose and throat of the baby find their way through the cracks on the nipple while breastfeeding. The stagnant milk provides an ideal conducive environment for growth and proliferation of bacteria causing infection, thereby leading to abscess formation. Lactational mastitis and abscess can cause sepsis and systemic symptoms such as generalized malaise, nausea, fever, tachycardia, or painful lump with redness of overlying skin and fluctuant swelling. Treatment should never be delayed. If necessary, intravenous antibiotics should be administered immediately. Only 5–10% of patients with lactational mastitis will form an abscess. Rarely, infection from methicillin-resistant *S aureus* can prove fatal. Hence, a sample of pus aspirate must always be sent for culture sensitivity examination [9, 10].

An ultrasound (US) exam will confirm if there is any underlying abscess. Features such as tissue edema, enlarged reactive axillary nodes, and pus collection confirm infective etiology. However, breast cancer can mimic mastitis and must be ruled out in these scenarios by triple assessment.

If US confirms pus collection/abscess, then this pus can be aspirated under US guidance using local anesthesia. The aspirate should be tested for culture and sensitivity. This procedure should be repeated until symptoms resolve; patient's progress must be carefully monitored.

Because the patient is still breastfeeding, the choice of antibiotics should be carefully considered. Patients can continue to breastfeed during the treatment of lactational mastitis and even breast abscess. Stagnant milk in the breast provides ideal conditions for bacteria to grow. Patients should be encouraged to breastfeed or use milk pump while the treatment is going on. Any potential delay in treatment can lead to sepsis and local complications such as skin necrosis.

If the pus is thick and cannot be aspirated, or if the abscess is very superficial, then mini-incision and drainage can be considered. The abscess cavity should not be packed after incision and drainage. This can lead to persistent cavity and milk fistula (see also Chap. 13). In addition, putting in drains should be avoided, however large the cavity might be. The mini-incision should be made at a non-dependent area of the abscess cavity, and it should be as small as possible [11].

Wherever possible, a small tissue sample should be collected for histology to rule out any possibility of underlying malignancy. The abscess cavity can be irrigated with local anesthetics. Re-aspiration may be required after every 2–3 days until there is no more pus.

Malignancy must be ruled out if the patient's condition does not improve. A tissue biopsy is necessary. Furthermore, US of the axillary nodes can provide insight, and it is a useful tool to differentiate between mastitis and inflammatory breast cancer (see also Chap.7).

Granulomatous lobular mastitis is an uncommon variant of breast inflammation that mimics breast cancer. It is associated with pregnancy and lactation and occasionally develops within few years of pregnancy. Although its etiology remains unknown, *Corynebacterium* has been associated with this condition. Even the imaging can sometimes fail to differentiate between lobular mastitis and cancer [12–16]. Patients usually present with a firm growing mass in the breast with occasional redness of overlying skin. In the later stages, multiple superficial abscesses are common. The final diagnosis is based on histology, and it classically shows non-caseating granulomas with chronic inflammatory changes. Antibiotics and steroids are prescribed, along with incision and drainage/aspiration of associated abscesses. Surgical excision of the mass is not required routinely (see also Chap. 7).

Breasts are expected to grow in size during pregnancy and lactation. However, if the growth is rapid, disproportionate, enormous, and causes physical and psychological symptoms, it is termed gigantomastia. It is probably an autoimmune condition, but hormonal changes during pregnancy are also thought to play a significant role. Cases have been reported in the literature in patients with systemic *lupus* erythematosus, and lupus supports autoimmune etiology [17–23]. Some definitions of gigantomastia are based on the amount of tissue that might need excision during reduction surgery. These vary from 800 to 1800 g or more of tissue. Usually the onset of symptoms occurs during puberty, pregnancy or lactation; it can affect both breasts or rarely one breast. The breasts can rapidly enlarge, becoming enormous within a short time span. Although benign, it can lead to skin ulceration, sepsis, infarction, and bleeding. Diagnosis of gigantomastia is largely clinical, although random core biopsies can be performed. It requires surgical intervention such as breast reduction surgery or even mastectomy which can be done after delivery, or better after end of lactation if the patient can wait (see also Chap. 6).

5.3 Nipple Discharge

Blood-stained nipple discharge is not common during pregnancy, although slightly hemoglobin-positive (on a dipstick), multiple-duct, bilateral, nonspontaneous discharge is common. But unilateral, spontaneous, blood-stained single duct discharge is suspicious. Any local nipple pathology such as rash, ulcer, or crack should be ruled out as the cause on examination [24] (see also Chap. 2). Intraductal papilloma can occur during pregnancy and lactation. The presenting feature is usually blood-stained nipple discharge. In absence of any atypia on histopathology and benign features on imaging, intraductal papillomas can be managed conservatively during pregnancy and lactation. However, their progress should be monitored, and surgery should be considered after completion of breastfeeding (see also Chap. 8).

Bilateral, multiple-duct (serous yellow, white, or green) discharge, on the other hand, is mostly physiologic and related to the pregnancy-induced changes in the breast tissue. Sometimes the discharge is related to some medications (hormonal, H_2-receptor antagonists, antihypertensives, or antidepressants) the patient consumes.

US, Doppler, mammography, magnetic resonance imaging (MRI) without contrast, magnetic resonance ductography, nipple discharge cytology, and even ductoscopy have been described as investigation modalities. Real-time color Doppler gray-scale US is the diagnostic procedure of choice in this group of young patients who are either pregnant or breastfeeding. It probably has slightly higher sensitivity than mammography because of the density of breast tissue. Mammography also should be undertaken carefully using abdominal shielding. MRI scan is not considered very useful because there is a controversy surrounding gadolinium contrast in pregnant and lactating patients. The rapid uptake of gadolinium by lactating tissue also reduces sensitivity. Finally, it is difficult to perform an MRI in pregnant patients because of issues with prone position (see Chap. 3).

Cytology of nipple discharge or galactography are not very sensitive tests and can give false-positive results (see also Chap. 4).

Breast duct endoscopy or breast ductoscopy offers a minimally invasive approach in the investigation of pathological nipple discharge. However, its role during pregnancy or lactation is not yet fully explored [25–28].

If no cause is found, total duct excision procedure is the last option for diagnosis in selected few patients where malignancy is suspected. This generally does not need to be performed during pregnancy and can await delivery, except in highly suspicious cases which should be approached very soon.

5.4 Palpable Breast Mass

During pregnancy, the breast tissue undergoes changes as a result of increased levels of hormones. The changes include growing vascularity, increased ductal and glandular tissue, and reduction in supporting stromal tissue (see also Chap. 1). This gives rise to extremely dense breast tissue, which in turn reduces sensitivity of clinical examination (see also Chap. 2). This is also the reason for late presentation and manifestation of breast cancer. It also poses challenges for imaging and reduces the sensitivity of conventional imaging techniques (see also Chap. 3).

Many preexisting breast lumps enlarge during pregnancy because of the effects of hormonal changes and become more apparent. The common lumps found during pregnancy and lactation are fibroadenomas, cysts, galactoceles, and lactating adenomas (see also Chap. 6). It is not necessary to excise biopsy-proven benign lumps during pregnancy, and although a US may show benign features such as anechoic lesions with posterior enhancement (in case it is a cyst or a galactocele) or well-circumscribed isoechoic lobulated lesions in favor of fibroadenoma, core biopsy is recommended for new solid lesions [29, 30].

Galactocele (see also Chaps. 3 and 6) is a retention cyst with pent-up milk as a result of blocked milk duct. It can appear at any time during the third trimester of pregnancy and lactation, as well as during weaning of breastfeeding. It could be associated with inflammation and fat necrosis; patients usually present with a tender lump, usually at a central location. US shows a unilocular or multiloculated fluid-filled, thin-walled structure. Aspiration provides relief and galactoceles usually do not re-fill after aspiration [31].

Fibroadenomas (see also Chaps. 3 and 6) are hormone sensitive, and hence, they can grow during pregnancy. US features include a well-circumscribed lobulated lump with isoechoic texture and posterior acoustic shadowing. Owing to a sudden increase in size, these lumps can have cystic spaces, increased vascularity, and even central infarction. Final diagnosis is made after conducting a core biopsy. Fibroadenomas are known to enlarge during pregnancy because of the effect of hormonal changes (estrogen) and regress in size after pregnancy. If a patient has a biopsy-proven fibroadenoma from before pregnancy, follow-up scans should be undertaken if it grows in size, and progress should be monitored.

Lactating adenomas (see also Chap. 6) typically develop during the third trimester (either as a single lump or occasionally multiple lumps) and decrease in size or even disappear after delivery. On US, lactating adenomas mimic fibroadenomas. Core biopsy is essential [30].

Occasionally, the fibroadenoma or lactating adenoma can undergo infarction, leading to sudden increase in size and associated pain. Core biopsy usually will help diagnose infarction.

Phyllodes tumors (see also Chap. 19) are rare during pregnancy, and malignant phyllodes with metastasis during pregnancy or lactation are extremely rare. Their growth is supposedly not affected by pregnancy or lactation because there are no hormone receptors. However, it is very difficult to differentiate fibroadenomas from phyllodes on imaging, and hence, biopsy must be performed. Furthermore, if the biopsy suggests a phyllodes tumor, its benign or malignant nature cannot be confirmed on mere core biopsy. So, in such situations, the tumors must be surgically excised with adequate surrounding margins to avoid recurrence.

Fat necrosis can occur as a result of trauma or enlargement of breasts during pregnancy. Patients usually present with a painful tender lump, sometimes with associated redness of overlying skin. Oil cysts can form as a result of fat necrosis. US-guided core biopsy is the diagnostic procedure of choice. FNA cytology is difficult to interpret because of physiologic changes in the breast tissue.

Core biopsy can lead to complications such as bleeding from the nipple, hematoma, and milk fistula. Lactating mothers should be warned of these complications. The local anesthesia administered takes about four hours to wear off; some of it can get into the milk, and hence, patients should be advised not to breastfeed from the affected side for about 4 hours. Patients should then preferably use a milk pump and discard the milk before resuming breastfeeding from that side. Milk fistula, although uncommon, is very difficult to treat. It may continue until the patient stops breastfeeding (see also Chap. 13).

It is helpful to keep a record of the preexisting lumps in the breast such as fibroadenomas, in the form of US images and biopsies. It helps with comparison and monitoring during pregnancy and lactation (see also Chap. 2).

5.5 Pregnancy-Associated Breast Cancer

Most of the breast disorders during pregnancy are benign in nature, but they have similar presentation as pregnancy-associated breast cancer (PABC). Hence, clinical examination followed by appropriate imaging is of paramount importance during pregnancy and lactation.

Any cancer diagnosed during pregnancy or within one year of pregnancy or while breastfeeding is termed PABC [32–37]. The average age of patients with PABC is 32–38 years, and it is associated with about 1 in 3000 to 10,000 pregnancies (see also Chap. 9).

Unlike patients undergoing breast cancer screening, PABC is diagnosed after manifestation of physical signs such as breast lump, lymphadenopathy, change of breast shape, skin changes, etc. (See Chap. 11). The most common presentation is the appearance of a painless mass. It is usually associated with higher grade tumors and advanced stage at presentation. The outcome of PABC is the same when compared with non-pregnant women. However, the diagnosis is often delayed. As a result, the outcome and prognosis remain poor. Patients with PABC are likely to have large tumors, vascular invasion, and even distant metastases at presentation.

The changes in hormonal levels during pregnancy also affect the breast tissue. The breast tissue increases in volume and density. This makes the clinical examination difficult and less reliable (see also Chap. 2). Nevertheless, a thorough examination must be conducted alongside imaging. The radiologic features of PABC are same as other breast cancers on US as well as mammography (see Chap. 3).

The carriers of *BRCA* 1 and 2 mutations are at a higher risk of developing PABC, and these patients should undergo imaging, including US and MRI before and after pregnancy. Their symptoms should be assessed during pregnancy with triple assessment whenever required, and they should be under strict surveillance throughout their pregnancy [38] (see also Chap. 17).

5.6 Other Breast Symptoms during Pregnancy and Lactation

Symptoms associated with nipple and areola that are observed during pregnancy, and especially during lactation include eczematous rash, ulceration, and bleeding. Careful evaluation by the breast surgeon and even a dermatologist is prudent. If the symptoms do not subside with routine treatment such as good hygiene and the use of antibacterial soap, moisturizer, antibiotics and steroid cream, then a punch biopsy of the lesion is necessary.

Raynaud phenomenon was described by Maurice Raynaud in 1862. It is one of the reasons of painful breastfeeding. If not treated appropriately, many women discontinue breastfeeding on account of excruciating pain. It is characterized by episodic spasms of the arterial wall muscles in small arteries of mainly the digits, followed by vasodilatation. Similar spasms can also affect the nipple and areola arteries during pregnancy and especially lactation. It is often mistaken for a *Candida albicans* infection. There has to be a precipitation of symptoms because of exposure to cold, biphasic or triphasic color change, and association of symptoms with pregnancy and breastfeeding to confirm the diagnosis of Raynaud phenomenon. Patients experience extreme pain while breastfeeding and notice "blanching" of nipple, which later turns dark bluish-purple before returning to normal [39–42]. Patients with Raynaud phenomenon are more likely to develop hypertension during pregnancy and are susceptible to preterm labor, miscarriage, and fetal complications. Symptoms can be managed by avoiding cold and any vasoconstrictive drugs and nicotine. However, a calcium channel blocker like nifedipine is also recommended. The amount of nifedipine excreted in breast milk is very small, and hence it is considered safe to use during breastfeeding.

Accessory breast tissue is the aberrant excess breast tissue along the embryonal milk line. It occurs in 0.4–0.6% of population. It is mostly located in the axilla. It can be unilateral or bilateral and can grow in size during pregnancy (see also Chap. 2). Patients usually present with axillary lump, pain, and cosmetic issues. Accessory tissue in axilla can cause discomfort and pain, even restricted shoulder joint movement, and it can be cosmetically unacceptable. It can be treated with excision or liposuction. But it is better to delay surgery until after stopping breastfeeding. Because it is a part of normal breast tissue in an abnormal location, it can develop benign breast diseases as well as malignancy [43–45].

Accessory nipple can become prominent and uncomfortable during pregnancy and lactation. Some patients have accessory third nipple along the milk line, typically close to the inframammary fold. This is even known to produce milk during lactation. Accessory nipple excision can be offered if patients are symptomatic, but is better be deferred until after delivery.

References

1. Beesley R, Johnson JV (2008) Global library women's medicine. The breast during pregnancy and lactation
2. Yu JH, Kim MJ, Cho H, Liu HJ, Han SJ, Ahn TG (2013) Breast diseases during pregnancy and lactation. Obstet Gynecol Sci 56(3):143–159
3. Lee SS, Hartman HJ, Kuzmiak CM, Crosby KL (2013) The management of breast symptoms in the pregnant and lactating patient. Curr Obstet Gynecol Rep 2(1):53–58
4. Bell H, Peters G, Lynch A, Harle R (2013) Breast disorders during pregnancy and lactation: the differential diagnoses. J Clin Gynecol Obstet 2(2):47–50
5. Scott-Conner CE, Schorr SJ (1995) The diagnosis and management of breast problems during pregnancy and lactation. Am J Surg 170(4):401–405
6. Joshi S, Dialani V, Marotti J, Mehta TS, Slanetz PJ (2013) Breast disease in the pregnant and lactating patient: radiological-pathological correlation. Insights Imaging 4(5):527–538
7. Kuo K, Caughey AB (2018) Management Strategy for Breast Cancer in Pregnancy. Obstet Gynecol 132(1):122–125
8. Parker S, Saettele M, Morgan M, Stein M, Winkler N (2017) Spectrum of pregnancy-and lactation-related benign breast findings. Curr Probl Diagn Radiol 46(6):432–440
9. Dixon JM (1994) Breast infection. BMJ 309(6959):946–950
10. Iddon J, Dixon JM (2013) Mastalgia. BMJ 347:f3288
11. Dixon JM (2007) Breast abscess. Br J Hosp Med 68(6):315–320
12. Galea MH, Robertson JF, Ellis IO, Elston CW, Blamey RW (1989) Granulomatous lobular mastitis. Aust NZ J Surg 59(7):547–550
13. Dixon JM, Chetty U, Salam IM, Sim AJ (1995) Diagnosis and treatment of granulomatous mastitis. Br J Surg 82(8):1143–1144
14. Barreto DS, Sedgwick EL, Nagi CS, Benveniste AP (2018) Granulomatous mastitis: etiology, imaging, pathology, treatment, and clinical findings. Breast Cancer Res Treat 171(3):527–534
15. Omranipour R, Mohammadi SF, Samimi P (2013) Idiopathic granulomatous lobular mastitis-report of 43 cases from Iran; introducing a preliminary clinical practice guideline. Breast Care 8(6):439–443
16. Zhou F, Yu LX, Ma ZB, Yu ZG (2016) Granulomatous lobular mastitis. Chronic Dis Transl Med 2(1):17–21

17. Türkan H, Gökgöz MŞ, Taşdelen İ, Dündar HZ (2016) Gestational gigantomastia. J Breast Health 12(2):86
18. Ezem BU, Osuagwu CC, Opara KA (2011) Gestational gigantomastia with complete resolution in a Nigerian woman. BMJ Case Rep 2011:bcr0120102632
19. Dancey A, Khan M, Dawson J, Peart F (2008) Gigantomastia–a classification and review of the literature. J Plast Reconstr Aesthet Surg 61(5):493–502
20. Kulkarni D, Beechey-Newman N, Hamed H, Fentiman IS (2006) Gigantomastia: a problem of local recurrence. Breast 15(1):100–102
21. Vinicki JP, Gonzalez CN, Dubinsky D, Nasswetter G, Cardinal LH, Hojman J (2015) Gestational gigantomastia in autoimmune diseases. J Clin Rheumatol 21(2):110–112
22. Shoma A, Elbassiony L, Amin M, Zalata K, Megahed N, Elkhiary M et al (2011) "Gestational gigantomastia": a review article and case presentation of a new surgical management option. Surg Innov 18(1):94–101
23. Swelstad MR, Swelstad BB, Rao VK, Gutowski KA (2006) Management of gestational gigantomastia. Plast Reconstr Surg 118(4):840–848
24. Stone K, Wheeler A (2015) A review of anatomy, physiology, and benign pathology of the nipple. Ann Surg Oncol 22(10):3236–3240
25. Han Y, Li J, Han S, Jia S, Zhang Y, Zhang W (2017) Diagnostic value of endoscopic appearance during ductoscopy in patients with pathological nipple discharge. BMC Cancer 17(1):300
26. Waaijer L, Simons JM, Borel Rinkes IH, van Diest PJ, Verkooijen HM, Witkamp AJ (2016) Systematic review and meta-analysis of the diagnostic accuracy of ductoscopy in patients with pathological nipple discharge. Br J Surg 103(6):632–643
27. Yang X, Li H, Gou J, Tan Q, Wang L, Lin X et al (2014) The role of breast ductoscopy in evaluation of nipple discharge: a Chinese experience of 419 patients. Breast J 20(4):388–393
28. Dooley WC (2009) Breast ductoscopy and the evolution of the intra-ductal approach to breast cancer. Breast J 15:S90–S94
29. Langer A, Mohallem M, Berment H, Ferreira F, Gog A, Khalifa D et al (2015) Breast lumps in pregnant women. Diagn Interv Imaging 96(10):1077–1087
30. Magno S, Terribile D, Franceschini G, Fabbri C, Chiesa F, Di Leone A et al (2009) Early onset lactating adenoma and the role of breast MRI: a case report. J Med Case Rep 3(1):43
31. Muttarak M, Padungchaichote W (2003) Clinics in diagnostic imaging (84). Singapore Med J 44(4):211–215
32. Ruiz R, Herrero C, Strasser-Weippl K, Touya D, Louis JS, Bukowski A et al (2017) Epidemiology and pathophysiology of pregnancy-associated breast cancer: a review. Breast 35:136–141
33. Amant F, Loibl S, Neven P, Van Calsteren K (2012) Breast cancer in pregnancy. Lancet 379(9815):570–579
34. Jones AL (2006) Management of pregnancy-associated breast cancer. Breast 15:S47–S52
35. Boudy AS, Naoura I, Selleret L, Zilberman S, Gligorov J, Richard S et al (2018) Propensity score to evaluate prognosis in pregnancy-associated breast cancer: analysis from a French cancer network. Breast 40:10–15
36. Bae SY, Jung SP, Jung ES, Park SM, Lee SK, Yu JH et al (2018) Clinical characteristics and prognosis of pregnancy-associated breast cancer: poor survival of luminal b subtype. Oncology 95(3):163–169
37. Taylor D, Lazberger J, Ives A, Wylie E, Saunders C (2011) Reducing delay in the diagnosis of pregnancy-associated breast cancer: How imaging can help us. J Med Imaging Radiat Oncol 55(1):33–42
38. Johannsson O, Loman N, Borg A, Olsson H (1998) Pregnancy-associated breast cancer in BRCA1 and BRCA2 germline mutation carriers. Lancet 352(9137):1359–1360
39. Anderson JE, Held N, Wright K (2004) Raynaud's phenomenon of the nipple: a treatable cause of painful breastfeeding. Pediatrics 113(4):e360–e364
40. Kahl LE, Blair C, Ramsey-Goldman R, Steen VD (1990) Pregnancy outcomes in women with primary Raynaud's phenomenon. Arthritis Rheum 33(8):1249–1255
41. Wu M, Chason R, Wong M (2012) Raynaud's phenomenon of the nipple. Obstet Gynecol 119:447–449
42. Arnold KC, White DE, Flint CJ (2015) Association between Raynaud's phenomenon and pregnancy complications [266]. Obstet Gynecol 125:85S–86S
43. Aydogan F, Baghaki S, Celik V, Kocael A, Gokcal F, Cetinkale O et al (2010) Surgical treatment of axillary accessory breasts. Am Surg 76(3):270–272
44. Down S, Barr L, Baildam AD, Bundred N (2003) Management of accessory breast tissue in the axilla. Br J Surg 90(10):1213–1214
45. Lesavoy MA, Gomez-Garcia A, Nejdl R, Yospur G, Syiau TJ, Chang P (1995) Axillary breast tissue: clinical presentation and surgical treatment. Ann Plast Surg 35(4):356–360

Part II

Benign and Premalignant Disorders of the Breast in Pregnancy and Lactation

Benign Disorders of the Breast in Pregnancy and Lactation

6

Nur Aishah Taib and Kartini Rahmat

Abstract

Benign cystic or solid lumps frequently occur in the breasts of young women, and consequently can also be seen during pregnancy and lactation. Simple cysts do not increase the risk of malignancy. The current management is routine follow-up. Complex cysts are thick walled or contain a mass, and should be followed by a US-guided biopsy and then treated similar to any non-gravid, non-lactating patient.

Galactoceles can be detected during the last trimester of pregnancy and during or after stopping lactation. Aspiration can be done to confirm the content. Co-existence of galactocele and malignancy is extremely rare, and the key is to follow up until it resolves.

Fibroadenoma is the most frequent lesion found during pregnancy and lactation. Management is usually conservative after triple assessment. Surgery is usually not recommended in pregnant and lactating women unless rapid increase in size occurs or there is discordance in the triple assessment.

Lactating adenomas are sometimes interpreted as a variant of fibroadenoma. They can naturally disappear at the end of pregnancy or lactation. Management is usually conservative, and an excisional biopsy is only mandated if it is rapidly enlarging or if there is discordance in the triple assessment.

Gestational gigantomastia is a rare condition consisting of diffuse severe hypertrophy of both breasts during pregnancy. Mastectomy and reconstruction may rarely be required in such cases.

Keywords

Complex cysts · Fibroadenoma · Galactocele · Gestational gigantomastia · Lactating adenomas · Lactation · Pregnancy · Simple cysts

N. A. Taib (✉)
University of Malaya, Cancer Institute, Kuala Lumpur, Malaysia

Department of Surgery, Faculty of Medicine, University of Malaya, Kuala Lumpur, Malaysia
e-mail: naisha@um.edu.my

K. Rahmat
Department of Biomedical Imaging, Faculty of Medicine, University of Malaya, Kuala Lumpur, Malaysia

S. Alipour, R. Omranipour (eds.), *Diseases of the Breast during Pregnancy and Lactation*, Advances in Experimental Medicine and Biology 1252,
https://doi.org/10.1007/978-3-030-41596-9_6

6.1 Cystic Masses

6.1.1 Fibrocystic Changes

Fibrocystic change (FCC) of the breast is a common benign condition that occurs between the ages of 35 and 50 years. Symptoms include cyclical mastalgia and tender nodularity of the breasts. It often occurs bilaterally, mainly in the upper outer quadrants, and may consist of physiological changes arising in response to hormonal fluctuations that take place during the menstrual cycle [1] (see also Chap. 4). Swelling, pain, and tenderness of FCC occur during ovulation and decrease after menstruation begins. It may be characterized by sensation of pain or itchiness in the nipples.

Although the pathophysiology is not well defined, the close association of FCC with the menstrual cycle and the fact that these changes diminish after the onset of menopause suggest a link between cyclical hormonal changes and breast tissue alterations. This chronic fluctuation in hormone levels leads to the development of small cysts and areas of fibrosis.

Clinical findings range from nodularity to discrete diffuse masses dispersed throughout the breast, but mostly located in the upper outer quadrants of the breasts. The masses may wax and wane, or remain static at times. The nodularity of the breast may render detection of breast cancer difficult. Rarely, cancers such as lobular carcinoma may present with a non-lump nodularity of the breast and must be ruled out if there is persistent nodularity.

Diagnosis requires a comprehensive history to ascertain risk factors, and diagnostic imaging, with a mammogram recommended for those aged 35 years and older and an ultrasound (US) for any age. The findings of FCC on US are nonspecific and include multiple cysts of varying sizes, dilated ducts, and echogenic foci representing fibrous tissue that may cause posterior sound attenuation [2].

Dupont has classified FCC into nonproliferative lesions, proliferative lesions without atypia, and proliferative lesions with atypia (atypical hyperplasia) [3]. The great majority of breast biopsies (up to 70%) show nonproliferative lesions [1].

Adenosis is a common proliferative lesion without atypia that is characterized by glandular structures and maintain normally arranged epithelial and myoepithelial components. Adenosis causes lobular-based lesions with increased numbers of acini or tubular structures that may vary from case to case [4]. Sclerosing adenosis (SA) is a compact proliferation of acinar or tubular structures with pronounced intralobular fibrous connective tissue [4]. Occasionally, SA may be associated with atypical lobular hyperplasia, lobular carcinoma in situ, and ductal carcinoma in situ. SA increases risk of cancer slightly, similar to usual ductal hyperplasia [4].

6.1.1.1 Concerns in Pregnancy and Lactation

Pregnancy provides a period of sustained presence of high levels of estrogen and progesterone, hence providing a proliferative environment in the breast. During this time, the breast is tender, swollen, and nodular, which in turn can mask palpable masses (see also Chaps. 1 and 2).

The diagnosis of FCC is one of exclusion; thus, close clinical follow-up to ensure resolution or stability of palpable masses is required to ensure that cancers are not left undetected. Although Magnetic Resonance Imaging (MRI) can be done when US is normal and the European Committee on Radiation Risk has also reported that gadolinium-based contrast could be safely used during pregnancy, because it would not only be less absorbed into placenta but also be rapidly excreted to kidney [3], scarce safety information is available to make it a routine investigation in pregnancy. However, the contrast medium in MRI, gadolinium, is considered to be relatively safe for use during lactation as it is excreted in small amounts in breast milk [5] (see also Chap. 3). Image-guided core biopsy is suggested for these masses if there are corresponding US-detected lesions because they may be diffuse, and freehand biopsy may miss the lesion.

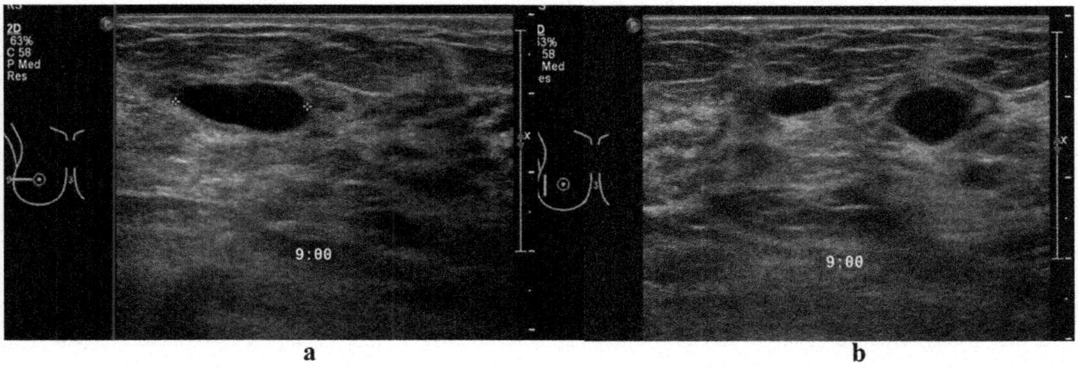

Fig. 6.1 (**a**) Anechoic thin walled cysts with posterior acoustic enhancement. (**b**) Multiple simple cysts

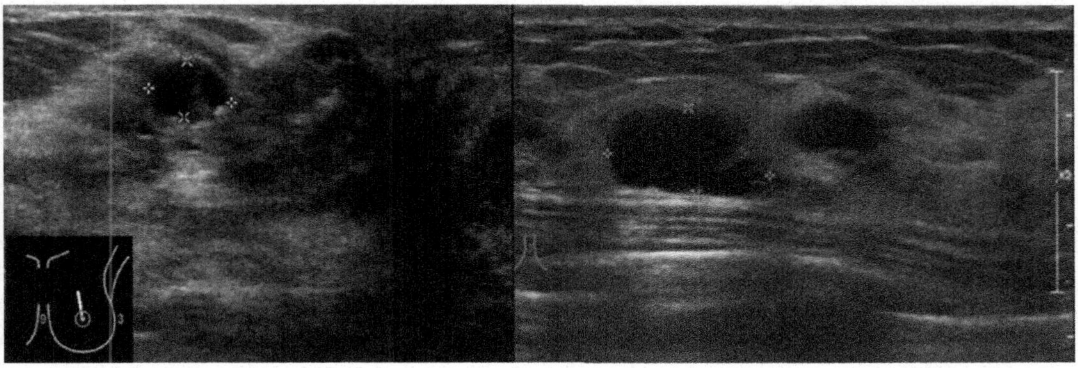

Fig. 6.2 Complicated cyst with septations and irregular walls

Treatment of FCC is conservative, and excision should be reserved for cases with very high suspicion of missing a diagnosis of malignancy.

6.1.2 Simple Cysts

Simple cysts are macrocysts that form because of involution of the breast and obstruction of the ducts. These fluid-filled cysts develop from dilatation of the terminal ductal lobular unit (TDLU) [2]. In most cysts, the epithelial lining is either flattened or totally absent. In only a small number of cysts, an apocrine epithelial lining is observed.

These cysts occur between the ages of 30 and 55 years. Cystic lumps may appear overnight and can be painful. These are thin-walled and appear as a well-circumscribed anechoic mass with posterior acoustic enhancement (Fig. 6.1) with absent vascularity on US [1]. On a mammogram, these can be found as a classical radiolucent halo around a well-circumscribed mass. Cysts are not associated with an increased risk of carcinoma development. The current consensus on the management of simple cysts is routine follow-up of the patient, without further therapy. Aspiration can be done for symptomatic tender palpable cysts.

6.1.3 Complex Cysts

Complicated cysts are cysts with septa showing low echoes in US (Fig. 6.2), whereas complex cysts are cysts that are thick walled or have masses within the walls of the cysts. Changing the position of the patient during US can help differentiate debris from intracystic masses, that is, debris move with change in position [2]. Incidence for malignancy in complex cysts is less

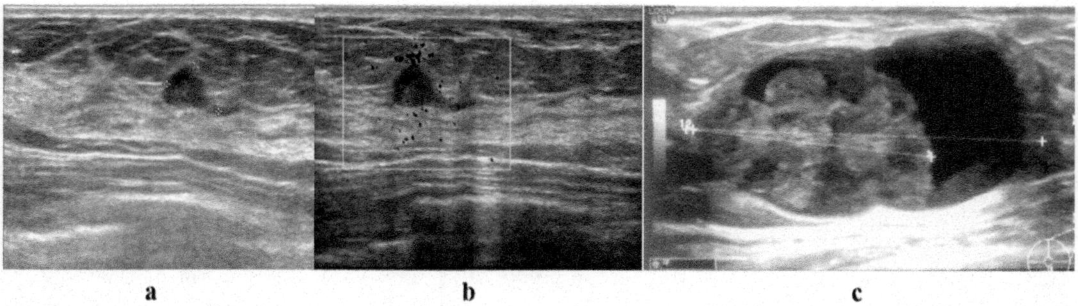

Fig. 6.3 Bloody nipple discharge – papillary lesion/ papilloma. (**a, b**) Periareolar complex cyst- intraductal papilloma (**c**) Complex cyst- intracystic papillary carcinoma

than 2% [6]. However, if the lesion also includes an intracystic mass, it should be regarded as "suspicious for neoplasm" and managed as a solid lesion; a US-guided core needle biopsy would be indicated.

Papillary lesions may present as cystic masses. Intraductal papillary lesions (Fig. 6.3a,b) mostly present with bloody nipple discharge, can be benign (intraductal papilloma), and are associated with high-risk lesions, or malignancy. Papillomas may obstruct ducts and secrete fluid, thereby forming cysts. Central papillomas originate from large ducts usually in the subareolar region with no TDLU involvement, whereas peripheral papillomas usually begin in the TDLU and extend into the ducts [6]. Risk of malignancy is higher in peripheral papillomatosis (see also Chap. 8).

Intracystic masses (Fig. 6.3) may also be due to intraductal proliferative lesions, some of which are classified as high-risk lesions. These pathologies include the following: atypical ductal hyperplasia, atypical lobular hyperplasia, and lobular carcinoma in situ (LCIS) [5] (see also Chap. 8). These are frequently managed with complete surgical excision after a US-guided biopsy because histopathologic assessment of the entire lesion may show associated malignancy. Rarely, a favorable-prognosis malignancy can present as a papillary lesion within a cyst. This is called intracystic papillary carcinoma or encapsulated papillary carcinoma(Fig. 6.3c). It makes up only 2% of breast carcinomas and occurs in older women, but it can also be found in younger women, thus affecting women from 34 to 92 years of age [6].

6.1.3.1 Concerns in Pregnancy and Lactation

Cystic lesions in the third trimester of pregnancy and in lactating women are usually caused by galactocele. Very rarely simple cysts and complex cysts may coexist in the older gravid patient. If complex cysts are detected during pregnancy or in lactating women, it should be followed by a US-guided biopsy; the patient should be treated similar to any non-gravid or non-lactating patient. Larger bore core biopsies using 12 G vacuum-assisted systems and surgical excision are better avoided in lactating women to prevent a milk fistula (see also Chap. 13).

6.1.4 Galactoceles

Galactoceles are benign lumps that can be detected during the last trimester of pregnancy and during lactation, especially after stopping breastfeeding. These lumps are painless and usually cystic in nature. Galactoceles are focal dilations of the ductal system resulting from distal duct obstruction of the TDLU [5]. They contain fluid with differing amounts of protein, fat, and lactose [2]. Initially, the content is liquid, thus appearing cystic and over time, as milk curdles, it may result in solid components [5]. On aspira-

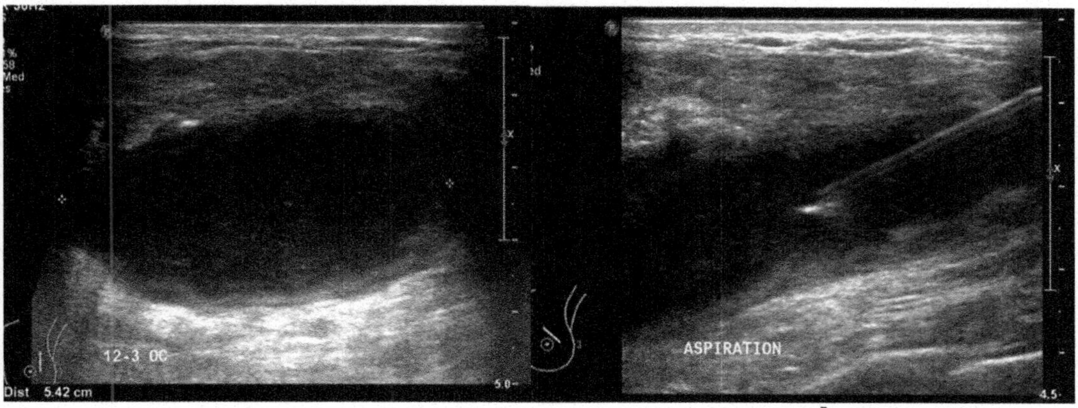

a b

Fig. 6.4 Galactocele (**a**) Well defined cystic mass, anechoic, with fat-fluid collection within it. (**b**) Needle in lesion, during aspiration of contents

tion, milky fluid or sticky material may be observed.

Any new palpable lump must be investigated with triple assessment including breast exam, imaging, and cytologic or histologic assessment when needed. US is an important part of the diagnosis workup and should be conducted to examine these lesions (see also Chap. 3). The fat-fluid composition of the content shows high echogenicity of lipid and low echogenicity of fluid that gives a characteristic speckled appearance, which in turn moves with movement or aspiration (Fig. 6.4).

Any masses within complex cysts will need US-guided biopsy; however, aspiration can be done to confirm the milky contents. Co-existence of galactocele and malignancy is extremely rare. The key is to follow up until the lesion resolves.

6.2 Solid Masses

6.2.1 Fibroadenomas

Fibroadenoma is the most common lesion in the breast that presents in adolescents and young women, but it can be found at any age [4]. It is asymptomatic in 25% of women [7]. Although most frequently unilateral, in 20% of cases, multiple lesions occur in the same breast or bilaterally [1]. They can cause cyclical mastalgia. Clinical examination may find a firm, highly mobile lump that is not attached to the surrounding breast tissue, which is classically known as the breast mouse.

Fibroadenoma consists of proliferation of the TDLU with both epithelial and stromal components [4]. The stroma proliferates around tubular glands (pericanalicular growth) or compressed cleft-like ducts (intracanalicular growth) [1]. The lesion is a hormone-dependent neoplasm that undergoes lactational changes during pregnancy and involutes along with the rest of the breast in perimenopause. Juvenile fibroadenomas that occur in adolescents are characterized by a more predominant stromal proliferation and resemble epithelial changes in gynecomastia; some refer to it as giant fibroadenoma when it is larger than 5 cm [1]. Complex fibroadenomas occur in older patients and have a slightly higher risk of cancer; they contain cysts, SA, epithelial calcification, or papillary apocrine changes [4]. Biopsy may sometimes mimic benign phyllodes tumor; hence, rapidly enlarging masses are best classified after excisional biopsy [1] (see also Chap. 4).

In US, fibroadenomas are oval in shape with smooth regular margins, homogeneous internal echoes, and posterior enhancement (Fig. 6.5) (see also Chap. 3). Atypical fibroadenomas can present with inhomogeneous internal echotexture and

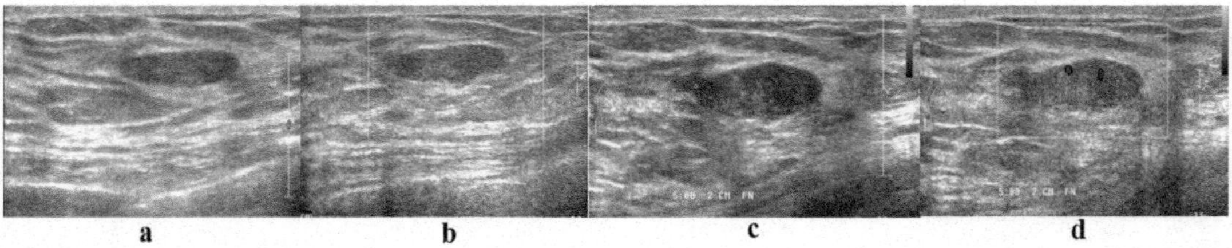

Fig. 6.5 Fibroadenoma (**a, b**) B- mode ultrasound: images showing well defined oval solid hypoechoic lesions with smooth margins. (**c, d**) With color Doppler signal showing absence and presence of internal vascularity

posterior shadowing. Unfortunately, triple-negative breast cancer, medullary, and mucinous cancer may appear the same [8–10]. Posterior shadows occur with associated macrocalcifications, which classically are referred to as popcorn macrocalcifications on a mammogram.

6.2.1.1 Concerns in Pregnancy and Lactation

Fibroadenomas can be classified into growing fibroadenoma, fibroadenoma with infarction, and fibroadenoma with lactational change in pregnancy or lactation [5]. It is the most frequent lesion found during this period. During pregnancy, hormone-sensitive fibroadenomas are stimulated by pregnancy hormones leading to secretory hyperplasia [7, 11]. Spontaneous necrosis of fibroadenomas is very rare, but it is sometimes observed during pregnancy or lactation. It can be presumed when there is a sudden pain during the third trimester or during birth; it is caused by embolism within the vessels [11, 12]. Imaging findings of fibroadenoma are not very different during pregnancy; however, findings such as dilation of lactiferous ducts and increase in blood flow are very similar to those of complex fibroadenoma; posterior shadows depend on degree of necrosis. When interpreting fine-needle aspiration for diagnosing palpable fibroadenoma, one must consider the physiological changes of cells during pregnancy [14] (see also Chaps. 1, 3, and 4).

Management is usually conservative after triple assessment. Excisional biopsy is usually not recommended in pregnant and lactating women to avoid hemorrhage, milk fistula, and infection, unless rapid increase in size occurs (see also Chap. 13). However, if there is discordance in the triple assessment, surgical excision is required.

6.2.2 Lactating Adenomas

Lactating adenoma is sometimes interpreted as a variant of fibroadenoma, tubular adenoma, or lobular hyperplasia, which are also caused by physiological changes [4, 5] (see also Chap. 4). Compared with fibroadenomas that have mixed stromal and epithelial components, lactating adenomas only consist of epithelial components [5, 11]. Lactating adenoma can naturally disappear at the end of pregnancy or lactation. It may also disappear by necrosis, similar to a fibroadenoma [5]. Approximately 5% of lactating adenomas are reported to undergo infarction [15].

The usual presentation is a painless, soft, mobile breast mass found in the third trimester of pregnancy or breastfeeding period. They may be multiple and bilateral, or may develop in ectopic breast tissue along the milk line from the axilla to the groin [13, 16].

However, acute infarction can cause significant enlargement and moderate to severe pain, thereby making it difficult to be differentiated from an infected galactocele, abscess, or a high-grade malignancy with or without cystic degeneration [13, 16].

It may be difficult to distinguish between a lactating adenoma and a fibroadenoma by imaging. A radiolucent or hyperechoic area, which indicates fat content of breast milk or lactation hyperplasia, can be seen on US (Fig. 6.6) [5]. Occasionally, the differential diagnosis of malignant lesions is difficult mainly because of unclear margins, microlobulation, and posterior acoustic shadowing and heterogeneity, especially with the necrosis of lactating adenomas [5] (see also Chap. 3). Moreover, radiological diagnosis is often challenging owing to lactational changes. Fine-needle aspiration cytology during

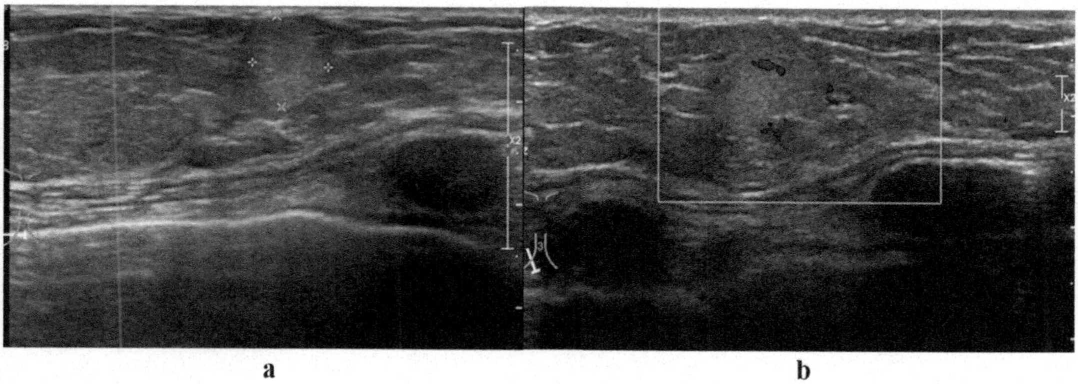

Fig. 6.6 Lactating adenoma. (**a, b**) Well defined oval hypoechoic lesion with smooth regular margins showing internal vascularity on color Doppler

pregnancy and lactation can be confused with malignant change [14]. Hence, a core biopsy would be useful. Management is usually conservative, and an excisional biopsy is only mandated if it is rapidly enlarging or if there is discordance in the triple assessment.

Lactating adenoma usually does not relapse after complete resection. Associated malignant lesions are very rare, and it has no increased risk of cancer [4].

6.3 Gestational Gigantomastia

Gestational gigantomastia or gravid macromastia is a rare condition that is characterized by rapid and diffuse hypertrophy of both breasts during pregnancy, and is rarely unilateral [17] (see also Chap. 5). Approximately about 100 cases have been reported in the literature [17]. It is usually reported in the second to fourth decade of life and in the second and third trimesters, and rarely after pregnancy. It is usually reported in the African and Asian populations. It complicates between 1:28,000 and 1:118,000 deliveries [19]. Proposed etiologies include hormonal and/or autoimmune mechanisms [20, 21] and placenta-breast axis mediated by parathyroid hormone–related protein (PTHrP), calcitriol, and human placental lactogen [22, 23], but it is mostly unknown.

Although benign, this condition causes severe psychological, emotional, and physical distress because of rapid and dramatic enlargement causing back pain and skin changes, such as peau d'orange and necrosis that may ulcerate. The weight of the breast and stretching of the skin cause great discomfort, and reports of significant weight gain up to 30 kg that affect the quality of life of the patient have been reported (Fig. 6.7) [19].

The features on imaging are non-specific. Ultrasound generally reveals edematous skin and non-specific heterogenous stromal changes without any mass lesion (Fig. 6.8).

A biopsy of the lesion is important to differentiate from other differential diagnoses that include phyllodes tumour, non-Hodgkin's lymphoma and lymphoblastic lymphoma [24]. Biopsy will show changes that are consistent with pregnancy, including extensive lobular hyperplasia, dilated ducts, and pseudoangiomatous hyperplasia. Some areas may also show interstitial edema and lymphoplasmocytes in the stroma and foci of increased fat and connective tissue [24].

Increased prolactin levels beyond pregnancy ranges [20] and associated raised PTHrP and hypercalcemia have been reported [22, 23] but are not diagnostic.

This condition is treated best with dopamine agonists [17–28]. Although bromocriptine has been more widely reported as treatment for gestational gigantomastia, cabergoline has also been mentioned in pregnancy albeit in smaller numbers. The literature shows that the risk of malformations with either medicine is not greater than

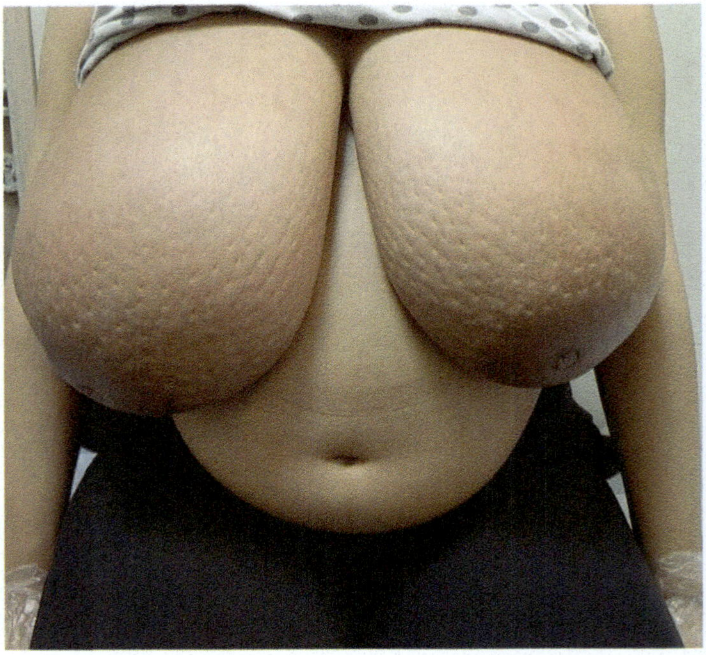

Fig. 6.7 Gestational gigantomastia in a 27 years old Iranian woman, 20 weeks of gestational age. (Courtesy of Dr. Sadaf Alipour)

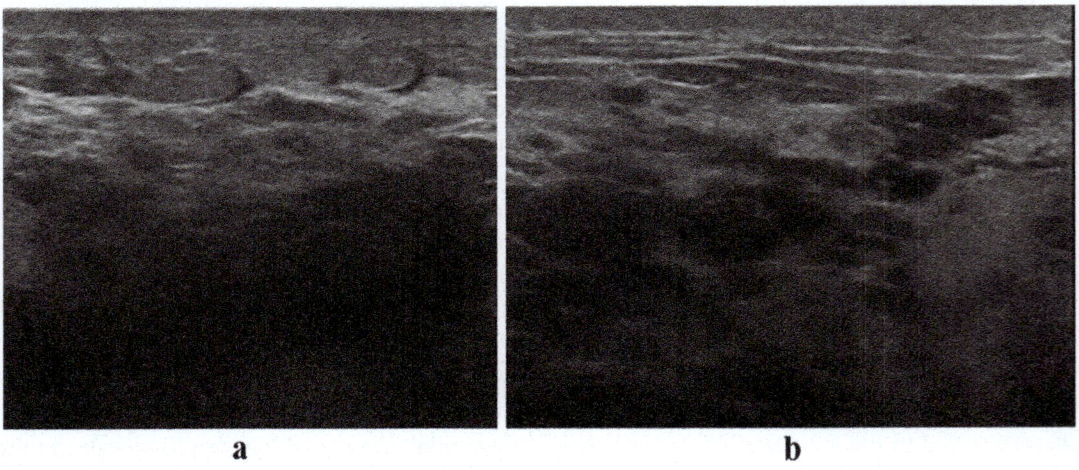

Fig. 6.8 (**a**, **b**) Gestational gigantomastia: ultrasound shows non-specific stromal changes

what was found in the general population [29]. The drugs would be continued to term and delivery of the baby. Termination of pregnancy is rarely required unless there are other medical indications or the gestation is near-term. Psychosocial aspects of this condition have been reported, for example, depression that will require holistic care.

Although spontaneous resolution of the breast enlargement occurs in most cases, secondary infection on the skin ulcers complicate matters [17–28], and mastectomy and reconstruction may be required in such cases. Hence, early diagnosis, before severe skin complications set in, becomes important.

The rapid growth may complicate anesthesia on account of reduced chest compliance and lung volumes. Hence, administration of regional anesthesia has been recommended for Cesarean deliveries [19] (see also Chap. 14). A multidisciplinary approach involving an endocrinologist, obstetrician, and breast surgeon would be required to manage the patient.

References

1. Guray M, Sahin AA (2006) Benign breast diseases: classification, diagnosis, and management. Oncologist 11(5):435–449
2. Vachhani PG, Shah A, Fabrega-Foster K, Harvey S (2017) Cysts with masses and masses with cysts: an imaging review of cystic breast masses. Appl Radiol 46(10):8–18
3. Dupont WD, Page DL (1985) Risk factors for breast cancer in women with proliferative breast disease. N Engl J Med 312(3):146–151
4. Lakhani SR, Ellis IO, Schnitt SJ, Tan PH, Van de Vijver MJ (2012) WHO classification of tumours of the breast. IARC, Lyon
5. Yu JH, Kim MJ, Cho H, Liu HJ, Han SJ, Ahn TG (2013) Breast diseases during pregnancy and lactation. Obstet Gynecol Sci 56(3):143–159
6. D'Orsi CJ, Sickles EA, Mendelson EB et al (2013) ACR BI-RADS Atlas: breast imaging reporting and data system. American College of Radiology, Reston
7. El-Wakeel H, Umpleby HC (2003) Systematic review of fibroadenoma as a risk factor for breast cancer. Breast 12(5):302–307
8. Yoon GY, Cha JH, Kim HH, Shin HJ, Chae EY, Choi WJ (2018) Sonographic features that can be used to differentiate between small triple-negative breast cancer and fibroadenoma. Ultrasonography 37(2):149–156
9. Souza P, Matsumoto R, de Barros N, Oliveira FA (2017) Imaging aspects of a rare invasive breast carcinoma: the medullary subtype. Eur Congress Radiol
10. Memis A, Ozdemir N, Parildar M, Ustun EE, Erhan Y (2000) Mucinous (colloid) breast cancer: mammographic and US features with histologic correlation. Eur J Radiol 35(1):39–43
11. Majmudar B, Rosales-Quintana S (1975) Infarction of breast fibroadenomas during pregnancy. JAMA 231(9):963–964
12. Singh SK, Gupta P, Arora R, Singla S, Mishra B, Singh K (2017) Giant fibroadenoma with massive infarction masquerading as malignancy: a case report. Int Cancer Conf J 6(4):167–170
13. Manipadam MT, Jacob A, Rajnikanth J (2010) Giant lactating adenoma of the breast. J Surg Case Rep 2010(9):8
14. Novotny DB, Maygarden SJ, Shermer RW, Frable WJ (1991) Fine needle aspiration of benign and malignant breast masses associated with pregnancy. Acta Cytol 35(6):676–686
15. Baker TP, Lenert JT, Parker J, Kemp B, Kushwaha A, Evans G et al (2001) Lactating adenoma: a diagnosis of exclusion. Breast J 7(5):354–357
16. Szabo J, Garcia D, Ciomek N, Margolies L (2017) Spuriously aggressive features of a lactating adenoma prompting repeated biopsies. Radiol Case Rep 12(2):215–218
17. Kuhn-Beck F, Foessel L, Bretz-Grenier MF, Akladios CY, Mathelin C (2014) Unilateral gigantomastia of pregnancy: case-report of a giant breast hamartoma. Gynecol Obstet Fertil 42(6):444–447
18. Rezai S, Nakagawa JT, Tedesco J, Chadee A, Gottimukkala S, Mercado R et al (2015) Gestational gigantomastia complicating pregnancy: A case report and review of the literature. Case Rep Obstet Gynecol 2015:892369
19. Gomez-Rios MA, Nieto-Serradilla L, Kuczkowski KM, Couceiro NE (2012) Gestational gigantomastia and anesthesia. Anesthesiology 116(1):193
20. Vinicki JP, Gonzalez CN, Dubinsky D, Nasswetter G, Cardinal LH, Hojman J (2015) Gestational gigantomastia in autoimmune diseases. J Clin Rheumatol 21(2):110–112
21. Mangla M, Chhatwal J, Nautiyal R, Prasad D (2019) Gestational gigantomastia in the setting of myasthenia gravis. J Obstet Gynaecol India 69(Suppl 1):84–87
22. Modarressi T, Levine MA, Khan AN (2019) Response to Letter to the Editor: Gestational gigantomastia complicated by PTHrP-mediated hypercalcemia. J Clin Endocrinol Metab 104(11):5100–5101
23. Modarressi T, Levine MA, Tchou J, Khan AN (2018) Gestational gigantomastia complicated by PTHrP-mediated hypercalcemia. J Clin Endocrinol Metab 103(9):3124–3130
24. Ezem BU, Osuagwu CC, Opara KA (2011) Gestational gigantomastia with complete resolution in a Nigerian woman. BMJ Case Rep 2011:bcr0120102632
25. Moazzami B, Chaichian S, Farahvash MR, Taheri S, Ahmadi SA, Mokhtari M et al (2016) A rare case of gestational gigantomastia with hypercalcemia: the challenges of management and follow up. J Reprod Infertil 17(4):243–246
26. Hayes MM, Konstantinova AM, Kacerovska D, Michal M, Kreuzberg B, Suvova B et al (2016) Bilateral gigantomastia, multiple synchronous nodular pseudoangiomatous stromal hyperplasia involving breast and bilateral axillary accessory breast tissue, and perianal mammary-type hamartoma of anogenital mammary-like glands: a case report. Am J Dermatopathol 38(5):374–383
27. Eler dos Reis P, Blunck Santos NQ, Barbosa Pagio FA, Chambo F, Chambo D, Chambo Filho A (2014) Management and follow-up of a case of gestational gigantomastia in a brazilian hospital. Case Rep Obstet Gynecol 2014:610363
28. Bukhari SS, Manan H, Khan MM, Raza SS (2018) Resolution of gestational gigantomastia with termination of pregnancy. J Ayub Med Coll Abbottabad 30(2):298–300
29. Molitch ME (2015) Endocrinology in pregnancy: management of the pregnant patient with a prolactinoma. Eur J Endocrinol 172(5):R205–R213

7 Mastitis, Breast Abscess, and Granulomatous Mastitis

Ramesh Omranipour and Mahtab Vasigh

Abstract

Breastfeeding is immunoprotective and World Health Organization recommends exclusive breastfeeding for about six months with continuation of breastfeeding for one year or longer as mutually desired by mother and infant. But the target for duration of exclusive breastfeeding has not been reached in a significant number of women. It may be due to inflammatory breast disease such as milk stasis or lactational mastitis.

In this chapter we discuss the most common complications of breastfeeding including milk stasis, mastitis, and breast abscess. Also idiopathic granulomatous mastitis, a less common condition, is discussed due to its confusing characteristics and not universally-accepted treatment strategies.

Breastfeeding mastitis is inflammation of the breast that can be infectious or non-infectious. With proper diagnosis and treatment of this condition, more severe complications like breast abscess could be avoided, so that breastfeeding could be continued in some circumstances.

Keywords

Breast abscess · Breastfeeding · Granulomatous mastitis · Mastitis · Pregnancy

7.1 Overview

The benefits of breastfeeding are well recognized, and the World Health Organization (WHO) recommends exclusive breastfeeding for the first 6 months after birth, continuing for up to one year and beyond [1]. Only 50% of women worldwide reach the 6-month period of breastfeeding [2]. It may be due to inflammatory breast disease such as milk stasis or puerperal mastitis in some instances, which necessitates adding supplements or completely ceasing breastfeeding [3].

R. Omranipour (✉)
Breast Disease Research Center (BDRC), Tehran University of Medical Sciences, Tehran, Iran

Department of Surgical Oncology, Cancer Institute, Tehran University of Medical Sciences, Tehran, Iran
e-mail: omranipour@tums.ac.ir

M. Vasigh
Breast Disease Research Center (BDRC), Tehran University of Medical Sciences, Tehran, Iran

S. Alipour, R. Omranipour (eds.), *Diseases of the Breast during Pregnancy and Lactation*, Advances in Experimental Medicine and Biology 1252,
https://doi.org/10.1007/978-3-030-41596-9_7

7.2 Mastitis

Mastitis is an inflammatory condition of the breast that is usually associated with lactation and is less common during pregnancy. It can progress from the non-infective stage to infective mastitis. According to variations in the definition and length of follow up in the postpartum period, the incidence of mastitis in lactating women is 3–20% with a wide variation among the studies. The time of occurrence is important in estimating the incidence. Breastfeeding-associated inflammatory breast diseases appear mostly during the first 12 weeks postpartum [4, 5].

7.2.1 Clinical Presentation

Mastitis during pregnancy and breastfeeding must be differentiated from severe engorgement, breast abscess, plugged duct, galactocele, and inflammatory breast cancer. The clinical characteristics of mastitis include tenderness, swelling, and a warm wedge-shaped area over the breast, associated with fever (>38.5 °C), fatigue and flu-like symptoms (see also Chap. 5). It may or may not be accompanied by an infection [5]. The presentation can be subtle with few clinical signs in the early stages (Fig. 7.1). A large area of breast swelling with overlying skin erythema can be recognized in patients with an advanced infection. Reactive lymphadenopathy may be associated with axillary pain and swelling. There are some predisposing factors including primiparity, obesity, smoking, maternal malnutrition, illness of mother or baby, poor positioning of the baby, cessation of breastfeeding, cracked and sore nipples [6–8].

7.2.2 Pathophysiology and Bacteriology

The primary cause is milk stasis [3, 9, 10]. The suckling of the infant causes erosions leading to painful nipple and areola. The pain leads to incomplete emptying of the breast and milk stasis in the mammary alveoli. Intraductal pressure rises owing to milk stasis, and milk penetrates into the connective tissue. This penetration opens intercellular junctions of the ductal epithelium into the connective tissue, creating a primary sterile inflammation. It is usually followed by a secondary bacterial infection [10]. The pathogenesis of inflammatory breast diseases seems to be associated with stress [3]. Hypothetically, a change in the intramammary cytokine profile (e.g. enhanced concentration of proinflammatory Th-1-cytokines) occurs because of an elevated amount of stress around birth that can lead to breast infections during the puerperal period. An increase in inflammatory cytokines is accompanied by a decrease in anti-inflammatory cytokines and local immunodeficiency [11].

Nasopharangeal organism from the newborn babies, such as *Staphylococcus aureus* and

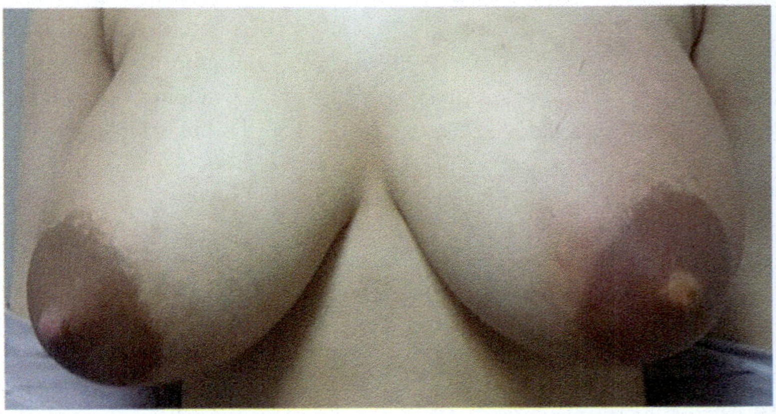

Fig. 7.1 Mastitis during pregnancy. Mild erythema and edema is seen. (Courtesy of Dr. Sadaf Alipour)

Streptococcus, may infect the breast via the damaged epithelial cells of the nipple-areola complex during breastfeeding. Milk stasis in itself would be a good culture medium and cause symptoms [12, 13].

7.2.3 Diagnosis

The diagnosis of mastitis is based on clinical manifestations, laboratory tests are not routinely needed [5, 7, 14, 15]. Although differential diagnosis of mastitis from milk stasis is possible by quantifying the leucocytes and pathogenic bacteria in the breast milk, in practice infectious mastitis is usually treated without this test if clinical symptoms of the patient do not improve after 24 hours of conservative management. However, a culture of the breast milk is useful to guide the selection of antibiotics; this is particularly important in the setting of infection that is severe, hospital acquired, recurrent, or unresponsive to initial antibiotics [7, 16, 17]. Imaging is useful if lactational mastitis does not respond to supportive care and antibiotics. Ultrasound (US) exam can help in differentiating mastitis from breast abscess and also from lactational phlegmon which can occur in this inflammatory spectrum [18]. Owing to the thickening of the skin and fibrous tissue in mastitis, mammography should not be performed, unless there are suspicious malignant findings. In addition, it is rare to discover other severe abnormalities in mammography [12].

7.2.4 Treatment

Conservative management includes continued breastfeeding and draining the milk frequently [5, 19]. Other measures include supportive care, rest and adequate fluid intake, non-steroidal anti-inflammatory drug (NSAID) consumption, warm or cold compress, and analgesics. Antibiotics are recommended if symptoms have not improved [20] although a Cochrane systematic review found insufficient evidence, owing to a lack of studies, to confirm when to use antibiotics in the treatment of mastitis [21]. *S. aureus* infections result in severe clinical symptoms from the beginning, whereas infections caused by *Streptococci* are diffuse and cause local abscess only in the advanced stage of infection. Infectious mastitis can be treated efficiently using antibiotics, especially amoxicillin-clavulanate 875 mg taken twice daily orally for 10–14 days. Cephalexin or dicloxacillin 500 mg taken every 6 h, orally for 10–14 days, is an alternative empiric therapeutic regimen [7]. It is quite rare, but puerperal mastitis by methicillin-resistant *S aureus* (MRSA) can be very fatal [7].

In the setting of severe infection (e.g. hemodynamic instability or progressive erythema), empiric inpatient therapy with vancomycin (15–20 mg/kg per dose every 8–12 h, not to exceed 2 g per dose) should be initiated; then, it is tailored based on culture and sensitivity results [21].

7.3 Breast Abscess

The prevalence of breast abscess during pregnancy and lactation has been reported from 0.4% to 11% [8, 22]. It could be related to malnutrition, low immunity, diabetes mellitus, obesity, and poor hygiene of skin and overlying clothing. The nipple and skin are usually the primary sources of infection. The predisposing factors leading to breast abscess include overlying skin diseases, minor cracks, and various forms of trauma to a bulky breast [23].

A breast abscess is determined as a localized accumulation of infected fluid in the breast tissue. A hard, tender, and sometimes fluctuant mass with overlying erythema of the skin is the most common presentation of a breast abscess [24] (Fig. 7.2).

Breast abscesses are usually associated with lactation (puerperal) but can be non-puerperal [8]. Pregnancy over the age of 30 years, first pregnancies, gestational age ≥ 41 weeks, and mastitis are considered as risk factors for development of lactational breast abscesses [25] (see also Chap. 5).

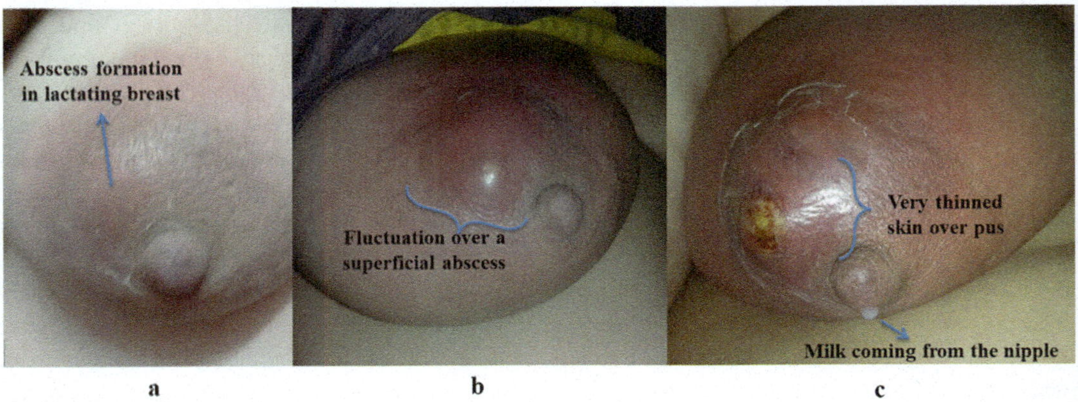

Fig. 7.2 Lactating breast abscess in a 40 years-old nursing woman 16 months after delivery (first two photographs taken by the patient). (**a**) Two days after beginning of symptoms. (**b**) After 3 days, fluctuation over the abscess. (**c**) After 4 days, impending rupture. The abscess was drained surgically. (Courtesy of Dr. Sadaf Alipour)

7.3.1 Bacteriology

S. aureus is the most common causative organism [1], other organisms like *Streptococcus* or *Escherichia coli* are less common [26]. MRSA has also been reported as a causative organism in several studies [26–28].

7.3.2 Diagnosis

If abscess formation is suspected, US is required for diagnosis and treatment (see also Chap. 3). Irregular boundaries, hypo-echoic or anechoic mass, thick irregular walls, posterior acoustic enhancement and liquid debris (fluid-debris) shades can be observed in the abscess. Sometimes, the air in the abscess can cause a bright reflection. The floating hyper-echoic dots help in the differentiation from malignancy [12]. When the patient is resistant to treatment and satisfactory recovery is not observed after 1 week of different therapeutic modalities, US-guided tissue sampling and blood tests for HIV should be considered [19].

7.3.3 Treatment

The goal of any of the interventions performed in treating an abscess is to remove the infected fluid as soon as possible. Thereby, pain and discomfort is reduced, allowing the woman to continue breastfeeding her infant with little or no interruption. Maintaining the integrity of the breast is also important, that is, the procedure should cause minimal or preferably no scarring and should preserve the function of the breast [28]. Recently, the treatment of lactational abscesses with single or serial needle aspiration has been favored in several studies. This is considered to be effective and less invasive [6, 29, 30]. But currently it is not clear whether needle aspiration is a more effective option to incision and drainage for treatment of breast abscess [6]. There are reports of insertion of a drain in the cavity of large abscesses after aspiration to inhibit early re-accumulation of pus. The results are promising, and this can replace open drainage with a lower rate of complications [31, 32]. Incision and drainage (I&D) is recommended if the abscess is subareolar, the skin over the abscess is thin and shiny, or the abscess appears as if it will burst [33]. I&D is also recommended when the abscess is large (>3 cm) or if there are multiple abscesses that fail to respond to aspiration [6, 12]. In order to allow continued breastfeeding, incision should not be made in the areola, and breastfeeding from the affected breast is recommended even if a drain is inserted in the abscess cavity. I&D will involve hospitalization and regular dressings. This is thought to cause considerable distress to both mother and baby during what is already a difficult time [17]. Daily washing out of the wound may be necessary until secretions decrease or become clear. Usually, the abscess cavity granulates and closes within four weeks [28].

Antibiotics are also recommended following either a needle aspiration or I&D [28]. Antibiotics of choice, such as dicloxacillin or flucloxacillin 500 mg 4 times daily orally or first generation cephalosporins may be prescribed. Erythromycin or clindamycin may be prescribed for women who are allergic to penicillin. In cases of MRSA, a breast milk culture and an assay of antibiotic sensitivities should be undertaken. MRSA strains are mostly sensitive to vancomycin or trimethoprim/sulfamethoxazole and less to rifampin. It should be presumed that regardless of susceptibility test results, MRSA is resistant to treatment with macrolides and quinolones [26].

Continuing breastfeeding helps improve inflammation and promote drainage, and can be safely performed during antibiotic therapy [13, 26]. If there is no improvement despite these treatments and if there is any suspicion of malignancy, a cytology test and biopsy should be performed [26] (Fig. 7.3).

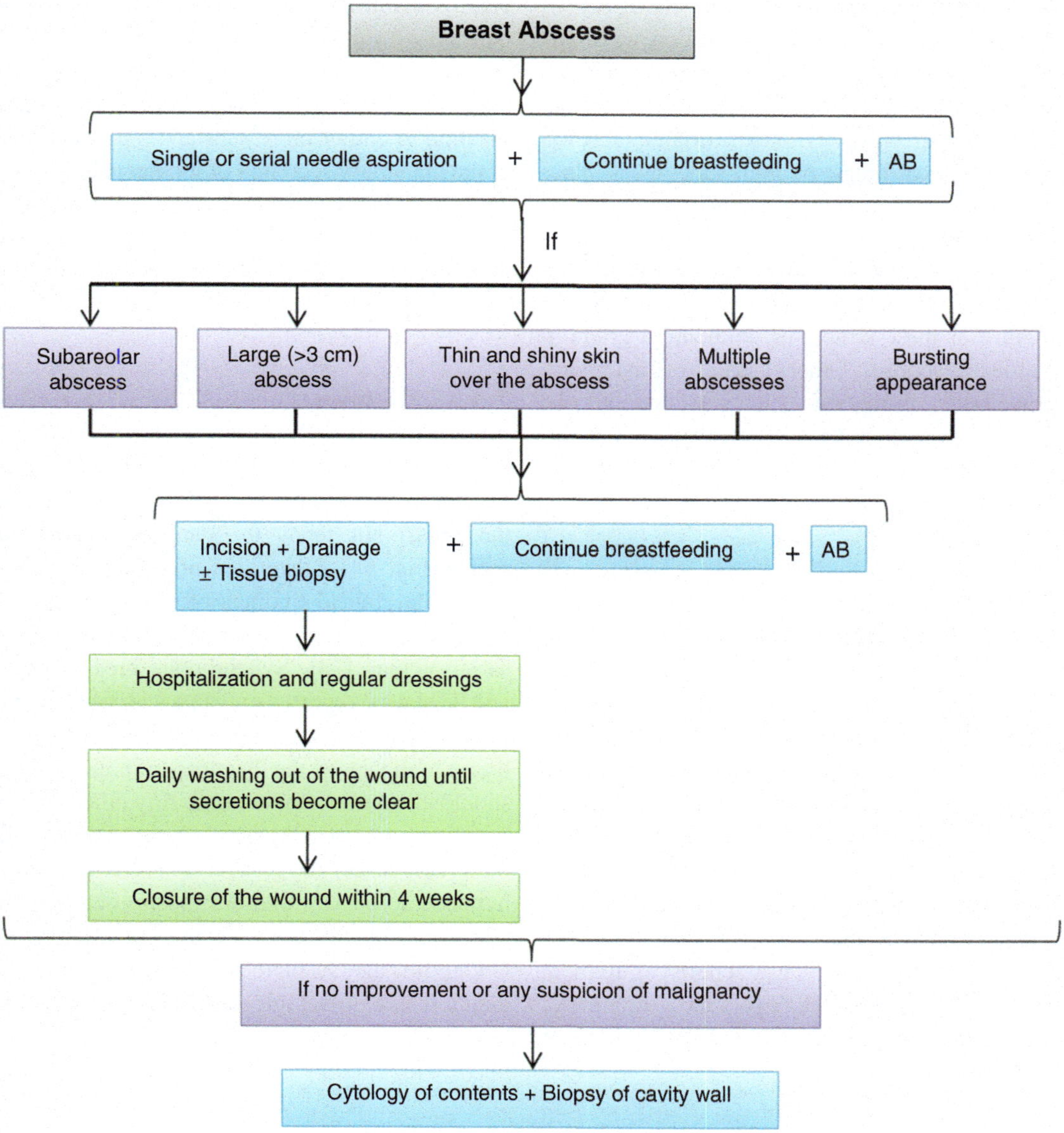

Fig. 7.3 Treatment of breast abscess during pregnancy or lactation. *AB* antibiotic therapy

Delayed, inappropriate, or even inadequate treatment of a breast abscess may result in permanent tissue damage, disfigurement, and more extensive lesions, which in about 10% of women could affect future lactation [6].

7.4 Idiopathic Granulomatous Mastitis

Numerous pathologic processes responsible for inflammation of the breast, can be included under the generic heading of granulomatous mastitis. Idiopathic granulomatous mastitis (IGM) is considered to be an idiopathic condition with currently no universally accepted treatment. Several etiologies have been considered for IGM. They include hypersensitivity to extravasated lactation products, local breast trauma, subclinical infection, autoimmune process, recent history of pregnancy, lactation, and use of oral contraceptives [34–38]. High levels of serum prolactin and distension of the acini and ducts may result in rupture of these structures, inducing a granulomatous response. Other causes of mammary granuloma formation such as sarcoidosis, Wegener granulomatosis, tuberculosis, and a fungal infection must be excluded before a confirmative diagnosis [39].

7.4.1 Concerns in Pregnancy and Lactation

IGM is unusual in pregnancy, although it usually occurs few years after lactation [40] (see also Chap. 5). In a retrospective study of 25 women diagnosed with IGM from Malaysia, 1 patient was 25 weeks pregnant and 5 patients were lactating at the time of presentation [41]. In addition, in a report of 24 patients from Jordan, 4 were pregnant at the time of presentation [42]. There were few other case reports of IGM at 11 weeks [43], 17 weeks [44], and 7 months of pregnancy [45]. The first two cases of IGM were treated with corticosteroids, and postpartum recurrences responded to steroid therapy, as well.

7.4.2 Pathogenesis

Kessler and Wolloch drew attention to the distinction between granulomatous and plasma cell mastitis. Stains and cultures for bacteria, acid-fast organisms, and fungi are typically negative [46]. Although the role of *Corynebacterium* species in the pathogenesis of IGM has not been clearly confirmed, remarkable supporting evidence has been reported by Taylor et al. [47]. The perilobular distribution and granulomatous character of the inflammation suggests a cell-mediated reaction to one or more substances concentrated in the mammary secretion of lobular cells, but no specific antigen has been identified [34]. The lesion usually appears after, rather than during pregnancy [48]. Coexistent erythema nodosum has been reported [49].

7.4.3 Diagnosis

Women with IGM typically present with a distinct, firm-to-hard mass that involves any part of the breast but tends to spare the subareolar region (Fig. 7.4). Bilateral involvement is uncommon. The clinical findings often suggest carcinoma, and mammography may also be described as suspicious [34]. Synchronous breast cancer and IGM were reported only in a few case reports. Although IGM is not the underlying cause of breast malignancy, the diagnosis of breast cancer should always be kept in mind. Any other lesions in the presence of IGM should be assessed to rule out breast cancer [50]. The lesions are frequently hypoechoic on US, and are mostly detected by this modality rather than by mammography. Owing to the suspicious inherent differential diagnoses, a histologic diagnosis with needle biopsy is necessary.

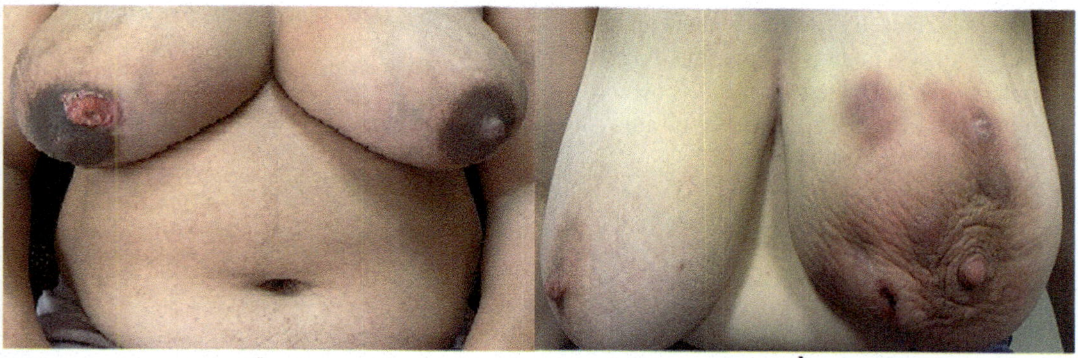

Fig. 7.4 Idiopathic Granulomatous Mastitis (**a**) In the right breast of a pregnant woman (26 weeks of gestational age); (**b**) In the left breast of a 31 years-old woman during lactation

Persistence or recurrence of the inflammatory process has been described after biopsy, which may lead to skin ulceration [51].

7.4.4 Treatment

There is no universally accepted treatment for IGM [34–37]. The most commonly used therapies include surgical excision of the granuloma, drainage of the wound, and concomitant steroid therapy [34–37]. Others suggest the use of antibiotics, wide surgical resection, mastectomy, and use of immunosuppressants [34–37]. Several studies have proved that recurrences after surgical excision are frequent, even after bilateral mastectomy [42, 53]. The rare coexistence of breast cancer with IGM supports the argument against operative management of this benign condition [55]. Currently, surgical management has become less prominent in treating IGM, and non-surgical recommendations have become more common including medications (oral or topical steroids, methotrexate, azathioprine, and anti-tuberculosis medications) or close observation [42, 53, 56–58]. There are significant reports in the literature indicating that treatment with methotrexate is effective, can prevent complications, and can limit adverse effects associated with corticosteroid use [53].

Corticosteroids have been effective in resolving the lesions after a specific infectious etiology has been ruled out [59]. In 1980, DeHertogh et al. were the first to investigate the efficacy of different treatments, and they concluded that corticosteroid is an appropriate option for the treatment of the disease [59]. Furthermore, Sakurai et al. sought this pharmaceutical approach, and they authenticated that corticosteroids could be efficient in 87% of patients without any relapse [37]. The results were validated by Su et al. while their research implied that low doses of corticosteroids were efficient [60]. Some authors concluded that steroid therapy is effective and resolution can be obtained without surgery [37]. Other researchers found that surgical excision and antibiotics should be the primary treatment modalities [38, 54].

There are reports demonstrating that clinical observation of the patient associated with education and reassurance can be an effective strategy to manage IGM with resolution after an average of about 7 months [55].

With such wide variations in treatment patterns, it is not surprising that published recurrence rates can approach 50% [51]. In fact, current treatment methods are considered suboptimal as all therapies can have significant adverse effects [34–36, 54].

Despite the relatively high incidence of the disease in Iran, the authors have only had the experience of 4 cases of IGM during pregnancy in around 25 years of surgical practice. All were controlled with low-dose prednisolone (15 mg per day during pregnancy and postpartum periods). Of these, 2 stopped breastfeeding because of IGM by themselves, whereas the remaining 2 had more than 6 months of breastfeeding. Their symptoms and lesions are controlled by corticosteroid therapy and NSAIDs, and they are visited every 6 months for follow-up.

References

1. World Health Organization (2003) Nutrition. www.who.int/nutrition/topics/exclusive_breastfeeding/en (accessed 2019) 2012 July 29
2. Gartner LM, Black LS (1997) Breastfeeding and the use of human milk. Pediatrics 100(6):1035–1039
3. Wockel A, Beggel A, Gensch M, Abou-Dakn M (2007) Psychological stress and puerperal mastitis-possible pathophysiological mechanisms. Curr Wom Health Rev 3(2):123–127
4. Kinlay JR, O'Connell DL, Kinlay S (1998) Incidence of mastitis in breastfeeding women during the six months after delivery: a prospective cohort study. Med J Australia 169(6):310–312
5. Amir LH, Committee AoBMP (2014) ABM clinical protocol# 4: mastitis, revised March 2014. Breastfeed Med 9(5):239–243
6. Irusen H, Rohwer AC, Steyn DW, Young T (2015) Treatments for breast abscesses in breastfeeding women. Cochrane Database Syst Rev 8:CD010490
7. Spencer JP (2008) Management of mastitis in breastfeeding women. Am Fam Physician 78(6):727–731
8. Kataria K, Srivastava A, Dhar A (2013) Management of lactational mastitis and breast abscesses: review of current knowledge and practice. Indian J Surg 75(6):430–435
9. Scott-Conner CE, Schorr SJ (1995) The diagnosis and management of breast problems during pregnancy and lactation. Am J Surg 170(4):401–405
10. Wöckel A, Abou-Dakn M, Beggel A, Arck P (2008) Inflammatory breast diseases during lactation: health effects on the newborn-a literature review. Mediators Inflamm 2008:298760
11. Wöckel A, Beggel A, Rücke M, Abou-Dakn M, Arck P (2010, Jan) Predictors of inflammatory breast diseases during lactation–results of a cohort study. Am J Reprod Immunol 63(1):28–37
12. Yu JH, Kim MJ, Cho H, Liu HJ, Han S-J, Ahn T-G (2013) Breast diseases during pregnancy and lactation. Obstet Gynecol Sci 56(3):143–159
13. Scott-Conner C (1997) Diagnosing and managing breast disease during pregnancy and lactation. Medscape Womens Health 2(5):1
14. Wambach KA (2003) Lactation mastitis: a descriptive study of the experience. J Hum Lact 19(1): 24–34
15. Osterman KL, Rahm V-A (2000) Lactation mastitis: bacterial cultivation of breast milk, symptoms, treatment, and outcome. J Hum Lact 16(4):297–302
16. Reddy P, Qi C, Zembower T, Noskin GA, Bolon M (2007) Postpartum mastitis and community-acquired methicillin-resistant Staphylococcus aureus. Emerg Infect Dis 13(2):298–301
17. World Health Organization (2000) Mastitis: causes and management. World Health Organization, Geneva. Publication Number. WHO/FCH/CAH/00.13
18. Johnson HM, Mitchell KB (2020) Lactational phlegmon: a distinct clinical entity affecting breastfeeding women within the mastitis-abscess spectrum. Breast J 26(2):149–154
19. Marchant DJ (2002) Inflammation of the breast. Obstet Gynecol Clin 29(1):89–102
20. Lawrence RA, Lawrence RM (2010) Breastfeeding e-book: a guide for the medical professional: Elsevier Health Sciences
21. Jahanfar S, Ng CJ, Teng CL (2009) Antibiotics for mastitis in breastfeeding women. Cochrane Database Syst Rev 1:CD005458
22. Son EJ, Oh KK, Kim EK (2006) Pregnancy-associated breast disease: radiologic features and diagnostic dilemmas. Yonsei Med J 47(1):34–42
23. Sangri AM, Shaikh AG, Unar F (2017) Benign breast diseases in pregnancy. Pak J Surg 22(4):125–128
24. Barbosa-Cesnik C, Schwartz K, Foxman B (2003) Lactation mastitis. JAMA 289(13):1609–1612
25. Kvist LJ, Rydhstroem H (2005) Factors related to breast abscess after delivery: a population-based study. BJOG 112(8):1070–1074
26. Stafford I, Hernandez J, Laibl V, Sheffield J, Roberts S, Wendel G (2008) Community-acquired methicillin-resistant Staphylococcus aureus among patients with puerperal mastitis requiring hospitalization. Obstet Gynecol 112(3):533–537
27. Branch-Elliman W, Golen TH, Gold HS, Yassa DS, Baldini LM, Wright SB (2011) Risk factors for Staphylococcus aureus postpartum breast abscess. Clin Infect Dis 54(1):71–77
28. Abou-Dakn M, Richardt A, Schaefer-Graf U, Wöckel A (2010) Inflammatory breast diseases during lactation: milk stasis, puerperal mastitis, abscesses of the breast, and malignant tumors–current and evidence-based strategies for diagnosis and therapy. Breast Care 5(1):33–37
29. Dixon JM (1988) Repeated aspiration of breast abscesses in lactating women. Br Med J 297(6662):1517–1518
30. Schwarz RJ, Shrestha R (2001) Needle aspiration of breast abscesses. Am J Surg 182(2):117–119
31. Tewari M, Shukla H (2006) An effective method of drainage of puerperal breast abscess by percutaneous placement of suction drain. Indian J Surg 68(6):330–333
32. Kousar N, Durrani TA, Ghafoor T, Qazi W (2018) Large lactational breast abscess: primary closure with drain versus conventional incision and drainage. JSZMC 9(3):1439–1442
33. Dirbas F, Scott-Conner C (2011) Breast surgical techniques and interdisciplinary management. Springer Science & Business Media

34. Al-Khaffaf B, Knox F, Bundred NJ (2008) Idiopathic granulomatous mastitis: a 25-year experience. J Am Coll Surg 206(2):269–273
35. Ocal K, Dag A, Turkmenoglu O, Kara T, Seyit H, Konca K (2010) Granulomatous mastitis: clinical, pathological features, and management. Breast J 16(2):176–182
36. Omranipour R, Mohammadi S-F, Samimi P (2013) Idiopathic granulomatous lobular mastitis-report of 43 cases from Iran; introducing a preliminary clinical practice guideline. Breast Care 8(6):439–443
37. Sakurai K, Fujisaki S, Enomoto K, Amano S, Sugitani M (2011) Evaluation of follow-up strategies for corticosteroid therapy of idiopathic granulomatous mastitis. Surg Today 41(3):333–337
38. Yau FM, Macadam SA, Kuusk U, Nimmo M, Van Laeken N (2010) The surgical management of granulomatous mastitis. Ann Plast Surg 64(1):9–16
39. Seo HR, Na KY, Yim HE, Kim TH, Kang DK, Oh KK, Kang SY, An YS, Chun M, Kim W, Park RW (2012, Mar 1). Differential diagnosis in idiopathic granulomatous mastitis and tuberculous mastitis. J Breast Cancer 15(1):111–118
40. Kaviani A, Vasigh M, Omranipour R, Mahmoudzadeh H, Elahi A, Farivar L et al (2019) Idiopathic granulomatous mastitis: looking for the most effective therapy with the least side effects according to the severity of the disease in 374 patients in Iran. Breast J 25(4):672–677
41. Azlina AF, Ariza Z, Arni T, Hisham AN (2003) Chronic granulomatous mastitis: diagnostic and therapeutic considerations. World J Surg 27(5):515–518
42. Bani-Hani KE, Yaghan RJ, Matalka II, Shatnawi NJ (2004) Idiopathic granulomatous mastitis: time to avoid unnecessary mastectomies. Breast J 10(4):318–322
43. Garcia-Rodiguez JA, Pattullo A (2013) Idiopathic granulomatous mastitis: a mimicking disease in a pregnant woman: a case report. BMC Res Notes 6:95
44. Goldberg J, Baute L, Storey L, Park P (2000) Granulomatous mastitis in pregnancy. Obstet Gynecol 95(5 Pt 2):813–815
45. Poniecka AW, Krasuski P, Gal E, Lubin J, Howard L, Poppiti RJ (2001) Granulomatous inflammation of the breast in a pregnant woman. Acta Cytol 45(5):797–801
46. Kessler E, Wolloch Y (1972) Granulomatous mastitis: a lesion clinically simulating carcinoma. Am J Clin Pathol 58(6):642–646
47. Taylor GB, Paviour SD, Musaad S, Jones WO, Holland DJ (2003) A clinicopathological review of 34 cases of inflammatory breast disease showing an association between corynebacteria infection and granulomatous mastitis. Pathology 35(2):109–119
48. Going JJ, Anderson TJ, Wilkinson S, Chetty U (1987) Granulomatous lobular mastitis. J Clin Pathol 40(5):535–540
49. Lucas R, Gussman D, Polis RL, Rattigan MI, Matulewicz TJ (2014) Idiopathic granulomatous mastitis with erythema nodosum simulating breast abscess in pregnancy: a case report. Obstet Med 7(1):37–39
50. Kaviani A, Zand S, Karbaksh M, Ardalan FA (2017) Synchronous idiopathic granulomatosis mastitis and breast cancer: a case report and review of literature. Arch Breast Cancer 6:32–36
51. Aghajanzadeh M, Hassanzadeh R, Sefat SA, Alavi A, Hemmati H, Delshad MS, Alavi CE, Rimaz S, Geranmayeh S, Ashtiani MN, Habibzadeh SM (2015, Aug 1) Granulomatous mastitis: presentations, diagnosis, treatment and outcome in 206 patients from the north of Iran. Breast 24(4):456–460
52. Azizi A, Prasath V, Canner J, Gharib M, Sadat Fattahi A, Naser Forghani M, Sajjadi S, Farhadi E, Vasigh M, Kaviani A, Omranipour R (2020, Apr 5) Idiopathic granulomatous mastitis: management and predictors of recurrence in 474 patients. Breast J
53. Akbulut S, Arikanoglu Z, Senol A, Sogutcu N, Basbug M, Yeniaras E et al (2011) Is methotrexate an acceptable treatment in the management of idiopathic granulomatous mastitis? Arch Gynecol Obstet 284(5):1189–1195
54. Ma X, Min X, Yao C (2020) Different treatments for granulomatous lobular mastitis: a systematic review and meta-analysis. Breast Care 5(1):54–60
55. Bouton ME, Jayaram L, O'Neill PJ, Hsu CH, Komenaka IK (2015) Management of idiopathic granulomatous mastitis with observation. Am J Surg 210(2):258–262
56. Lai EC, Chan WC, Ma TK, Tang AP, Poon CS, Leong HT (2005) The role of conservative treatment in idiopathic granulomatous mastitis. Breast J 11(6):454–456
57. Joseph KA, Luu X, Mor A (2014) Granulomatous mastitis: a New York public hospital experience. Ann Surg Oncol 21(13):4159–4163
58. DeHertogh DA, Rossof AH, Harris AA, Economou SG (1980) Prednisone management of granulomatous mastitis. NEJM 303(14):799–800
59. Su FH, Liu SC, Suen JH, Chen DS, Sister Mary Ann Lou (2005) Idiopathic granulomatous mastitis: a case successfully treated with a minimum dose of a steroid. Chang Gung Med J 28(6):431–435

Premalignant Disorders of the Breast in Pregnancy and Lactation

8

Ramesh Omranipour, Sadaf Alipour, Fereshteh Ensani, and Faina Nakhlis

Abstract

Papillomas, atypical hyperplasias, and lobular carcinoma in situ of the breast are not malignant tumors, but present serious management challenges when they are diagnosed in a breast biopsy. Upgrading after excision and increased possibility of future cancer are risks that accompany these lesions. While some features have been defined as high-risk for upgrading, many practitioners now recommend conservative non-surgical treatment and vacuum-assisted biopsy. However, the challenge gets worse when the patient is pregnant or breast-feeding because of the limitations in imaging and treatment in relation to the fetus. This chapter deals with these problems, although the best management strategy cannot be defined because of lack of evidence at present.

Keywords

Atypical hyperplasia · Breast cancer · Lactation · Lobular carcinoma in situ · Papilloma · Pregnancy · Premalignant lesion

8.1 Papilloma

8.1.1 Overview

Papillomas are benign proliferative lesions of the breast comprised of a fibrovascular core covered by inner myoepithelial and outer epithelial layers (see also Chap. 4). They usually grow in lactiferous ducts as solitary or multiple lesions. They may present as palpable masses, image-only detected lesions, and/or as nipple discharge.

Solitary papillomas are usually centrally located and can produce clear or bloody nipple

R. Omranipour
Breast Disease Research Center (BDRC), Tehran University of Medical Sciences, Tehran, Iran

Department of Surgical Oncology, Cancer Institute, Tehran University of Medical Sciences, Tehran, Iran
e-mail: omranipour@sina.tums.ac.ir

S. Alipour (✉)
Breast Disease Research Center (BDRC), Tehran University of Medical Sciences, Tehran, Iran

Department of Surgery, Arash Women's Hospital, Tehran University of Medical Sciences, Tehran, Iran
e-mail: salipour@tums.ac.ir

F. Ensani
Department of Pathology, Tehran University of Medical Sciences, Tehran, Iran

F. Nakhlis
Brigham and Women's Hospital, Department of Surgery, Harvard Medical School, Boston, MA, USA
e-mail: fnakhlis1@bwh.harvard.edu

S. Alipour, R. Omranipour (eds.), *Diseases of the Breast during Pregnancy and Lactation*, Advances in Experimental Medicine and Biology 1252,
https://doi.org/10.1007/978-3-030-41596-9_8

discharge. Multiple papillomas are typically peripherally located, less often cause nipple discharge, and may have a higher likelihood of being associated with adjacent premalignant and malignant lesions [1]. Papillomas may develop in women of all ages. Solitary papillomas are often seen in women of 30–50 years of age while multiple intraductal papillomas occur at a younger age [2].

8.1.2 Presentation and Diagnosis

Papillomas are commonly asymptomatic, but they can present as pathologic (bloody or clear, single duct) nipple discharge and, less commonly, as a mass. Bloody nipple discharge is more commonly observed in high risk papillary lesions rather than benign papillomas [3].

The ultrasound (US) presentation of papillomas consists of a solid intracystic lesion or a dilated duct with or without an associated intraductal mass, but it may sometimes present as a solid nodule with irregular margins [4] which is not distinguishable from other breast lesions. Color Doppler imaging may detect the flow from an afferent vascular feeding pedicle. In mammography, papillomas have no specific sign; but larger lesions may appear as round or oval masses or as nonspecific densities. Up to 25% of solitary papillomas are associated with benign-appearing calcifications [5], as can be the case with multiple peripheral papillomas [6]. Larger papillomas may appear as enhancing nodules on MRI, which may have a higher sensitivity in defining the number and extent of the papillary lesions than mammography and ultrasonography [7]. Ductography and ductoscopy are other assessment tools that are rarely used for evaluation of nipple discharge. Cytology smear can confirm the presence of erythrocytes and give information about exfoliated cells, which is rarely helpful, as a tissue diagnosis is still necessary.

The diagnosis is often made by image guided percutaneous needle biopsy, although core needle biopsy (CNB) may not be sufficient to distinguish benign and malignant papillary lesions, partly due to fragmentation of the tissues (see also Chap. 4).

8.1.3 Management

If the result of CNB shows atypia or suspected malignancy in the papillary lesion, surgical excision is the standard treatment. However, in benign papillary lesions without atypia, both options of follow-up by imaging and surgical excision have been recommended.

One of the main concerns with papillomas is the risk of an occult existing malignancy in the lesion. This happens when a lesion diagnosed as benign papilloma in CNB is upgraded to malignancy in the histological exam after surgical resection.

Overall, the rate of upgrading to carcinoma of a benign needle biopsy proven papillary lesion without atypia during surgical excision has been reported as high as 41% [8] and 30% [9], to as low as 2.3% [10], 2.12% [11], and 0.8% [12]. In a retrospective review of 97 intraductal papillomas detected in core biopsy between 2005 and 2013, rate of upgrade to cancer on excision was 21% for intraductal papillomas with atypia and 6% for intraductal papillomas without atypia [13]. In a meta-analysis by Wen and Cheng [14], in the thirty-four studies including 2236 breast papillary lesions, the median percentage of underestimation of malignant papillary lesions at CNB was 15.7% (95% CI:12.8–18.5%). Seely et al. [15] compared the upgrade rates to malignancy and atypia in two types of percutaneous biopsy techniques: vacuum-assisted biopsy (VAB) and 14-gauge CNB. This study showed that the upgrade rate to malignancy was 5.5 times higher with a 14 G CNB than with 10–12 G VAB, attributing this difference to the lesion being under-sampled with a smaller biopsy device, retrieving smaller core samples. In a similar study [16] a non-vacuum-assisted CNB approach was associated with a much higher upgrade rate than that of VAB (10.2% versus 0%).

It has also been shown that when an initial diagnosis of breast papilloma is made with VAB, the upgrade rate to malignancy tends to be low, and long-term imaging follow-up can safely replace surgical excision [11, 16–18]. Kuehner et al. [19] reported a cohort study of benign papillary lesions diagnosed by image-guided CNB between 2012 and 2013. Among 407 patients in

Table 8.1 Probable risk factors for upgrading of biopsy-detected papilloma to malignancy in surgery

Patient and lesion characteristics	Advanced patient age [21], or age > 50 years [19][a]
	Presence of bloody nipple discharge [20]
	Palpable mass [19, 20]
	Lesion size > 15 [20], or > 10 [19], or > 5 [21] mm[a]
	Peripheral location [20], or located > 5 cm from nipple [19]
Imaging features	Presence of mammographic calcifications [21][a]
	BI-RADS > 4b in mammography [20]
Biopsy details	Diagnosis by core needle biopsy[b] [15, 16]
	Presence of atypia in specimen [13]

a. In another study, age, lesion size and radiologic features were not associated with increased risk of upgrading [22]; b. Versus vacuum-assisted biopsy. *BI-RADS* Breast Imaging-Reporting and Data System

this study, 327 (80%) underwent surgical excision and 61 (15%) had imaging follow-up. The upgrade rate to cancer in the surgical excision group was low (3.4% were ductal carcinoma in situ, 2.4% were invasive cancers) and upgrades were more common among women older than 50, in those with palpable masses, lesions greater than 1 cm, and lesions located more than 5 cm from the nipple. No cancers were diagnosed in the imaging follow-up group. In a study by Ahn et al. [20] similar factors, such as peripheral location and palpability, were found to be associated with a higher risk of an adjacent carcinoma. Additionally, the study showed that the presence of bloody nipple discharge, lesion size on imaging greater than 15 mm and Breast Imaging-Reporting and Data System (BI-RADS) classification greater than 4b are associated with higher upgrade rates to malignancy. Similarly, in a study of 182 patients with papillary lesions who underwent surgical excision, upgrade rate was 12%, and advanced patient age, tumor larger than 0.5 cm and presence of mammographic calcification within the lesion were associated with higher upgrade rates [21]. However, in another study on 123 papillary lesions age, race, lesion size and radiologic features were not associated with an increased risk of upgrade to carcinoma on excision: the upgrade rate was 8.3% for papillary lesions and 12.5% for papillary lesions with hyperplasia [22]. Factors that might increase the rate of upgrading of a biopsy-detected papilloma to malignancy are shown in Table 8.1.

8.1.4 Concerns in Pregnancy and Lactation

Pregnancy produces changes within the lactiferous ducts that can cause bloody discharge by formation of delicate intraductal spurs which can be traumatized and shed into the lumen of a duct [23]. This usually happens when the vascularity is increased in the late second trimester and in the third trimester of pregnancy, and often ceases after delivery. Occult blood has been detected in up to 20% of pregnant women with non-bloody nipple discharge and in 15% of lactating women [24] (see also Chaps. 2, 4 and 5).

However, both solitary and multiple (peripherally-located) papillomas can occur in pregnant and lactating women, and can be the cause of the discharge. US remains the main modality for assessment of nipple discharge in pregnancy. Mammography is relatively safe during pregnancy and lactation but is often requested only if there is a high suspicion of malignancy. As well, although MRI is an acceptable and useful modality for assessing papillomas in non-pregnant women, routine use of gadolinium in the evaluation of pregnant patients is not appropriate (see also Chap. 3). The use of breast MRI in lactation also can be of limited value because breast parenchyma in a lactating patient may show rapid enhancement following contrast injection, followed by an early plateau of enhancement making interpretation of the findings difficult [8].

As with non-pregnant women, the best method of definitive diagnosis when an intraductal lesion is detected in the US breast exam of a pregnant or lactating woman with or without nipple discharge is percutaneous US-guided needle biopsy. As well, like non-pregnant cases, detection of atypical or malignant cells should lead to surgical excision as standard treatment despite higher risk of bleeding and milk fistula. When surgery is planned, excision via a subareolar incision is often feasible for most central papillomas. Surgery of non-palpable papillomas can be performed with guide of nipple discharge (steel probe duct cannulation from nipple surface, injection of dye or water into the duct) or wire localization under imaging techniques. In these circumstances, more peripheral placement of the incision might lower the incidence of milk fistula if the surgery is performed in late pregnancy or during breastfeeding (see also Chap. 13).

Although a number of studies have reported high upgrade rates [8, 13, 21], follow-up by imaging can be helpful in the management of benign intraductal papilloma in gravid and breastfeeding women, albeit data regarding the management of papillary lesions in these patients are scant.

Since pregnancy-associated breast cancer (PABC) can rarely present only as bloody nipple discharge, if the result of physical exam, breast ultrasonography (US) and cytology smear are normal in a pregnant woman with bloody nipple discharge, she can be followed for several months after delivery [24] (see also Chaps. 2 and 5).

Juvenile papillomatosis of the breast is a rare entity in children which occurs less commonly in young women. In a report of 18 patients with juvenile papillomatosis, five were diagnosed during pregnancy [8]. The treatment is surgical excision, and obtaining negative margins is recommended to prevent recurrence. If the mass is diagnosed in the third trimester of pregnancy, excision can be postponed until after delivery. In the first and second trimesters, it is safe to do complete excision after performing needle biopsy.

8.2 Atypical Hyperplasias and Lobular Carcinoma In-Situ

8.2.1 Overview

Atypical ductal hyperplasia (ADH) is recognized as a borderline lesion due to the overlap of its histologic characteristics with Usual Ductal Hyperplasia (UDH) and low-grade Ductal Carcinoma in Situ (DCIS), and harbors some but not all of the diagnostic criteria of the latter. It is composed of epithelial atypical cells within the ductal–lobular system of the breast [25, 26].

Lobular neoplasia (LN) is a term comprising both atypical lobular hyperplasia (ALH) and lobular carcinoma in situ (LCIS). LN is characterized by proliferation of atypical epithelial cells within the terminal duct lobular units (TDLUs); but the degree of TDLU involvement by neoplastic cells differs in ALH and LCIS [27, 28]. From the standpoint of cellular appearance, classical LCIS and ALH are essentially indistinguishable but there are some quantitative differences between the two entities: in classical LCIS, at least half of the acini in a lobule are filled and distended with uniform, small, loosely cohesive cells, and sometimes signet ring cells while the nucleoli are small or absent. However, if fewer than a half of the acini are involved, the lesion would be called ALH [27, 28]. LN is multifocal and multicentric in 50–85% of cases, and bilateral in 25–40% [27, 29–31].

8.2.2 Histologic and Molecular Characteristics

The cytologic features of ADH consist of evenly spaced, uniform small atypical cells with rounded nuclei and well-defined borders. The size and extent of the lesion is also important in the diagnosis of ADH, and it is defined as partial involvement of ducts or complete involvement of fewer than 2 duct profiles [25, 26]. The histologic differentiation of ADH and DCIS is not very objective, and Tavassoli has suggested a classification system using the term ductal intraepithelial neoplasia (DIN), ranging from DIN 1 to 3 based on nuclear grade; in this classification carcinoma is used for invasive tumors only [29, 30].

In LN, the E-cadherin stain is typically negative and this immunostain is often helpful in differentiating LN from DCIS, because the latter is expectedly characterized by E-cadherin positivity [27, 28]. The proliferative rate of the cells constituting ALH is low. These characteristically show strong expression of estrogen receptors (ER) and no expression of high-molecular-weight cytokeratins with CK5/6 immunostains12 [26].

Non-classical or variant forms of LCIS such as pleomorphic LCIS (pLCIS) and florid LCIS (fLCIS) harbor some additional features such as discohesion, central necrosis, calcifications, large nuclei, and prominent nucleoli [27, 28]. The term lobular intraepithelial neoplasia (LIN) was suggested by Tavassoli to introduce a three-level classification for LN, ranging from LIN1 to LIN3 based on the degree and extent of pathologic changes in cells and acini [28, 33].

8.2.3 Epidemiology

Since the establishment of mammographic screening, atypical hyperplasia (AH) has been diagnosed more frequently. It is seen in 0.5 to 17% of benign breast biopsy results depending on method of biopsy (surgical, vacuum-assisted, or core needle biopsy), means of radiologic assistance (no imaging assistance, ultrasound, mammography, or MRI guidance), type of biopsied lesion (from palpable mass to mammographic microcalcifications), [34–43] and definition of diagnostic criteria [42]. The true incidence of LN is unknown, because it is considered to be an incidental finding without any specific imaging or clinical features, largely associated with other lesions in the histologic specimen; and diagnosis may vary based on pathologic definition.

The mean age at diagnosis of ADH is 46 years [40]. LN is detected in all ages and has been said to be more common in pre- and perimenopausal women, [27, 29, 44] the most frequent age of detection being between 40 and 55 years [27, 30, 31, 40, 43, 44]. The incidence of LN has been reported to range from 0.1% for pure ALH to 0.5–7.8% for LCIS with or without ALH [27–31, 33, 40, 44]. Surveillance, Epidemiology and End Results (SEER) has reported a substantial increase in age-adjusted incidence rates of LN from 1978 to 1998 [29, 30, 43, 44]. This rise was mainly noted among women over 40 years old, particularly those between 50–59 years of age; and may be related to routine mammographic screening and an increasing access to image-guided biopsy technology, improved accuracy histologic interpretation, and widespread use of hormone replacement therapy in peri- and post-menopausal women at the reported time period [28, 30, 33].

8.2.4 Presentation and Diagnosis

ADH most commonly presents as mammographic microcalcifications, which are similar in appearance to those associated with DCIS or invasive breast cancers. It also seldom presents as a palpable or an image-detected mass [25, 40, 45–48] and it may rarely be seen on a core biopsy of mammographic distortion [45] or of a non-mass enhancement in MRI [48].

LN is an incidental finding detected on core or excisional biopsies of breast imaging abnormalities. The most common imaging findings are clustered mammographic microcalcifications, and these are more common in pLCIS [27–31, 33, 34, 44, 48]. LN may also present as a "shadowing, avascular, irregular, hypoechoic mass" on ultrasound; and a "heterogeneous non–mass-like enhancement with persistent enhancement kinetics" on MRI [31].

8.2.5 Clinical Course and Prognosis

There are actually two main concerns regarding AH and LCIS. The first issue is the synchronous presence of cancer in a biopsied lesion, which can be detected as an upgrade of the first diagnosis when surgical resection of the biopsied lesion is undertaken.

The rate of upgrade of ADH, diagnosed on CNB, to in situ or invasive cancer in further excision is 0–65%, [25, 26, 29, 34, 38, 49, 50] with a preponderance for DCIS [41, 46]. The risk of upgrade is highest when ADH has been detected in a lesion biopsied by a 14-gauge needle without radiologic assistance or under ultrasound guidance and lowest when a stereotactic VAB of a mammographically-detected lesion is performed with a 9- or 11- gauge needle [34]. In addition, the most common suggested prognosticators of upgrade are multiple ADH foci, presence of necrosis and marked atypia, high estimated percent of residual lesion in imaging, no mass lesions on mammography, initial radiological large size, and less well trained pathologist [40, 46, 51]. It is remarkable that ADH detected in MRI has a higher rate of upgrade to cancer than ADH diagnosed by mammographic sampling, may be because MRI is primarily used in higher risk women [32]. Factors that might be associated with an increased rate of upgrading of biopsy-detected ADH to malignancy are summarized in Table 8.2.

The rate of upgrade during histologic exam of the surgically excised samples of biopsy-proven LN has been from 9–33%, but this falls to 1–3% with radiological–pathologic concordance [34, 52].

The second concern about AH and LCIS is the future probability of cancer occurrence in a woman who has been diagnosed with ADH or LN.

Overall, AH carries a relative risk of around three– to four-fold increase in future cancer, and the risk is greater for women whose lesion is diagnosed at younger age [35, 37, 53–55]. The cancer may occur in the same or opposite breast, [53] and is more frequently invasive rather than in situ [55]. It is noteworthy that AH was not present in most of the benign breast biopsies that preceded a breast cancer [42].

ADH is considered both a risk factor for, and a potential non-obligate precursor of breast cancer; with a three- to five-fold increased risk in the ipsi-(50–75%) or contralateral (25–50%) side [25, 26, 32, 36, 38–41, 47, 48, 50], and no specific histology or grade [47].

Because cancers that follow ALH and LCIS are both ductal and lobular invasive carcinomas and are seen in both breasts, LN has been considered a risk factor for future cancer development. However, a greater risk of LCIS than the general population and a higher frequency of ipsilateral rather than contralateral cancer suggest the possibility of LN being also a non- obligate breast cancer precursor. The overall risk is higher for

Table 8.2 Probable risk factors for upgrading of biopsy-detected ADH to malignancy in surgery [32, 34, 40, 46, 51]

Imaging characteristics of the lesion	No mass on mammography
	Radiological large size
	Detected in MRI [a]
	Significant residue in imaging
Biopsy details [b]	By 14-gauge needle
	Without radiologic assistance
	Under US guidance
Histological assessment	Interpretation by less experienced pathologist
	Presence of multiple foci of ADH
	Presence of necrosis
	Presence of marked atypia

a. versus lesion detected in mammography; b. versus stereotactic vacuum-assisted biopsy with a 9- or 11- gauge needle. *ADH* Atypical Ductal Hyperplasia, *US* ultrasound

LCIS than ALH: eight- to ten-fold versus four- to five- fold [27, 29, 30, 33, 39, 40, 43, 44, 48, 53].

8.2.6 Management

Because of the significant upgrade rate, most clinicians recommend excision as standard treatment for a lesion diagnosed as ADH in biopsy [25, 26, 32, 35, 38, 41, 46–48, 50, 51, 56]. Observation without surgery has been proposed in selected cases of biopsy-proven ADH, including small lesions detected as mammographic microcalcifications and completely removed by VAB, micropapillary type, and few foci of ADH; however long-term safety has not yet been proved [32, 34, 48, 51, 56]. To address the elevated lifetime breast cancer risk associated with ADH, counseling about risk assessment and risk-reducing options should be provided. The Breast Cancer Prevention Trial [57] showed that tamoxifen lowered the risk of subsequent cancer in 86% of women harboring ADH; the problem is very low long-term adherence of women to the therapy [25, 26, 32, 38, 40, 41, 47, 50, 51]. Some patients choose to undergo bilateral risk-reducing mastectomies when diagnosed with ADH due to the increased future breast cancer risk, this approach may be considered excessive by most specialists [25, 41].

Ideal management of LN is still subject to controversy, fluctuating from only high-risk screening to radical surgery [29, 31]. Ipsilateral mastectomy and contralateral mirror-image breast biopsy, as well as bilateral mastectomy due to bilateral risk of breast cancer, have been carried out in the past for cases of LCIS [44]. LN detected in core needle biopsy in the presence of radiologic-pathologic concordance for lesions that are BI-RADS 4 or less does not require excision and the patient should only be counseled about her elevated future breast cancer risk [27, 28, 35, 43, 44, 48, 50, 52, 54]; although according to some clinicians, excisional biopsy is still a suitable option in these cases [31]. The consensus recommendations for LN diagnosed on needle biopsy in the "first International Consensus Conference on lesions of uncertain malignant potential in the breast (B3 lesions)" in January 2016 was to excise the lesion by VAB only, and then follow it by imaging for 5 years [56]. Surgery is more frequently done when an associated lesion in the biopsy specimen needs excision, there is pathological-mammographic discordance, ALH is diagnosed in biopsies taken under US or MRI guidance, necrosis is seen in classical LCIS, or LCIS constitutes a bulky mass, as well as in pLCIS [28, 43, 48, 50, 54].

Endocrine therapy for 5 years as chemoprevention is recommended in patients with LN [28, 31, 44, 48, 54] and can cause a more than 86% and 55% decrease in future breast cancer risk in ALH and LCIS, respectively [31, 50].

For women with genetic or additional risk factors or those who are excessively worried about plain observation or chemoprophylaxis, bilateral risk-reducing mastectomy with or without immediate reconstruction is an option. This procedure is associated with 95% breast cancer risk reduction, and the patient should be instructed about the remaining slight risk of breast cancer and the effect of the procedure on quality of life [30, 31, 44]. Variant LCIS, including pLCIS is often approached as DCIS from the management point of view; and is therefore treated by excision of the lesion with clear margins, followed by endocrine therapy if hormone receptors are positive [27, 28, 34, 44]. The role of adjuvant radiotherapy for non-classical LCIS is not well established.

Based on the above data, our recommendation for the management of core biopsy-proven cases of LN with radiological–pathologic concordance would be observation, with or without prophylactic endocrine therapy.

8.2.7 Follow Up

Women diagnosed with LCIS should be counseled about the lifetime risk of getting breast cancer, [44] and possible risk-reducing options. They are recommended to follow healthy dietary and lifestyle habits [54]. There is insufficient data to opt for or against breast cancer screening by MRI in women with ADH and LCIS. After diagnosis of these lesions and ruling out associated lesions, follow up screening by self- examination, clinical breast exam, and annual mammography should be undertaken [26, 30–32, 35, 41, 44, 48].

8.2.8 Concerns in Pregnancy and Lactation

The approach to AH and LCIS during pregnancy has not been discussed in the literature. One reason is that these lesions are most frequently diagnosed incidentally during biopsy taken from a mammographic lesion, and mammography is not done during pregnancy. Still, some lesions may be detected in a specimen biopsied under US guidance, or may be an associated finding during biopsy of a palpable lesion during gestation [31, 45].

In order to assess what surgeons would do in approaching this issue, we designed a multiple choice questionnaire which enclosed two sets of questions about the approach to ADH, ALH, or LCIS during each trimester of pregnancy or breastfeeding. The proposed treatment modules were observation, immediate VAB, immediate surgery, surgery in next trimester, and surgery after delivery or breastfeeding. The survey was sent out to 671 surgeons in 24 countries, including specialists from the International Network on Cancer, Infertility and Pregnancy (INCIP). Answers from the 101 responding surgeons showed that they preferred extraction of ADH by VAB during pregnancy and either by VAB or surgery during breastfeeding. Their approach to LN was mostly conservative, except for LCIS found in a mass lesion, where VAB for the first trimester, surgery during the second trimester or breastfeeding, and surgery after delivery for lesions discovered in the third trimester were more frequently chosen [58]. Our suggestion is to excise the lesions via immediate vacuum biopsy in this unusual circumstance.

8.2.9 Is Pregnancy Allowed in Women with ADH or LCIS?

The effects of pregnancy on AH or LCIS have not been specifically contemplated in the literature. This may however be a serious question for women in reproductive ages that have been diagnosed with one of these pathologies. The subject may be considered from the point of view of the influence of female sex hormones on the lesions, namely because of massive hormonal increases that occur during pregnancy.

Hormone replacement therapy (HRT) is very different from pregnancy, both from the perspective of type and amount of hormonal changes in the body, and the age of the patient. It is nonetheless a hormonal issue that has been pointed out in relation to AH and LCIS in the literature. It has been demonstrated that HRT had caused a rise in the incidence of AH when it was routinely prescribed, [59] and use of HRT in women having been diagnosed with AH is not recommended in first place. In addition, when these pathologies are detected in a woman already on HRT, discontinuation of the medicine should be considered; this rule also applies to oral contraceptive pills [40, 44, 54].

From its early descriptions, LCIS has been known as a disease of premenopausal women. Although confining the disease to that period has been challenged thereafter, it is still known to be more common in premenopausal age, probably due to higher levels of sex hormones. On the other hand, one of the possible etiologies for the increased incidence of LCIS between years 1978 and 1998 is common use of HRT by postmenopausal women [27, 33]. Also, it has been said that the risk of cancer following ADH and LN is higher in premenopausal patients; then again body hormonal milieu has been held responsible [48].

Given the uncertainty associated with the safety or not in patients with ADH and LN, there is still no strong basis to discourage pregnancy on these patients. The most reasonable suggestion seems to be guideline-concordant management of the lesions in question before a planned pregnancy, and careful observation of the pregnant and nursing mother by appropriate breast assessment during that period. Anti-estrogens taken after a diagnosis of AH or LCIS should be discontinued before planning for pregnancy because of the possible adverse effects on the child [25, 36, 40], whereas the prescription can be used again later.

References

1. Mulligan AM, O'Malley FP (2007) Papillary lesions of the breast: a review. Adv Anat Pathol 14(2):108–119
2. Al Sarakbi W, Worku D, Escobar PF, Mokbel K (2006) Breast papillomas: current management with a focus on a new diagnostic and therapeutic modality. Int Semin Surg Oncol 3(1):1
3. Wang LJ, Wu P, Li XX, Luo R, Wang DB, Guan WB (2018) Magnetic resonance imaging features for differentiating breast papilloma with high-risk or malignant lesions from benign papilloma: a retrospective study on 158 patients. World J Surg Oncol 16(1):234
4. Qian Y, Chang C, Zhang H (2019) Ultrasound imaging characteristics of breast lesions diagnosed during pregnancy and lactation. Breastfeed Med 14(10):712–717
5. Cardenosa G, Eklund GW (1991) Benign papillary neoplasms of the breast: mammographic findings. Radiology 181(3):751–755
6. Dabbs DJ (2016) Breast pathology, 2nd edn. Elsevier Health Sciences, Elsevier, Philadelphia
7. Eiada R, Chong J, Kulkarni S, Goldberg F, Muradali D (2012) Papillary lesion of the breast: MRI, ultrasound, and mammographic appearance. AJR Am J Roentgenol 198(2):264–271
8. Shiino S, Tsuda H, Yoshida M, Jimbo K, Asaga S, Hojo T et al (2015) Intraductal papillomas on core biopsy can be upgraded to malignancy on subsequent excisional biopsy regardless of the presence of atypical features. Pathol Int 65(6):293–300
9. Rasmussen BB, Balslev E, Christensen IJ, Lanng C, Bak A, Galatius H et al (2018) Diagnostic challenges in clinical, radiological and histopathological tests regarding papillomatous lesions of the breast. Breast 40:177–180
10. Pareja F, Corben AD, Brennan SB, Murray MP, Bowser ZL, Jakate K et al (2016) Breast intraductal papillomas without atypia in radiologic-pathologic concordant core-needle biopsies: rate of upgrade to carcinoma at excision. Cancer 122(18):2819–2827
11. Fuentes JA, Martínez CE, Casadiego AK, Freites VF, Marín VA, Castellano AC (2019) Papillary breast lesions diagnosed by percutaneous needle biopsy: management approach. Ecancermedicalscience 13
12. Han SH, Kim M, Chung YR, Yun BL, Jang M, Kim SM et al (2018) Benign intraductal papilloma without atypia on core needle biopsy has a low rate of upgrading to malignancy after excision. J Breast Cancer 21(1):80–86
13. Nakhlis F, Ahmadiyeh N, Lester S, Raza S, Lotfi P, Golshan M (2015) Papilloma on core biopsy: excision vs. observation. Ann Surg Oncol 22(5):1479–1482
14. Wen X, Cheng W (2013) Non-malignant breast papillary lesions at core-needle biopsy: a meta-analysis of underestimation and influencing factors. Ann Surgical Oncol 20(1):94–101
15. Seely JM, Verma R, Kielar A, Smyth KR, Hack K, Taljaard M et al (2017) Benign Papillomas of the breast diagnosed on large-gauge vacuum biopsy compared with 14 gauge Core needle biopsy–do they require surgical excision? Breast J 23(2):146–153
16. Kim MJ, Kim SI, Youk JH, Moon HJ, Kwak JY, Park BW, Kim EK (2011) The diagnosis of non-malignant papillary lesions of the breast: comparison of ultrasound-guided automated gun biopsy and vacuum-assisted removal. Clin Radiol 66(6):530–535
17. Yamaguchi R, Tanaka M, Tse GM, Yamaguchi M, Terasaki H, Hirai Y et al (2015) Management of breast papillary lesions diagnosed in ultrasound-guided vacuum-assisted and core needle biopsies. Histopathology 66(4):565–576
18. Hawley JR, Lawther H, Erdal BS, Yildiz VO, Carkaci S (2015) Outcomes of benign breast papillomas diagnosed at image-guided vacuum-assisted core needle biopsy. Clin Imaging 39(4):576–581
19. Kuehner G, Darbinian J, Habel L, Axelsson K, Butler S, Chang S et al (2019) Benign papillary breast mass lesions: favorable outcomes with surgical excision or imaging surveillance. ANN Surg Oncol 26(6):1695–1703
20. Kyung Ahn S, Han W, Moon HG, Kim MK, Noh DY, Jung BW et al (2018) Management of benign papilloma without atypia diagnosed at ultrasound-guided core needle biopsy: scoring system for predicting malignancy. Eur J Surg Oncol 44(1):53–58
21. Kupsik M, Perez C, Bargaje A (2019) Upstaging papillary lesions to carcinoma on surgical excision is not impacted by patient race. Breast Dis 38(2):67–72
22. MacColl C, Salehi A, Parpia S, Hodgson N, Ramonas M, Williams P (2019) Benign breast papillary lesions diagnosed on core biopsy: upgrade rate and risk factors associated with malignancy on surgical excision. Virchows Arch:1–7
23. Olsen CG, Gordon JR (1990) Breast disorders in nursing mothers. Am Fam Physician 41(5):1509–1516
24. Sabate JM, Clotet M, Torrubia S, Gomez A, Guerrero R, de Las HP et al (2007) Radiologic evaluation of breast disorders related to pregnancy and lactation. Radiographics 27(suppl_1):S101–S124
25. Allison KH, Jensen KC (2016) Intraductal proliferations (DCIS, ADH, and UDH). In: A comprehensive guide to core needle biopsies of the breast. Springer, Switzerland, pp 337–376
26. Schnitt SJ, Collins LC (2018) Intraductal proliferative lesions: usual ductal hyperplasia, atypical ductal hyperplasia, and ductal carcinoma in situ. In: Biopsy interpretation of the breast, 3rd edn. Wolters Kluwer, pp 96–168
27. Schnitt SJ, Collins LC (2018) Lobular carcinoma in situ (LCIS) and atypical lobular hyperplasia. In: Biopsy interpretation of the breast, 3rd edn. Wolters Kluwer, pp 208–263
28. Hwang H, Sahoo S (2016) Atypical lobular hyperplasia and lobular carcinoma in situ. In: A comprehensive guide to core needle biopsies of the breast. Springer, Switzerland, pp 561–594
29. Veronesi U, Goldhirsch A, Veronesi P, Gentilini OD, Leonardi MC. Breast Cancer: innovations in research and management: springer; 2017

30. King TA, Reis-Filho JS (2014) Lobular carcinoma in situ: biology and management. In: Diseases of the Breast, 5th edn. Wolter Kluver, Philadelphia, pp 324–336
31. Hoda SA. Lobular carcinoma in situ and atypical lobular hyperplasia. In:. Rosen's breast pathology. 4th ed. Wolters Kluwer; 2015:797–854
32. Chinyama CN (2013) Atypical Ductal Hyperplasia (ADH). In: Benign breast diseases: radiology-pathology-risk assessment. Springer, pp 159–167
33. Dabbs DJ, Oesterreich S (2017) Lobular Neoplasia and invasive lobular carcinoma. In: Breast pathology, 2nd edn. Elsevier, Philadelphia, pp 436–470
34. Calhoun BC (2018) Core needle biopsy of the breast: an evaluation of contemporary data. Surg Path Clin 11(1):1–16
35. Hartmann LC, Degnim AC, Santen RJ, Dupont WD, Ghosh K (2015) Atypical hyperplasia of the breast-risk assessment and management options. N Engl J Med 372(1):78–89
36. Laura C, Collins SJS (2014) Pathology of benign breast disorders. In: Diseases of the breast, 5th edn. Wolters Kluwer, Philadelphia, pp 71–88
37. Vogel VG (2018) Primary prevention of breast cancer. In: The breast, 5th edn. Elsevier, Philadelphia, pp 219–236
38. Kader T, Hill P, Rakha EA, Campbell IG, Gorringe KL (2018) Atypical ductal hyperplasia: update on diagnosis, management, and molecular landscape. Breast Cancer Res 20(1):39
39. Simpson JF, Schnitt SJ, Visscher D (2012) Atypical ductal hyperplasia. In: Vijver MJ, Ellis IO (eds) WHO classification of tumours of the breast, 4th edn. International Agency for Research on Cancer (IARC), France, pp 88–89
40. Dion L, Racin A, Brousse S, Beltjens F, Cauchois A, Levêque J et al (2016) Atypical epithelial hyperplasia of the breast: state of the art. Expert Rev Anticancer Ther 16(9):943–953
41. Hoda SA (2015) Ductal hyperplasia: usual and atypical. In: Rosen's breast pathology, 4th edn. Elsevier, Philadelphia, pp 271–307
42. Hoda SA (2015) Precarcinomatous breast disease: epidemiologic, pathologic, and clinical considerations. In: Rosen's breast pathology, 4th edn. Wolters Kluwer, pp 309–330
43. Lakhani SR, Schnitt SJ, O'Malley F, Vijver MJ, Palacios PTSJ (2012) Lobular Neoplasia. In: WHO classification of tumours of the breast, 4th edn. International Agency for Research on Cancer, pp 77–80
44. Aydiner A, İğci A, Soran A (2015) Breast Disease: diagnosis and pathology. Springer, Switzerland
45. Mesurolle B, Perez JCH, Azzumea F, Lemercier E, Xie X, Aldis A et al (2014) Atypical ductal hyperplasia diagnosed at sonographically guided core needle biopsy: frequency, final surgical outcome, and factors associated with underestimation. AJR Am J Roentgenol 202(6):1389–1394
46. Pena A, Shah SS, Fazzio RT, Hoskin TL, Brahmbhatt RD, Hieken TJ et al (2017) Multivariate model to identify women at low risk of cancer upgrade after a core needle biopsy diagnosis of atypical ductal hyperplasia. Breast Cancer Res Treat 164(2):295–304
47. Collins LC (2018) Precursor lesions of the low-grade breast Neoplasia pathway. Surg Pathol Clin 11(1):177–197
48. Clauser P, Marino MA, Baltzer PA, Bazzocchi M, Zuiani C (2016) Management of atypical lobular hyperplasia, atypical ductal hyperplasia, and lobular carcinoma in situ. Expert Rev Anticancer Ther 16(3):335–346
49. Kwong A, Shek T (2018) Factors affecting the underdiagnosis of atypical ductal hyperplasia diagnosed by core needle biopsies–a 10-year retrospective study and review of the literature. Int J Surg 49:27–31
50. Sasaki J, Geletzke A, Kass RB, Klimberg VS, Copeland EM, Bland KI (2018) Etiology and management of benign breast disease. In: The breast, 5th edn. Elsevier, Philadelphia, pp 79–92
51. Menen RS, Ganesan N, Bevers T, Ying J, Coyne R, Lane D et al (2017) Long-term safety of observation in selected women following core biopsy diagnosis of atypical ductal hyperplasia. Ann Surg Oncol 24(1):70–76
52. Nakhlis F, Gilmore L, Gelman R, Bedrosian I, Ludwig K, Hwang ES et al (2016) Incidence of adjacent synchronous invasive carcinoma and/or ductal carcinoma in-situ in patients with lobular neoplasia on core biopsy: results from a prospective multi-institutional registry (TBCRC 020). Ann Surg Oncol 23(3):722–728
53. Calhoun BC, Grobmyer SR, Simpson JF (2018) Benign, high-risk, and premalignant lesions of the breast. The Breast. Fifth ed. Elsevier, In, pp 116–129
54. Orr B, Kelley JL III (2016) Benign breast diseases: evaluation and management. Clin Obstet Gynecol 59(4):710–726
55. Danforth DN (2018) Molecular profile of atypical hyperplasia of the breast. Breast Cancer Res Treat 167(1):9–29
56. Rageth CJ, Am O'Flynn E, Comstock C, Kurtz C, Kubik R, Madjar H et al (2016) First international consensus conference on lesions of uncertain malignant potential in the breast (B3 lesions). Breast Cancer Res Treat 159(2):203–213
57. Vogel VG (2001) Follow-up of the breast cancer prevention trial and the future of breast cancer prevention efforts. Clin Cancer Res 7(12):4413s–4418s
58. Alipour S, Omranipour R, Amant F, Eslami B (2020) Atypical lesions of the breast and lobular carcinoma in situ in pregnancy- Practice of surgeons. Eur J Breast Health 16(1):16–21
59. Gayet A, Esteve J, Seradour B, Piana L, Jacquemier J (2003) Does hormone replacement therapy increase the frequency of breast atypical hyperplasia in postmenopausal women? Results from the Bouches du Rhone district screening campaign. Eur J Cancer 39(12):1738–1745

Part III

Breast Cancer in Pregnancy and Lactation

Epidemiology of Pregnancy-Associated Breast Cancer

9

Anna L. V. Johansson and Hanne Stensheim

Abstract

Breast cancer diagnosed during pregnancy or lactation up to 1 year post-partum is often referred to as *pregnancy-associated breast cancer* (PABC), although the definition varies with length of post-partum period. The incidence rate has been reported to range from 17.5 to 39.9 per 100,000 births, but the rate is substantially lower during pregnancy (ranging from 3.0 to 7.7) than during the post-partum period (ranging from 13.8 to 32.2). The PABC incidence rate is increasing in many populations, and higher maternal age at birth is a likely explanation. Linkable population-based data on pregnancies and cancer are required to obtain reliable estimates of PABC incidence. In studies comparing outcomes in women with PABC to other young breast cancer patients, it is crucial to adjust for age, since the age distribution of PABC depends both on age at pregnancy and age at breast cancer. Large studies have shown similar prognosis for women with PABC compared to other young women with breast cancer, when accounting for differences in age, stage and other tumour characteristics.

Keywords

Age confounding · Exposure · Incidence · Pregnancy risk window · Prognosis

9.1 Definition of Pregnancy-Associated Breast Cancer Risk Windows

Pregnancy-associated breast cancer (PABC) is commonly defined as a breast cancer during pregnancy and up to 1 year postpartum. Some authors only include the pregnancy window, while others also include the second year post-partum or up to 5 or 10 years postpartum in the PABC risk window. The different pregnancy and postpartum risk windows reflect short- and long-term exposures and effects of pregnancy on breast cancer detection and management, e.g. mammographic density, masking, potential detection delays, diagnostic workup, treatment and survival. For most outcomes it is of relevance to separate effects in the different risk windows during pregnancy (first, second, third trimesters)

A. L. V. Johansson (✉)
Department of Medical Epidemiology and Biostatistics, Karolinska Institute, Stockholm, Sweden
e-mail: anna.johansson@ki.se

H. Stensheim
Cancer Registry of Norway, Oslo, Norway
e-mail: hanne.stensheim@kreftregisteret.no

S. Alipour, R. Omranipour (eds.), *Diseases of the Breast during Pregnancy and Lactation*, Advances in Experimental Medicine and Biology 1252,
https://doi.org/10.1007/978-3-030-41596-9_9

and after delivery (0–6 months, 6–12 months, etc.). The pregnancy window is of particular interest for long-term follow-up of children exposed to cancer treatment in utero.

From an epidemiological point of view, it is important to realize that *pregnancy-association* is an exposure, and that PABC is a subgroup of young patients diagnosed with breast cancer, namely those cases diagnosed while exposed to pregnancy and lactation.

9.2 Incidence of Pregnancy-Associated Breast Cancer

Breast cancer, together with malignant melanoma and cervical cancer, are the most common malignancies diagnosed in pregnant or recently pregnant women [1–8]. Around 4% of women with breast cancer under age 45 are diagnosed during pregnancy or within the first year postpartum [2, 4, 9].

The incidence rate of PABC has been reported from several population-based studies (Table 9.1). There is a large variation in incidence across different populations and calendar periods. Reported estimates of PABC incidence rates range from 17.5 to 39.9 per 100,000 births; rates range from 3.0 to 7.7 during pregnancy, and from 13.8 to 32.2 during the first year postpartum [1, 3–7].

Hence, the risk pattern before and after delivery is strikingly different, with a lower incidence during pregnancy and an increasing incidence after delivery. This risk pattern was assessed in detail by Andersson et al. [1] who reported the relative risks of PABC as: 1st trimester: 0.05 (95% CI 0.02–0.11), 2nd trimester: 0.26 (0.18–0.36), 3rd trimester: 0.72 (0.59–0.87), 0–6 months postpartum: 0.59 (0.51–0.69), 6–12 months postpartum: 1.12 (1.01–1.24) and 12–24 months postpartum: 1.10 (1.03–1.18), compared to an age- and year-matched control population. This risk pattern could reflect a biologically lower risk during pregnancy, diagnostic delays or a healthy mother effect (reverse causation) during pregnancy [10].

Several studies have shown increasing incidence rates of PABC over calendar time [2–6]. The increasing incidence can to a large extent be explained by increasing maternal age and the ongoing trend of postponement of childbearing to ages where breast cancer is more common [2, 4, 5]. Other than age, risk factors of PABC are largely unknown.

There are several challenges when estimating PABC incidence rates. First, the estimates depend on which denominator has been utilized for calculating the incidence rate. Most commonly, PABC incidence is expressed as number of PABC cases per 100,000 deliveries or births, but also pregnancies (including elective and spontaneous abortions), live births, and person-time at risk have been used as denominator. The denominator should ideally capture the population at risk of PABC, namely pregnant (or recently pregnant) women. The total number of pregnant women may be difficult to ascertain in a population even with a birth registry, due to high rates of spontaneous abortion in the first trimester. Hence, births (deliveries) are a more stable measure of the pregnant population. Number of deliveries is also a good estimate of number of women at risk in the postpartum period, and using the same denominator before and after delivery makes the rates comparable.

Second, the incidence estimates may differ between studies due to the inclusion or exclusion of abortions (spontaneous or induced) and stillbirths in both the numerator and denominator of the incidence rate. Such hampered case ascertainment would lead to underestimation of PABC rates during pregnancy, in particular in early trimesters. Eibye et al. [4] reported that 81% of patients diagnosed with PABC in first trimester underwent elective abortion. However, in recent years the use of therapeutic abortion is likely to have decreased as more aggressive treatments are given during pregnancy.

Third, to obtain unbiased estimates of PABC incidence, population-based individual level data on both pregnancies and breast cancer is required. In many countries, population data is available in birth registries and in cancer registries. However, in order to classify a breast cancer as PABC these two registry databases must be linkable on an individual level, which may not be administratively possible in all countries.

Table 9.1 Population-based studies estimating incidence of pregnancy-associated breast cancer

First author, year	Country	Years covered by study	PABC definition	Proportion of PABC among all BC	Total PABC N (Py + 1y PP) N (Py) N (1y PP)	PABC incidence (per 100,000) Rate (Py + 1yPP) Rate (Py) Rate (1y PP)	Denominator in incidence	Incidence trend over calendar time
Cottreau, 2018 [3]	USA, 5 states	2001–2013	Py + 1 y	Not reported	208 56 152	26.8 7.2[a] 19.6[a]	775,709 births	Increasing trend
Parazzini, 2017 [6]	Lombardy, Italy	2002–2011	Py + 1 y	Not reported	479 93 386	39.9 7.7 32.2	1,200,263 Py	Not reported for PABC
Andersson, 2015 [1]	Sweden	1963–2007	Py + 1 y	Not reported	1486 139 1347	32.4[b] 3.0[b] 29.4[b]	4,580,005 live births	–
Eibye, 2013 [4]	Denmark	1997–2006	Py + 1 y	491/10963 = 4.4%	426 91 335[c]	17.5 3.7 13.8[c]	Induced abortions and live births	Increasing trend
Lee, 2012 [5]	New South Wales, Australia	1994–2008	Py + 1 y	Not reported	377 95 282	28.8 7.3 21.5	1,309,501 maternities (births)	Increasing trend
Smith, 2003 [7]	California	1991–1999	Py + 1 y	Not reported	935 266 669	19.3 5.4 13.8	4,846,505 births	–

BC breast cancer, *N* number, *PABC* pregnancy-associated breast cancer, *PP* postpartum, *Py* pregnancy, *y* year

[a]For Cottreau et al.: We have calculated the incidence for pregnancy window taking number of PABC divided by births at risk: 208/775,709 = 26.8, 56/775,709 = 7.2 and 152/775,709 = 19.6

[b]For Andersson et al.: Study used pregnancy+ 2 years PABC window, we have re-calculated numbers for pregnancy+ 1 year using numbers given in the tables in the article. We have also calculated incidence using the reported number of live deliveries (4,508,005). E.g. total incidence is calculated as 1486/4,508,005 = 32.4/100,000

[c]For Eibye et al.: Numbers and incidence 1 year postpartum was not reported in the article, but we have made calculations by taking total minus pregnancy window (426–91 = 335, and 17.5–3.7 = 13.8)

9.3 Age-Specific Incidence of Preganancy-Associated Breast Cancer and Maternal Age

Similar to the overall breast cancer incidence, the age-specific incidence rates of PABC increases over a woman's reproductive period and is highest above age 40 [2–4, 6]. However, the PABC incidence also depends on the age at childbirth (i.e. age at exposure). In most populations, the mean age at childbirth is below 30 years. Because the age distribution of PABC is the overlap between the age distributions of pregnancy (exposure) and of breast cancer (outcome), the absolute numbers of PABC are therefore highest in ages 30–34 years [2]. In this age group, pregnancy and breast cancer is most likely to co-occur. So, although the PABC incidence rate increases with age, the absolute numbers of PABC continuously decrease to zero at menopause after which women are no longer exposed to pregnancy and PABC does not occur.

In studies of PABC, women with PABC are often compared to other premenopausal women with breast cancer, often denoted "non-PABC". Non-PABC is usually defined as a breast cancer diagnosed in a nulliparous woman or in women more than 1 year after the latest childbirth. Women with non-PABC are thus similar to "premenopausal breast cancer" and will have an increasing age-distribution, since breast cancer becomes more common at higher ages. Hence, since the age of PABC women is shifted towards younger ages, while the age of non-PABC women is shifted towards higher ages, age at diagnosis is a very strong confounder in comparisons between PABC and non-PABC. Any comparison between PABC and non-PABC must therefore be thoroughly adjusted for differences in age at diagnosis to avoid age confounding. This can either be achieved via fine matching between PABC cases and non-PABC controls (e.g. 1-year age categories are often required) or via adjustments in the statistical analysis. Residual age confounding is a problem that many studies of PABC may have overlooked when using too broad age categories in the adjustment.

9.4 Prognosis of Pregnancy-Associated Breast Cancer

Several studies of varying sizes have assessed prognosis following breast cancer during pregnancy and lactation with somewhat conflicting results [8, 9, 11–20]. Two meta-analyses including both hospital-based and population-based studies found a worse prognosis in women with PABC compared to non-PABC, but the association was stronger in the postpartum period and weaker in women diagnosed during pregnancy [12, 16]. Women with PABC are more often diagnosed with advanced stage tumors and hormone receptor negative disease [8, 9, 11]. After adjustments for tumor stage and biology, the survival is similar between PABC and non-PABC, indicating that the worse prognosis reported in some studies can to a large extent be explained by adverse tumor characteristics [8, 9, 11].

Differences between studies include study setting (hospital-based, population-based), country, ages at diagnosis, calendar periods, treatments, postpartum windows, length of follow-up and study size. Population-based studies of prognosis following PABC often lack detailed information on clinical factors and treatment data, while institution-based materials often include those variables at high quality, but are at risk of selection bias. The poorer survival in cases diagnosed postpartum may partly be explained by delayed diagnosis, suboptimal treatment and lack of adjustment for several important clinical factors (see also Chap. 11).

References

1. Andersson TM, Johansson AL, Fredriksson I, Lambe M (2015) Cancer during pregnancy and the postpartum period: a population-based study. Cancer 121(12):2072–2077
2. Andersson TM, Johansson AL, Hsieh CC, Cnattingius S, Lambe M (2009) Increasing incidence of pregnancy-associated breast cancer in Sweden. Obstet Gynecol 114(3):568–572
3. Cottreau CM, Dashevsky I, Andrade SE, Li DK, Nekhlyudov L, Raebel MA et al (2018) Pregnancy-associated cancer: a US population-based study. J Women's Health 28(2):250–257

4. Eibye S, Kjær SK, Mellemkjær L (2013) Incidence of pregnancy-associated cancer in Denmark, 1977–2006. Obstet Gynecol 122(3):608–617
5. Lee YY, Roberts CL, Dobbins T, Stavrou E, Black K, Morris J et al (2012) Incidence and outcomes of pregnancy-associated cancer in Australia, 1994–2008: a population-based linkage study. BJOG 119(13):1572–1582
6. Parazzini F, Franchi M, Tavani A, Negri E, Peccatori FA (2017) Frequency of pregnancy related cancer: a population based linkage study in Lombardy, Italy. Int J Gynecol Cancer 27(3):613–619
7. Smith LH, Danielsen B, Allen ME, Cress R (2003) Cancer associated with obstetric delivery: results of linkage with the California cancer registry. Am J Obstet Gynecol 189(4):1128–1135
8. Stensheim H, Møller B, van Dijk T, Fosså SD (2009) Cause-specific survival for women diagnosed with cancer during pregnancy or lactation: a registry-based cohort study. J Clin Oncol 27(1):45–51
9. Johansson AL, Andersson TM, Hsieh CC, Jirström K, Cnattingius S, Fredriksson I et al (2018) Tumor characteristics and prognosis in women with pregnancy-associated breast cancer. Int J Cancer 142(7):1343–1354
10. Sankila R, Heinävaara S, Hakulinen T (1994) Survival of breast cancer patients after subsequent term pregnancy: "healthy mother effect". Am J Obstet Gynecol 170(3):818–823
11. Amant F, von Minckwitz G, Han SN, Bontenbal M, Ring AE, Giermek J et al (2013) Prognosis of women with primary breast cancer diagnosed during pregnancy: results from an international collaborative study. J Clin Oncol 31(20):2532–2539
12. Azim HA Jr, Santoro L, Russell-Edu W, Pentheroudakis G, Pavlidis N, Peccatori FA (2012) Prognosis of pregnancy-associated breast cancer: a meta-analysis of 30 studies. Cancer Treat Rev 38(7):834–842
13. Beadle BM, Woodward WA, Middleton LP, Tereffe W, Strom EA, Litton JK et al (2009) The impact of pregnancy on breast cancer outcomes in women≤ 35 years. Cancer 115(6):1174–1184
14. Bonnier P, Romain S, Dilhuydy JM, Bonichon F, Julien JP, Charpin C et al (1997) Influence of pregnancy on the outcome of breast cancer: a case-control study. Int J Cancer 72(5):720–727
15. Daling JR, Malone KE, Doody DR, Anderson BO, Porter PL (2002) The relation of reproductive factors to mortality from breast cancer. Cancer Epidemiol Biomark Prev 11(3):235–241
16. Hartman EK, Eslick GD (2016) The prognosis of women diagnosed with breast cancer before, during and after pregnancy: a meta-analysis. Breast Cancer Res Treat 160(2):347–360
17. Johansson AL, Andersson TM, Hsieh CC, Cnattingius S, Lambe M (2011) Increased mortality in women with breast cancer detected during pregnancy and different periods postpartum. Cancer Epidemiol Biomark Prev 20(9):1865–1872
18. Moreira WB, Brandão EC, Soares AN, Lucena CE, Antunes CM (2010) Prognosis for patients diagnosed with pregnancy-associated breast cancer: a paired case-control study. Sao Paulo Med J 128(3):119–124
19. Rodriguez AO, Chew H, Cress R, Xing G, McElvy S, Danielsen B et al (2008) Evidence of poorer survival in pregnancy-associated breast cancer. Obstet Gynecol 112(1):71–78
20. Whiteman MK, Hillis SD, Curtis KM, McDonald JA, Wingo PA, Marchbanks PA (2004) Reproductive history and mortality after breast cancer diagnosis. Obstet Gynecol 104(1):146–154

Histology of Pregnancy-Associated Breast Cancer

10

Behnaz Jahanbin and Vahid Soleimani

Abstract

Breasts are one of the most common sites of neoplastic lesions in women during pregnancy and lactation. This chapter reviews carcinomas of the breast during pregnancy and lactation while focusing on histologic features, biomarker profiles and some involved molecular pathways. Also, a brief review of previous studies on this field is performed.

Keywords

Breast · Carcinoma · Hormone receptors · Pregnancy · Lactation

10.1 Overview

Breast carcinoma during pregnancy is not a common finding (from 1 in 10,000 to 1 in 3000 pregnancies) (see also Chap. 9), but owing to diagnostic problems and the biological differences in this setting, notable attention should be given not to miss the diagnosis [1–3].

The definition of pregnancy-associated breast cancer (PABC) is evolving; it is defined as breast cancer that is diagnosed during pregnancy, in the first postpartum year or at any time during lactation [3, 4]; but some authors extend this time up to 10 years after delivery. In addition, some authors have proposed that PABC should be viewed as 2 distinct subsets: (1) cases diagnosed during pregnancy and (2) cases diagnosed in the postpartum period [5].

The age distribution of PABC seems to be changing owing to delayed pregnancy and the probable increase in breast cancer rate [6–8] (see also Chap. 9).

The clinical presentation of PABC is not very different from non-pregnant women, but because of the attribution of symptoms to physiologic alterations related to pregnancy, the fear of cancer in patients, and physician's inclination to be reassuring, delayed diagnosis may occur [9, 10]. PABC is mostly characterized by a palpable mass, but nipple discharge, skin redness and thickening, ulcers, and nipple retraction are not exceptional symptoms [9] (see also Chap. 11).

It is known that pregnancy exerts a bidirectional effect on breast cancer development with a short-term increase in risk (up to 5 or 10 years after gestation) and a protective effect afterwards [11, 12].

B. Jahanbin (✉) · V. Soleimani
Department of Pathology, Cancer Institute, Imam Khomeini Hospital Complex, Tehran University of Medical Sciences, Tehran, Iran
e-mail: b-jahanbin@sina.tums.ac.ir

S. Alipour, R. Omranipour (eds.), *Diseases of the Breast during Pregnancy and Lactation*, Advances in Experimental Medicine and Biology 1252,
https://doi.org/10.1007/978-3-030-41596-9_10

10.2 Histologic Findings

Notable studies have discussed histologic findings, prognostic markers, and survival of PABC compared with non-PABC. There are no obvious differences in histologic types of breast cancer among PABC and non-PABC. Invasive ductal carcinoma, not otherwise specified (NOS), is the most prevalent type of tumor in PABC; representing 78–88% of cases [13, 14] (Fig. 10.1). Invasive lobular carcinoma is reported with a very low frequency in some studies [15, 16]. Mucinous, medullary, and other types of carcinoma are also reported.

Ductal carcinoma in PABC reveals a higher histologic grade, a more aggressive profile, and a more advanced stage at diagnosis: a larger tumor size, a higher frequency of nodal involvement, a less frequent expression of estrogen receptors (ERs) and progesterone receptors (PRs), and a higher proportion of inflammatory breast cancer [13–18] (see also Chaps. 11 and 20).

In a systematic review of 14 case-control, age-matched studies by Marikakis et al. [19], it was observed that the frequency of ER- and PR-negative breast cancers in PABC is significantly higher than non-PABC.

Bae et al. [8] in an assessment of 2810 patients with breast cancer diagnosed at an age of less than 40 years, including 2770 non-pregnant young patients with breast cancer and 40 PABC cases, found that PABC had significantly lower ER and PR positivity and higher HER2 overexpression. The most common subtype of tumors in

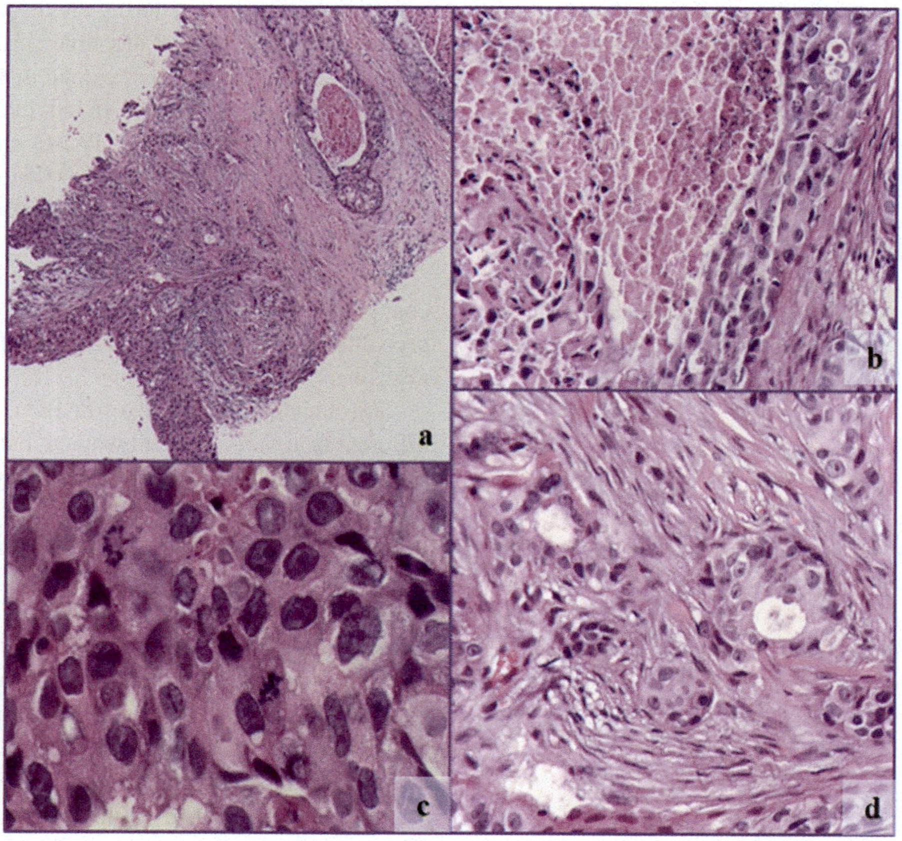

Fig. 10.1 Invasive ductal carcinoma, not otherwise specified (NOS) type in a 37-year-old woman, 7 months postpartum. (**a**) Predominantly solid nests of tumor cells with remarkable foci of necrosis. (**b**) High magnification view of tumor necrosis. (**c**) Presence of numerous mitotic figures including atypical forms. (**d**) Rarity of tubule formation

PABC was triple-negative, and luminal A was less prevalent. In univariate analysis, PABC had worse breast cancer–specific survival (BCSS) and disease-free survival (DFS) but in multivariate analysis, DFS showed no significant differences. In assessment of subtypes, luminal B subtype showed worse DFS and BCSS compared with young breast cancer, especially when Ki67 levels were higher; but no differences were seen in other subtypes.

In a retrospective matched case-control study, thirty-one PABC patients were analyzed by Madaras et al. [20], and pathologic features, including histopathology information on tumor type, grade, tumor size, lymph node involvement, lymphovascular invasion, Nottingham Prognostic Index, and associated in situ lesions and immunohistochemical characteristics (ER, PR, HER2, Ki67, and p53) were assessed and compared with the control group. It was found that the histopathologic features of the tumor were not significantly different between groups, but the incidence of high-grade extensive ductal carcinoma in situ was higher among patients with PABC. HER2 overexpression revealed no difference between PABC and the control group. Proliferative activity (Ki67) was above 15% in all PABC, and luminal B and triple-negative cases were the most common subtypes in PABC. Finally, this study demonstrated the inferior outcome of PABC, especially for those detected postpartum, in comparison with the control young non-PABC group.

In a retrospective chart review study done by Genin et al. [21], 41 cases of PABC were compared with 235 breast cancer patients younger than 45 years. They found that PABC was locally more advanced with a greater size but did not have a higher frequency of lymph node invasion. In comparison of prognostic profiles, PABC was associated with HER2 overexpression and hormone receptor negativity that were twice more frequent; triple-negative tumors were also more frequent in PABC. This poorer phenotype was largely related to postpartum tumors rather than to tumors during pregnancy.

In a case-control study by Azim et al. [22], 65 PABC patients (diagnosis of breast cancer during pregnancy) were compared with 130 age-matched controls and it was revealed that there was no difference in tumor features and subtypes between the PABC and control groups. Most tumors were ductal carcinoma, NOS, and poorly differentiated. A common subtype was HER2-negative luminal B tumors.

Relevant data of the above and several other pertinent studies are summarized in Table 10.1.

It is worth noticing that in multiple studies, postpartum breast cancers (PPBCs) have been evaluated as a unique entity, separate from breast cancers diagnosed during pregnancy (PBC).

Boudy et al. [23], in a retrospective study of 108 PABC cases, compared 51 patients with breast cancer during pregnancy with 57 patients diagnosed in the postpartum period. In this study, medium size, initial axillary pathology, histological type, and hormone receptors were similar between PBC and PPBC. HER2 overexpression was less frequent, and proliferative activity (Ki67) was lower in PBC than PPBC.

Another meta-analysis of 30 studies carried out by Azim et al. [24] not only confirmed that PABC was associated with poor prognosis, poor overall survival, and higher risk of recurrence even after adjustment for confounding factors; but also revealed that these findings were particularly obvious in patients diagnosed in the 1-year postpartum period rather than in PBC (see also Chap. 11).

In a meta-analysis of 41 studies by Hartman and Eslick [25], 4929 cases of PABC were compared with 61,041 patients of the control group, reporting an increased risk of death among patients with PABC compared with the non-pregnant control group. PPBC had the poorest overall survival in comparison with PBC and the control group. In addition, PPBC cases were the most at risk of disease progression or relapse. Overall, according to the death rate and DFS, the highest risk periods were during pregnancy and up to 1 year postpartum; however, an increased risk was evident up to 5 years postpartum.

In a multicenter cohort study performed by Goddard et al, 701 breast cancer patients aged 45 years or younger were analyzed. It was found that a diagnosis of PPBC within 10 years appears to be associated with an increased risk for metas-

Table 10.1 Some studies comparing cellular and molecular prognostic characteristics and survival in pregnancy- and non-pregnancy-associated breast cancer

First author, year	Number of PABC	Study type	Histologic features	Immunohistochemistry features	Survival
Ishida, 1992 [30]	192	Case-control	Larger T, more LN +, more LVI+	Less ER,PR +	Worse survival
Bonnier, 1997 [13]	154	Retrospective, multicenter	Larger T, more LN +, more IBC	Less ER,PR +	Worse OS and DFS
Middleton, 2003 [28]	39	Cross-sectional	Higher tumor grade, more advanced stage	Less ER,PR +; higher Ki67index	Similar
Halaska, 2009 [31]	32	Retrospective, case-control	Similar	More ER –	Similar
Azim, 2012 [22]	65	Case-control	Similar	Similar ER, PR, HER2, Ki67	Worse DFS, similar OS
Murphy, 2012 [17]	99	Retrospective, single institute	Higher tumor grade, more advanced stage	Less ER,PR +	Similar
Genin, 2012 [21]	41	Retrospective, chart review	More aggressive features, more advanced stage	Less ER,PR +, more HER2 +	–
Madaras, 2013 [20]	31	Retrospective, case-control	More high grade DCIS	Less ER,PR +, similar HER2, higher Ki67, more TN, more luminal B	Worse DFS and OS
Langer, 2014 [32]	117	Retrospective, single institution	Higher tumor grade, more LN +	Less ER,PR +	Debated
Johansson, 2018 [33]	778	Cohort, case-control	More advanced stage	Less ER,PR +, more HER2 +, more TN	Similar
Bae, 2018 [8]	40	Retrospective, database information	–	Less ER,PR +, more HER2+, more TN, less luminal A	Worse DFS and BCSS

BCSS breast cancer specific survival; *DFS* disease free survival; *ER* Estrogen receptor; *IBC* inflammatory breast cancer; *LVI* lymphovascular invasion; *OS* overall survival; *PABC* pregnancy-associated breast cancer; *PR* Progesterone receptor; *T* tumor size; *TN* triple negative tumor

tasis. This increased risk was highest in stages 1 and 2 at diagnosis and present in both groups of ER-negative and ER-positive patients. PPBC was not associated with increased proliferative activity (Ki67) but was associated with increased lymphovascular invasion and lymph node involvement compared with breast cancer in nulliparous young patients. The end point of analysis was distant metastasis-free survival. They concluded that diagnosis of breast cancer in ages 45 years or younger within 10 years after childbirth (PPBC) is as an independent, adverse prognostic factor and leads to a twofold increase in metastasis compared with a nulliparous patient [26].

Previous studies showed that breast cancer at a young age (<45 years) is associated with inferior survival and unfavorable clinicopathologic features. Breast cancer in young women illustrates larger tumor size, higher HER2 overexpression, lower hormone receptor positivity, higher grade of tumor, and more frequent lymph node involvement. The gene profile expression is also unique and different in young women [27]. According to these findings and the young age of patients with breast cancer during pregnancy or lactation, we can conclude that the poor prognosis and histopathologic features of PABC are similar to those of breast cancer in young women. Actually, PABC shares many histologic, prognostic, and gene profiling similarities with breast cancer in young women [28]. Some studies assessed the pathogenesis of PABC to explain the poor prognosis of these tumors.

Genin et al. [29] investigated the underlying reason for poor prognosis of PABC and evaluated angiogenesis and lymphangiogenesis as prognostic factors in breast cancer. They found that angiogenesis but not lymphatic angiogenesis was significantly increased in the tumor and healthy breast tissue of patients with PABC compared with controls.

Xu et al. [34] showed the absence of p63 and WT-1 expression in a vast majority of myoepithelial cells, cytoplasmic localization of p63 in the entire epithelial cell population of some lobules, and a substantially increased WT-1 expression in vascular structures of the invasive cancer component of PABC in comparison with that of non-PABC. All or nearly all epithelial cells with aberrant p63 and WT-1 expression lacked the expression of ER and PR, whereas they had a substantially higher proliferation index than their counterparts with p63 and WT-1 expression.

Harvell et al. [35], in an assessment of the cause of different behavior and prognosis of PABC compared with non-PABC, revealed that immunity-related genes were enriched in tumor-associated stroma of PABC. Compared with normal stroma, PABC-associated stroma overexpressed immune response genes, whereas genes involved in angiogenesis and extracellular matrix deposition were more commonly down-regulated. Moreover, immunomodulators and genes encoding cell proliferative factors, signaling, and cell death, were hormone-regulated in stroma.

Azim et al. [36] showed that 2 pathways were enriched in tumors diagnosed during pregnancy: the G protein–coupled receptor pathway and the serotonin receptor pathway. Tumors diagnosed during pregnancy had a higher expression of PD1, PDL1, and gene sets related to SRC, IGF1, and β-catenin in PABC in comparison with non-PABC. In another work, Azim et al. [37] showed that among PABC, those with a triple-negative profile revealed higher tumor-infiltrating lymphocytes (TIL), which is consistent with earlier studies in the non-pregnant setting.

In summary, PABC is a rare condition, and controversies about prognosis, histologic features, and biomarker profiles exist. In most studies, breast cancer diagnosed during pregnancy and postpartum is associated with a more advanced stage, a larger tumor size, and a higher histologic grade; and postpartum tumors show more adverse outcomes in comparison with tumors during pregnancy. Also, expression of hormone receptors is less frequent and triple-negative and luminal B subtypes are more frequent in PABC than in non-PABC.

References

1. Antonelli NM, Dotters DJ, Katz VL, Kuller JA (1996) Cancer in pregnancy: a review of the literature. Part I. Obstet Gynecol Surv 51(2):125–134
2. Loibl S, Von Minckwitz G, Gwyn K, Ellis P, Blohmer JU, Schlegelberger B et al (2006) Breast carcinoma during pregnancy: international recommendations from an expert meeting. Cancer 106(2):237–246
3. Rojas KE, Bilbro N, Manasseh DM, Borgen PI (2019) A review of pregnancy-associated breast cancer: diagnosis, local and systemic treatment, and prognosis. J Women's Health 28(6):778–784
4. Keyser CE, Staat MB, Fausett CM, Shields LC (2012) Pregnancy-associated breast cancer. Rev Obstet Gynecol 5(2):94
5. Borges VF, Schedin PJ (2012) Pregnancy-associated breast cancer: an entity needing refinement of the definition. Cancer 118(13):3226–3228
6. Stensheim H, Møller B, Van Dijk T, Fosså SD (2009) Cause-specific survival for women diagnosed with cancer during pregnancy or lactation: a registry-based cohort study. J Clin Oncol 27(1):45–51
7. Andersson TM, Johansson AL, Hsieh CC, Cnattingius S, Lambe M (2009) Increasing incidence of pregnancy-associated breast cancer in Sweden. Obstet Gynecol 114(3):568–572
8. Bae SY, Jung SP, Jung ES, Park SM, Lee SK, Yu JH et al (2018) Clinical characteristics and prognosis of pregnancy-associated breast cancer: poor survival of luminal b subtype. Oncology 95(3):163–169
9. Al-Amri AM (2015) Clinical presentation and causes of the delayed diagnosis of breast cancer in patients with pregnancy associated breast cancer. J Fam Community Med 22(2):96
10. Woo JC, Yu T, Hurd TC (2003) Breast cancer in pregnancy: a literature review. Arch Surg 138(1):91–98
11. Lambe M, Hsieh CC, Trichopoulos D, Ekbom A, Pavia M, Adami HO (1994) Transient increase in the risk of breast cancer after giving birth. NEJM 331(1):5–9
12. Rodriguez AO, Chew H, Cress R, Xing G, McElvy S, Danielsen B et al (2008) Evidence of poorer survival in pregnancy-associated breast cancer. Obstet Gynecol 112(1):71–78

13. Bonnier P, Romain S, Dilhuydy JM, Bonichon F, Julien JP, Charpin C et al (1997) Influence of pregnancy on the outcome of breast cancer: a case-control study. Int J Cancer 72(5):720–727
14. Parente JT, Amsel ME, Lerner R, Chinea FR (1988) Breast cancer associated with pregnancy. Obstet Gynecol 71(6 Pt 1):861–864
15. Gwyn K, Theriault R (2001) Breast cancer during pregnancy. Breast Cancer 15(1)
16. King RM, Welch JS, Martin JJ, Coulam CB (1985) Carcinoma of the breast associated with pregnancy. Surg Gynecol Obstet 160(3):228–232
17. Murphy CG, Mallam D, Stein S, Patil S, Howard J, Sklarin N et al (2012) Current or recent pregnancy is associated with adverse pathologic features but not impaired survival in early breast cancer. Cancer 118(13):3254–3259
18. Beadle BM, Woodward WA, Middleton LP, Tereffe W, Strom EA, Litton JK et al (2009) The impact of pregnancy on breast cancer outcomes in women≤ 35 years. Cancer 115(6):1174–1184
19. Marikakis N, Yiangou C, Agrawal A (2018) Hormone receptor expression in pregnancy-associated breast cancer: a systematic review of the literature. Eur J Surg Oncol 44(6):913
20. Madaras L, Kovács KA, Szász AM, Kenessey I, Tőkés AM, Székely B et al (2014) Clinicopathological features and prognosis of pregnancy associated breast cancer – a matched case control study. Pathol Oncol Res 20(3):581–590
21. Genin AS, Lesieur B, Gligorov J, Antoine M, Selleret L, Rouzier R (2012) Pregnancy-associated breast cancers: do they differ from other breast cancers in young women? Breast 21(4):550–555
22. Azim HA Jr, Botteri E, Renne G, Dell'Orto P, Rotmensz N, Gentilini O et al (2012) The biological features and prognosis of breast cancer diagnosed during pregnancy: a case-control study. Acta Oncol 51(5):653–661
23. Boudy AS, Naoura I, Zilberman S, Gligorov J, Chabbert-Buffet N, Ballester M et al (2017) Clues to differentiate pregnancy-associated breast cancer from those diagnosed in postpartum period: a monocentric experience of pregnancy-associated cancer network (CALG). Bull Cancer 104(6):574–584
24. Azim HA Jr, Santoro L, Russell-Edu W, Pentheroudakis G, Pavlidis N, Peccatori FA (2012) Prognosis of pregnancy-associated breast cancer: a meta-analysis of 30 studies. Cancer Treat Rev 38(7):834–842
25. Hartman EK, Eslick GD (2016) The prognosis of women diagnosed with breast cancer before, during and after pregnancy: a meta-analysis. Breast Cancer Res Treat 160(2):347–360
26. Goddard ET, Bassale S, Schedin T, Jindal S, Johnston J, Cabral E et al (2019) Association between postpartum breast cancer diagnosis and metastasis and the clinical features underlying risk. JAMA Net Open 2(1):e186997
27. Anders CK, Hsu DS, Broadwater G, Acharya CR, Foekens JA, Zhang Y et al (2008) Young age at diagnosis correlates with worse prognosis and defines a subset of breast cancers with shared patterns of gene expression. J Clin Oncol 26(20):3324–3330
28. Middleton LP, Amin M, Gwyn K, Theriault R, Sahin A (2003) Breast carcinoma in pregnant women: assessment of clinicopathologic and immunohistochemical features. Cancer 98(5):1055–1060
29. Genin AS, Antoine M, Aractingi S, Rouzier R (2014) Pregnancy stimulates tumor angiogenesis in breast carcinoma. Anticancer Res 34(1):125–131
30. Ishida T, Yokoe T, Kasumi F, Sakamoto G, Makita M, Tominaga T et al (1992) Clinicopathologic characteristics and prognosis of breast cancer patients associated with pregnancy and lactation: analysis of case-control study in Japan. Jpn J Cancer Res 83(11):1143–1149
31. Halaska MJ, Pentheroudakis G, Strnad P, Stankusova H, Chod J, Robova H et al (2009) Presentation, management and outcome of 32 patients with pregnancy-associated breast cancer: a matched controlled study. Breast J 15(5):461–467
32. Langer A, Mohallem M, Stevens D, Rouzier R, Lerebours F, Chérel P (2014) A single-institution study of 117 pregnancy-associated breast cancers (PABC): presentation, imaging, clinicopathological data and outcome. Diagn Interv Imaging 95(4):435–441
33. Johansson AL, Andersson TM, Hsieh CC, Jirström K, Cnattingius S, Fredriksson I et al (2018) Tumor characteristics and prognosis in women with pregnancy-associated breast cancer. Int J Cancer 142(7):1343–1354
34. Xu Z, Wang W, Deng CX, Man YG (2009) Aberrant p63 and WT-1 expression in myoepithelial cells of pregnancy-associated breast cancer: implications for tumor aggressiveness and invasiveness. Int J Biol Sci 5(1):82–96
35. Harvell DM, Kim J, O'Brien J, Tan AC, Borges VF, Schedin P et al (2013) Genomic signatures of pregnancy-associated breast cancer epithelia and stroma and their regulation by estrogens and progesterone. Horm Cancer 4(3):140–153
36. Azim HA Jr, Partridge AH (2014) Biology of breast cancer in young women. Breast Cancer Res 16(4):427
37. Azim HA Jr, Vingiani A, Peccatori F, Viale G, Loi S, Pruneri G (2015) Tumour infiltrating lymphocytes (TILs) in breast cancer during pregnancy. Breast 24(3):290–293

11 Clinical Presentation, Diagnosis and Prognosis of Pregnancy-Associated Breast Cancer

James Sun and Marie Catherine Lee

Abstract

Breast cancer in pregnancy is a rare entity generally presenting as a persistent breast mass, but is often a delayed finding due to the expected physiologic changes in the breast related to pregnancy and lactation. The preferred diagnostic workup of a persistent breast mass involves a combination of mammographic and ultrasonographic evaluation in addition to tissue diagnosis via core biopsy; breast MRI is not recommended. Surgical excision should be reserved for definitive treatment in order to minimize fetal exposure to anesthesia. Evaluation for distant metastatic spread can be performed using radiographs and ultrasound to limit fetal radiation exposure. Similar to the non-pregnant patient, prognosis is primarily driven by tumor biology, however, there is limited and conflicting data regarding the impact of pregnancy on breast cancer outcomes with a distinct difference in survival among patients with breast cancer during pregnancy compared to those diagnosed postpartum.

Keywords

Breast cancer in pregnancy · Breast imaging and pregnancy · Diagnosis · Presentation · Pregnancy and Breast Cancer Staging · PABC

J. Sun
Comprehensive Breast Program, Moffitt Cancer Center, Tampa, FL, USA

M. C. Lee (✉)
Comprehensive Breast Program, Moffitt Cancer Center, Tampa, FL, USA

Department of Surgery, University of South Florida, Tampa, FL, USA
e-mail: marie.lee@moffitt.org

11.1 Overview

In general, the differential diagnosis of a breast mass in premenopausal women is broad and encompasses benign entities such as abscess, fibroadenoma, phyllodes tumor, lipoma, fat necrosis, fibrocystic disease, galactocele or cysts; benign etiologies are far more common than malignancy in this age group. As this population does not routinely undergo breast screening, image-detected lesions without a palpable physical finding are extremely rare (see also Chaps. 4, 5 and 6).

Many benign breast entities can be diagnosed with imaging and do not require invasive intervention, such as fibrocystic breast disease or breast cysts. Fat necrosis is usually a post-traumatic or postoperative finding and frequently

S. Alipour, R. Omranipour (eds.), *Diseases of the Breast during Pregnancy and Lactation*, Advances in Experimental Medicine and Biology 1252,
https://doi.org/10.1007/978-3-030-41596-9_11

can be diagnosed with mammogram and sonogram in conjunction with a consistent clinical history. Breast abscesses present with skin erythema, induration, and/or fever, and can be associated with lactation, smoking, diabetes or other systemic illnesses; these may or may not require surgical intervention (see also Chap. 7). Solid lesions are often diagnosed by core needle biopsy or surgical excision, such as phyllodes tumors and symptomatic or enlarging fibroadenomas. Galactoceles generally develop months after discontinuation of lactation (see also Chaps. 6 and 19).

Breast cancer is an uncommon diagnosis in this age group, with only 0.3% of all breast cancers occurring in women between the ages of 20 and 29 [1] and has been associated with a significant delay in diagnosis [2]. As women more frequently delay child-bearing, breast cancer is now the most common malignancy diagnosed during pregnancy, with an estimated incidence of 1 in 3000 pregnancies in the United States [3, 4] (see also Chap. 9).

11.2 Clinical Presentation

The most common clinical presentation of breast cancer in the pregnant population is a persistent, palpable breast mass, often markedly increasing in size disproportionately to the surrounding and evolving breast tissue in a gravid woman. During pregnancy, accessory breast tissue may also swell and present as an enlarging mass in the axillary tail of the breast or in the axilla and may be considered for excision in the postpartum period.

Common presenting complaints in pregnancy-associated breast cancer (PABC) include a painless palpable mass, skin thickening, or asymmetric breast swelling. Patients and physicians may mistakenly attribute these findings to the normal physiologic changes of the breast in pregnancy. Normal physiologic changes, such as increased density of breast tissue, can also make palpation of a mass more difficult and contribute to a delay in diagnosis (see also Chaps. 1 and 2).

11.3 Diagnostic Evaluation

Diagnostic workup of a pregnant patient with a palpable breast mass includes a detailed history of both the presenting complaint as well as an obstetric history with particular attention to dates of conception and delivery. Assessment with physical exam, diagnostic breast imaging, and pathologic tissue sampling should follow promptly if there is any suspicion for malignancy. Focused breast imaging consists of mammography and ultrasound; digital mammography is preferred if available, especially in women whose breast density may be increased by both lactational changes and normal tissue (Fig. 11.1). Mammography is considered safe in pregnancy with appropriate abdominal shielding and can visualize calcifications, masses and architectural distortion with approximately 86% sensitivity [5]. A study evaluating mammography and ultrasonography for identification of breast cancer during pregnancy reported 23 gravid women with known cancers imaged prior to surgery [6]. Mammography identified malignancy in 18/20 (90%) of patients. Of the 20 women who underwent breast ultrasound, all breast malignancies were correctly identified. Axillary metastases were correctly identified in 15/18 (83%) of women who underwent nodal ultrasound. This study also demonstrated accurate assessment of chemotherapy response by ultrasound features. It concluded that mammography can identify breast cancer during pregnancy despite dense breast tissue but ultrasound may provide more information including nodal disease and response to neoadjuvant chemotherapy (see also Chap. 3).

Assessment of clinical axillary lymphadenopathy on physical exam is the most readily available method of clinical assessment of regional disease, however, physical examination of the axilla is known to greatly over- and underestimate clinical disease burden [7]. Ultrasound of the axilla as an adjunct may improve sensitivity of detecting axillary metastases [6]. Regional assessment of the axilla affects recommendations for chemotherapy, surgery and radiation therapy in patients with invasive breast cancer; percutaneous biopsy of an abnormal node may signifi-

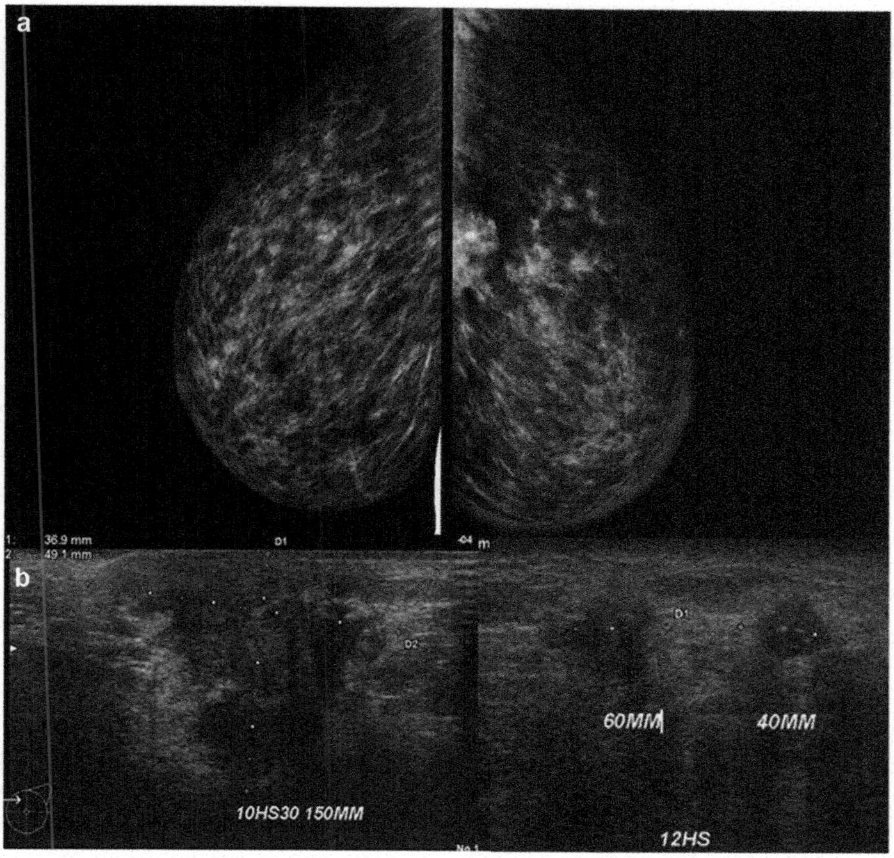

Fig. 11.1 Inflammatory breast cancer in a 32 year-old woman, 1 month pregnant. Multifocal lesions in (**a**) mammography, (**b**) ultrasound. (Courtesy of Dr. Adriana Langer)

cantly impact the treatment plan for a pregnant patient (see also Chap. 4).

Breast magnetic resonance imaging (MRI) with gadolinium contrast is often used in evaluating the extent of disease in non-pregnant patients with breast cancer, especially in those with dense breast tissue and lobular histology. A recent study demonstrated significant increase in neonatal rheumatological, inflammatory or infiltrative skin conditions, stillbirth or death associated with gadolinium exposure at any time during pregnancy [8]. Considering the prone position and the concerns regarding fetal exposure to gadolinium, breast MRI is not considered a safe modality for diagnostic workup of a breast mass in pregnant patients. In the postpartum setting, MRI can be performed but interpretation may be limited due to increased density and vascularity of lactating breast tissue (Fig. 11.2) (see also Chap. 3).

After appropriate diagnostic imaging, pathologic tissue assessment of a breast mass is critical. The current recommended modality for pathologic diagnosis is via image-guided, percutaneous core biopsy with clip placement [9]. This is most often done under ultrasonographic guidance but can be done under mammography if the abnormality is not detected with ultrasound. If the pathology result is discordant with diagnostic imaging, surgical excision should be considered. It is important to note that the proliferative lactational changes in normal breast tissue may be mistaken for atypia or even malignancy on cytology, so core needle biopsy is highly preferred over fine needle aspiration to minimize the possibility of a false positive diagnosis of malig-

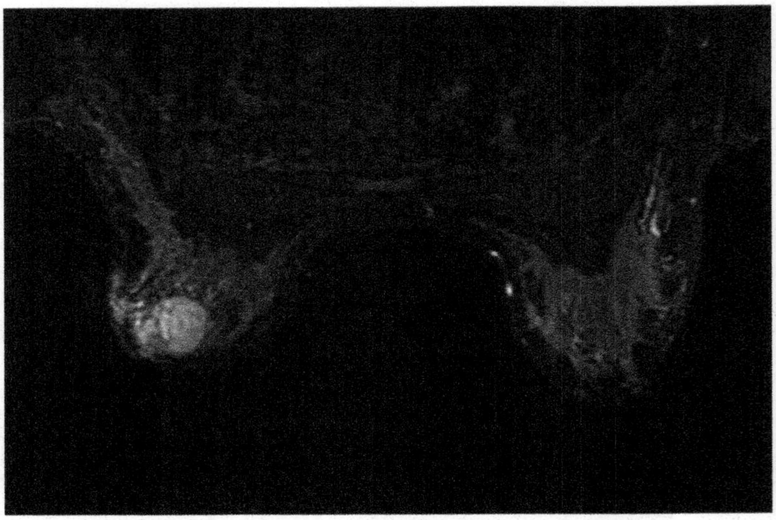

Fig. 11.2 T2-weighted breast MRI of a woman performed 10 days postpartum. A malignant phyllodes tumor distorts the left breast. Note diffusely dense lactating breast tissue bilaterally

nancy (see also Chap. 4). Surgical excision in lieu of percutaneous biopsy, though feasible, confers higher risk than a diagnostic core biopsy for initial pathologic tissue evaluation of any patient, pregnant or not, with a palpable breast mass [10]. Furthermore in the pregnant patient, every attempt should be made to establish the pathologic diagnosis percutaneously to avoid additional surgical procedures and anesthesia, thereby minimizing the risk to the pregnant patient as well as the fetus.

Further evaluation and treatment of benign breast findings in pregnant patients may be deferred until the postpartum period; however, a diagnosis of breast cancer during pregnancy warrants immediate attention. Invasive breast cancer is treated in a multidisciplinary approach, which may include surgery, chemotherapy or radiation. For the pregnant patient, the timing and order of treatment is often determined by gestational age as well as the stage of the cancer at diagnosis. Termination of the pregnancy does not improve outcomes and should not be recommended to patients in the context of breast cancer-related survival [2, 11]. Patients may safely undergo both local and systemic cancer treatment while maintaining their pregnancy, although radiation therapy is deferred until after delivery. Early planned delivery may be considered if felt to affect maternal oncologic outcomes (see also Chaps. 12 and 21).

Pregnant patients with an invasive breast cancer diagnosis, particularly those with biopsy-proven axillary disease, should also have staging studies performed. In the non-pregnant patient, either a whole body positron emission tomography (PET) scan or a computer tomography (CT) scan of the chest and abdomen with a bone scan are considered appropriate. However, these scans should be deferred until delivery in pregnant patients. Staging with a chest radiograph (with abdominal shielding) and an ultrasound of the liver is adequate for pregnant patients until the postpartum period. MRI of the brain without gadolinium contrast for patients with symptoms is considered safe. To evaluate for bone metastases when clinically indicated, MRI without contrast may be performed. Radionuclide bone scans are not recommended during pregnancy. However, modifications to radionuclide scans in gravid women have been described [12, 13]. Reduction in the dose of the tracer and an increase in imaging time appear to minimize radiation dose to the fetus [14]. Any lesion which is suspicious for distant metastatic disease on staging studies warrants a percutaneous biopsy, as patients with distant metastatic disease are not considered automatic surgical candidates. Central to the management of breast cancer in the pregnant patient is a multidisciplinary team consisting of surgical oncology, medical oncology, radiation oncology and maternal fetal medicine in developing a comprehensive treatment plan with the two-fold goal of maximizing breast cancer related outcomes while optimizing perinatal care for safe delivery of a healthy baby.

11.4 Prognosis

The impact of pregnancy on breast cancer outcomes is not clear and outcome data for this population is limited. PABC encompasses breast cancer diagnosed in pregnancy and up to 1 year postpartum, however, the most recent data suggests that outcomes are different between these two groups, with improved outcomes for women receiving treatment during pregnancy compared to women diagnosed in the postpartum setting (see also Chap. 10).

In general, PABC is less frequently hormone receptor positive and the frequency of HER2 expression varies in the literature, similar to non-pregnant breast cancer patients in this age group (Table 11.1). Multiple publications from institutions in the United States, Europe and Asia demonstrate significantly decreased overall survival among pregnant patients, in addition to presentation with more advanced disease, larger tumors and increased proportion of receptor negative cancers compared to age-matched non-pregnant patients [15–19].

Zhang et al. [20] found a trend to worse prognosis which did not reach statistical significance among women in China. A case-control study by Azim et al. [21] matched 65 patients with PABC to non-pregnant breast cancer patents in a 1:2 fashion. They found no differences in tumor characteristics, however PABC patients had inferior disease-free survival. PABC patients also had worse overall survival once patients who had not received neoadjuvant therapy were excluded. Also, A meta–analysis of 30 studies suggested an increased risk of death among women with PABC (Hazard Ratio; HR = 1.44). The authors found a trend to poorer prognosis if the cancer was diagnosed in the postpartum setting (HR = 1.81, p-value = 0.12) [22]. A more recent meta-analysis of 41 studies in 2016 also demonstrated an overall increased risk of death among pregnant patients compared to non-pregnant controls, again with the worst outcomes in women diagnosed postpartum [23] (see also Chap. 10).

In contrast, several case-control studies found no worsening of prognosis [24–27] and even potentially improved outcomes [28]. Amant et al. [29] described a series of patients with PABC, excluding those diagnosed in the postpartum setting. Adjusting for a variety of factors (age, stage, grade, hormone receptor status, HER2 status, histology, type of chemotherapy and postpartum therapies such as trastuzumab, radiotherapy and endocrine therapy), the hazard ratios were not statistically significant. Another large retrospective analysis was performed on 668 cases of breast cancer, 104 of whom were pregnancy-associated [30]. In this series, there were no statistically significant differences in outcomes

Table 11.1 Immunohistologic features of PABC

First author, year	Sample size	ER+ (%)	PR+ (%)	HER2 expression (%)
Elledge, 1993 [33]	12[a]	6 (50)	10 (83)	7 (58)
Middleton, 2003 [3]	25[a]	7 (28)	6 (24)	7 (28)
Reed, 2003 [34]	20	4 (22)	6 (33)	8 (44)
Beadle, 2009 [30]	104	36 (35)	30 (29)	Not reported
Halaska, 2009 [25]	30[a]	11 (37)	11 (37)	10 (33)
Azim, 2012 [21]	65	43 (66)	42 (65)	11 (17)
Murphy, 2012 [27]	99	39 (39)	26 (26)	20 (20)
Basaran, 2014 [4]	20	14 (70)[b]	14 (70)[b]	5 (25)
Madaras, 2014 [35]	31	13 (42)	4 (13)	6 (20)
Strasser-Weippl, 2015[c] [36]	109	58 (53)	39 (36)	12 (11)
Johansson, 2018[d] [37]	97	31 (32)	34 (35)	8 (8)

[a]Number of patients assessed with immunohistochemistry results
[b]ER or PR positivity reported together
[c]Significant number of hormone receptor statuses were unknown
[d]PABC diagnosed while pregnant, not including postpartum period

among the two groups and a trend to improved OS at 5-years (p-value = 0.068) for those who received some treatment during pregnancy. A population-based retrospective review of PABC in Sweden also did not identify a significant difference in breast cancer mortality between PABC and non-PABC patients when controlled for age [19].

Although the relationship between pregnancy and outcome is not clear, the data consistently underscore a pattern of later presentation and higher stage of PABC at presentation compared to non-pregnant cases. Not surprisingly, treatment delays can have a negative effect on outcome, although these delays likely occur in the diagnosis of PABC, rather than in initiation of treatment [31]. Another consideration is substandard administration of chemotherapy in patients prior to evidence of its safety during pregnancy, which would also lead to worse outcomes in PABC [32]. Given the variations in setting, timing and resources of these various relatively small studies, the difference in outcomes is attributed to the heterogeneity of data and paucity of cases overall.

In summary, any pregnant patient who presents with signs or symptoms suggestive of breast cancer must be evaluated promptly. Given the clinical finding of a persistent or enlarging breast mass in the setting of lactation-related breast changes, physicians should have a low threshold for imaging evaluation of a persistent breast finding in a pregnant patient. PABC is associated with a delay in diagnosis. A delayed or late presentation of PABC may contribute to the difference observed in outcomes among women receiving treatment during pregnancy compared to those receiving a PABC diagnosis in the postpartum setting. Our understanding of the true risk that PABC confers to a patient continues to evolve, with conflicting data as to the contribution of pregnancy to overall survival and other treatment outcomes.

References

1. Anders CK, Johnson R, Litton J, Phillips M, Bleyer A (2009) Breast cancer before age 40 years. Semin Oncol 36(3):237–249
2. Cardonick E (2014) Pregnancy-associated breast cancer: optimal treatment options. Int J Womens Health 6:935–943
3. Middleton LP, Amin M, Gwyn K, Theriault R, Sahin A (2003) Breast carcinoma in pregnant women: assessment of clinicopathologic and immunohistochemical features. Cancer 98(5):1055–1060
4. Basaran D, Turgal M, Beksac K, Ozyuncu O, Aran O, Beksac MS (2014) Pregnancy-associated breast cancer: clinicopathological characteristics of 20 cases with a focus on identifiable causes of diagnostic delay. Breast Care 9(5):355–359
5. Ahn BY, Kim HH, Moon WK, Pisano ED, Kim HS, Cha ES et al (2003) Pregnancy- and lactation-associated breast cancer: mammographic and sonographic findings. J Ultrasound Med 22(5):491–497
6. Yang WT, Dryden MJ, Gwyn K, Whitman GJ, Theriault R (2006) Imaging of breast cancer diagnosed and treated with chemotherapy during pregnancy. Radiology 239(1):52–60
7. Specht MC, Fey JV, Borgen PI, Cody HS III (2005) Is the clinically positive axilla in breast cancer really a contraindication to sentinel lymph node biopsy? J Am Coll Surg 200(1):10–14
8. Ray JG, Vermeulen MJ, Bharatha A, Montanera WJ, Park AL (2016) Association between MRI exposure during pregnancy and fetal and childhood outcomes. JAMA 316(9):952–961
9. Vashi R, Hooley R, Butler R, Geisel J, Philpotts L (2013) Breast imaging of the pregnant and lactating patient: imaging modalities and pregnancy-associated breast cancer. AJR Am J Roentgenol 200(2):321–328
10. Silverstein MJ, Recht A, Lagios MD, Bleiweiss IJ, Blumencranz PW, Gizienski T et al (2009) Special report: consensus conference III. Image-detected breast cancer: state-of-the-art diagnosis and treatment. J Am Coll Surg 209(4):504–520
11. Pavlidis N, Pentheroudakis G (2005) The pregnant mother with breast cancer: diagnostic and therapeutic management. Cancer Treat Rev 31(6):439–447
12. Loibl S, Von Minckwitz G, Gwyn K, Ellis P, Blohmer JU, Schlegelberger B et al (2006) Breast carcinoma during pregnancy: international recommendations from an expert meeting. Cancer 106(2):237–246
13. Baker J, Ali A, Groch MW, Fordham ER, Economou SG (1987) Bone scanning in pregnant patients with breast carcinoma. Clin Nucl Med 12(7):519–524
14. Bural GG, Laymon CM, Mountz JM (2012) Nuclear imaging of a pregnant patient: should we perform nuclear medicine procedures during pregnancy? Mol Imaging Radionucl Ther 21(1):1–5
15. Rodriguez AO, Chew H, Cress R, Xing G, McElvy S, Danielsen B et al (2008) Evidence of poorer survival in pregnancy-associated breast cancer. Obstet Gynecol 112(1):71–78
16. Bonnier P, Romain S, Dilhuydy JM, Bonichon F, Julien JP, Charpin C et al (1997) Influence of pregnancy on the outcome of breast cancer: a case-control study. Int J Cancer 72(5):720–727
17. Ali SA, Gupta S, Sehgal R, Vogel V (2012) Survival outcomes in pregnancy associated breast cancer: a retrospective case control study. Breast J 18(2):139–144

18. Yang YL, Chan KA, Hsieh FJ, Chang LY, Wang MY (2014) Pregnancy-associated breast cancer in Taiwanese women: potential treatment delay and impact on survival. PLoS One 9(11):e111934
19. Johansson AL, Andersson TM, Hsieh CC, Jirström K, Dickman P, Cnattingius S et al (2013) Stage at diagnosis and mortality in women with pregnancy-associated breast cancer (PABC). Breast Cancer Res Treat 139(1):183–192
20. Zhang J, Liu G, Wu J, Lu JS, Shen KW, Han QX et al (2003) Pregnancy-associated breast cancer: a case control and long-term follow-up study in China. J Exp Clin Cancer Res 22(1):23–27
21. Azim HA Jr, Botteri E, Renne G, Dell'Orto P, Rotmensz N, Gentilini O et al (2012) The biological features and prognosis of breast cancer diagnosed during pregnancy: a case-control study. Acta Oncol 51(5):653–661
22. Azim HA Jr, Santoro L, Russell-Edu W, Pentheroudakis G, Pavlidis N, Peccatori FA (2012) Prognosis of pregnancy-associated breast cancer: a meta-analysis of 30 studies. Cancer Treat Rev 38(7):834–842
23. Hartman EK, Eslick GD (2016) The prognosis of women diagnosed with breast cancer before, during and after pregnancy: a meta-analysis. Breast Cancer Res Treat 160(2):347–360
24. Petrek JA, Dukoff R, Rogatko A (1991) Prognosis of pregnancy-associated breast cancer. Cancer 67(4):869–872
25. Halaska MJ, Pentheroudakis G, Strnad P, Stankusova H, Chod J, Robova H et al (2009) Presentation, management and outcome of 32 patients with pregnancy-associated breast cancer: a matched controlled study. Breast J 15(5):461–467
26. Ibrahim EM, Ezzat AA, Baloush A, Hussain ZH, Mohammed GH (2000) Pregnancy-associated breast cancer: a case—control study in a young population with a high-fertility rate. Med Oncol 17(4):293–300
27. Murphy CG, Mallam D, Stein S, Patil S, Howard J, Sklarin N et al (2012) Current or recent pregnancy is associated with adverse pathologic features but not impaired survival in early breast cancer. Cancer 118(13):3254–3259
28. Litton JK, Warneke CL, Hahn KM, Palla SL, Kuerer HM, Perkins GH et al (2013) Case control study of women treated with chemotherapy for breast cancer during pregnancy as compared with non-pregnant patients with breast cancer. Oncologist 18(4):369–376
29. Amant F, von Minckwitz G, Han SN, Bontenbal M, Ring AE, Giermek J et al (2013) Prognosis of women with primary breast cancer diagnosed during pregnancy: results from an international collaborative study. J Clin Oncol 31(20):2532–2539
30. Beadle BM, Woodward WA, Middleton LP, Tereffe W, Strom EA, Litton JK et al (2009) The impact of pregnancy on breast cancer outcomes in women ≤ 35 years. Cancer 115(6):1174–1184
31. Johansson AL, Weibull CE, Fredriksson I, Lambe M (2019) Diagnostic pathways and management in women with pregnancy-associated breast cancer (PABC): no evidence of treatment delays following a first healthcare contact. Breast Cancer Res Treat 174(2):489–503
32. Shachar SS, Gallagher K, McGuire K, Zagar TM, Faso A, Muss HB et al (2017) Multidisciplinary management of breast cancer during pregnancy. Oncologist 22(3):324–334
33. Elledge RM, Ciocca DR, Langone G, McGuire WL (1993) Estrogen receptor, progesterone receptor, and HER-2/neu protein in breast cancers from pregnant patients. Cancer 71(8):2499–2506
34. Reed W, Hannisdal E, Skovlund E, Thoresen S, Lilleng P, Nesland JM (2003) Pregnancy and breast cancer: a population-based study. Virchows Arch 443(1):44–50
35. Madaras L, Kovács KA, Szász AM, Kenessey I, Tőkés AM, Székely B et al (2014) Clinicopathological features and prognosis of pregnancy associated breast cancer–a matched case control study. Pathol Oncol Res 20(3):581–590
36. Strasser-Weippl K, Ramchandani R, Fan L, Li J, Hurlbert M, Finkelstein D et al (2015) Pregnancy-associated breast cancer in women from Shanghai: risk and prognosis. Breast Cancer Res Treat 149(1):255–261
37. Johansson AL, Andersson TM, Hsieh CC, Jirström K, Cnattingius S, Fredriksson I et al (2018) Tumor characteristics and prognosis in women with pregnancy-associated breast cancer. Int J Cancer 142(7):1343–1354

Surgery for Pregnancy-Associated Breast Cancer

12

Ramesh Omranipour

Abstract

Surgery in the form of both mastectomy and breast conservation is the main step in the treatment of breast cancer. Numerous studies have shown an equivalent long-term survival for breast conserving surgery (BCS) and mastectomy. Patients desire and tumor characteristics, especially size and multicentricity, are the key factors that affect the decision between these two types of surgery. Patients with any contraindication for radiotherapy or previous history of radiation to the breast field are not suitable for BCS. There are few absolute contraindications for BCS, and early pregnancy is listed among them; mastectomy is preferred in the first trimester of pregnancy to avoid the impact of delaying radiation therapy on outcome of the cancer.

Keywords

Axillary dissection · Breast conserving surgery · Pregnancy-associated breast cancer · Mastectomy · Sentinel lymph node · Surgery

R. Omranipour (✉)
Breast Disease Research Center (BDRC), Tehran University of Medical Sciences, Tehran, Iran

Department of Surgical Oncology, Cancer Institute, Tehran University of Medical Sciences, Tehran, Iran
e-mail: omranipour@tums.ac.ir

12.1 Surgery of the Breast-Mastectomy versus Breast Conserving Surgery

Total mastectomy still remains the standard local treatment of the breast in Pregnancy-Associated Breast Cancer (PABC) in the first trimester of pregnancy when the patients desire to continue their pregnancy. Because breast tissue conserved during surgery needs to undergo radiation therapy, and this modality can be carried out in the postpartum period only, breast conserving surgery (BCS) can merely be offered to women who are not very far from delivery. However, radiation can be administrated 3–6 weeks after completing systemic therapy in women who need adjuvant chemotherapy, so that systemic therapy in the second and third trimesters can fill the period between BCS and postpartum radiation, allowing BCS to be considered as a suitable technique in this group. Total mastectomy should not be favored only because of pregnancy and possible delay of radiation when another treatment (systemic therapy) is considered in the gap, or when the interval between surgery and radiation is acceptable. The suitable options for breast surgery in PABC are demonstrated in Fig. 12.1.

In a study comparing medical records of 273 patients with PABC and 273 non-PABC controls from a Swedish database, the rate of mastectomy was significantly higher in the PABC group from 1992 to 1997 (74% versus 46%), but not in the

S. Alipour, R. Omranipour (eds.), *Diseases of the Breast during Pregnancy and Lactation*, Advances in Experimental Medicine and Biology 1252,
https://doi.org/10.1007/978-3-030-41596-9_12

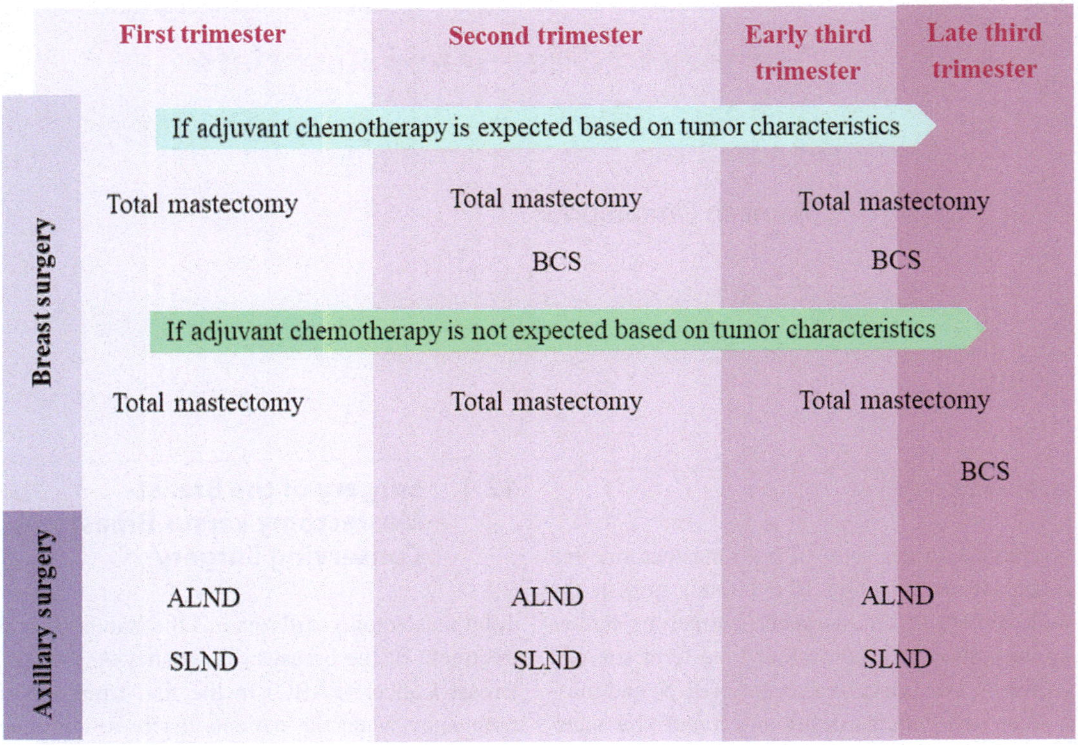

Fig. 12.1 Type of surgery of the breast and axilla for pregnancy-associated breast cancer in different trimesters; when two types of surgery are allowed, the choice depends on tumor characteristics and patient preference. Blue dye should not be used for sentinel lymph node dissection in pregnancy. *ALND* Axillary Lymph Node Dissection, *BCS* Breast Conserving Surgery, *SLND* Sentinel Lymph Node Dissection

period from 2004 to 2009 (69% versus 63%) [1]; which demonstrates a trend toward more frequent treatment of PABC with BCS.

Similar survivals has been reported with BCS and mastectomy in PABC [2, 3]. Kuerer et al. [2] reported no difference in disease-free and overall survival comparing BCS and modified radical mastectomy in stages I and II of PABC in 1996. Rodrigez et al. [3] reported similar survival between these 2 types of surgery in PABC after controlling for age, race, tumor size, hormone receptor status, and stage.

12.2 Surgery of the Axilla- Lymph Node Dissection versus Sentinel Lymph Node Biopsy

Axillary lymph node dissection was traditionally the standard of care in PABC because the rate of axillary lymph node involvement was reported to be higher than 50% at the time of diagnosis [4, 5]. Presently, due to the increasing number of node–negative cases of PABC, a considerable number of patients might benefit from sentinel lymph node biopsy (SLNB). The suitable options for axillary surgery in PABC are demonstrated in Fig. 12.1.

Considering that SLNB necessitates in-breast injection of radioisotope-labeled colloids or blue dyes, performing this procedure has some limitations during pregnancy.

Injection of isosulfan blue is not recommended because of the potential risk of teratogenicity, but in 30 pregnant patients with breast cancer or melanoma who received blue dye for SLNB, all gave birth to a healthy child, except for one who had electively terminated her pregnancy [6]. Both isosulfan blue and methylene blue are Food and Drug Administration (FDA) category C drugs during pregnancy, and anaphylactic shock has been reported with the use of isosulfan blue dye [7, 8]. As well, intestinal atresia and fetal demise have been reported with the use of methylene blue dye in gravid women, and fetal methemoglobinemia was reported only with the use of an intra-amniotic cavity injection. In the 1980s, methylene blue used to be injected in the amniotic cavity for diagnosis of premature membrane rupture; severe adverse effects such as fetal intestinal atresia and respiratory distress were reported consequently [9]. However, its circulating level and potential dosing to the fetus following subareolar injection had been examined by Pruthi et al., and they showed that as much as 5% of the administered dose of methylene blue in a subareolar injection could reach the fetus; so if the pregnant patient refuses a radioisotope injection or sentinel nodes fail to be mapped with a radioisotope, the injection of dilute methylene blue with a low risk of fetal complications has been proposed [10].

Khera et al. used patent blue on 8 pregnant women and found sentinel node(s) in all of them without complications [11]. Similarly, Gropper et al. used methylene blue in 7 gravid mothers and found sentinel node(s) in all of them without any obstetric complication; only 1 child was born with a cleft palate, but the mother was a smoker and had methadone consumption during pregnancy [6]. Despite this evidence, guidelines by the European Society for Medical Oncology (ESMO) discourage the use of blue dye during pregnancy because of the risk of anaphylaxis [12].

Injection of radioisotopes for SLNB raises the concern of fetal radiation exposure. Keleher et al. found the maximum dose of fetal radiation to be 0.43 mGy, which was lower than the recommended threshold and much lower than other nuclear medicine procedures that could be performed during pregnancy [13].

Pandit-Taskar et al. found that the maximal dose of radioactivity to which a fetus was exposed during SLNB was 0.014 mGy for an 18.5 MBq injection. As well, the effective dose to the whole body was estimated at 0.245 mSv in non-pregnant women of childbearing age, much below the 50 mSv threshold set by the National Council on Radiation Protection for pregnant women [14].

Gentilini et al. estimated the maximal absorbed dose of radiation from SLNB by placing a dosimeter around the abdomen of 26 non-pregnant women. A single peritumoral injection of 0.2 mL Tc- labelled human albumin colloid was administered. In 23 patients, the absorbed dose was lower than the sensitivity of the dosimeter. In the remaining 3 patients, the absorbed dose to the epigastrium, umbilicus, and hypogastrium was below the threshold (40–320, 120–150, and 30–40 GY, respectively) [15]. Results of SLNB in 12 pregnant women by the same author showed the safety of the procedure. In total, 11 normal weight babies were born without complication, and 1 baby had heart failure because of a ventricular septal defect which had been detected by echocardiography performed at week 21 of gestation before SLNB was done at 26 week of uterine life [16].

In a study by Cardonick et al. on 30 pregnant women who underwent SLNB, 9 (30%) had adverse pregnancy events including 2 miscarriages in the first trimester, 3 low birth weight, 2 prematurity complications, and 2 malformations [17].

In a large recent cohort from seven different countries, 145 women who underwent SLNB during pregnancy were identified from the International Network on Cancer, Infertility and Pregnancy; the German Breast Group; and the Cancer and Pregnancy Registry [18]. Twelve patients out of these were from Italy and had been reported in 2010 as well [16]. Mapping was unsuccessful in only 1 patient (0.7%, 1/145) and axillary recurrence was very low after 48 months (0.7%); positive sentinel nodes were found in 43 patients (29.7%), and locoregional recurrence occurred in 11 cases. The high identification rate and the low axillary recurrence in this international cohort suggest oncologic safety of SLNB for the mother, and the authors recommended SLNB in pregnant women with the same indication as non-pregnant cases, using radioactive colloid and single-day protocol.

In addition to the type of injection used for SLNB in pregnant women, the appropriate protocol regarding timing of the injection should be contemplated, too. The same day or 1-day protocol involves injecting the radioactive-labelled colloid on the same day of the operation; whereas in the 2-day protocol, injection is carried out the day before surgery. In SLNB performed in pregnancy, the one-day protocol is preferred to reduce the time between lymphoscintigraphy and surgery. Because the radioactive tracer is localized at the injection site and in the sentinel node that is subsequently removed by surgery, this protocol will reduce the risk of fetal exposure.

12.3 Breast Reconstruction after Mastectomy – Immediate versus Delayed Reconstruction

Delayed reconstruction is preferred in patients with PABC to decrease the time of anesthesia and complications of surgery, this also improves the cosmetic results of reconstruction (see also Chap. 26). Therefore, autologous tissue reconstruction should be delayed, but placement of a tissue expander is possible; this has been reported in 2 case series.

In the first series, Lohsiriwat et al. in 2013 reported 2-stage reconstructions by tissue expander in 12 patients with PABC and 1 with a definitive implant, without major complications or any fetal malformation [19].

In the second report by Caragacianu et al. in 2016, 10 of 29 patients with PABC had immediate reconstruction by tissue expander, and although the mean time of operation increased from 157 to 198 min, it did not lead to obstetrical and fetal complications [20].

Tissue expander insertion at the time of mastectomy seems to be safe during pregnancy and does not increase the morbidity of the surgery.

References

1. Johansson ALV, Weibull CE, Fredriksson I, Lambe M (2019) Diagnostic pathways and management in women with pregnancy-associated breast cancer (PABC): no evidence of treatment delays following a first healthcare contact. Breast Cancer Res Treat 174(2):489–503
2. Kuerer HM, Cunningham JD, Brower ST, Tartter PI (1997) Breast carcinoma associated with pregnancy and lactation. Surg Oncol 6(2):93–98
3. Rodriguez AO, Chew H, Cress R, Xing G, McElvy S, Danielsen B et al (2008) Evidence of poorer survival in pregnancy-associated breast cancer. Obstet Gynecol 112(1):71–78
4. Langer A, Mohallem M, Stevens D, Rouzier R, Lerebours F, Chérel P (2014) A single-institution study of 117 pregnancy-associated breast cancers (PABC): presentation, imaging, clinicopathological data and outcome. Diagn Interv Imaging 95(4):435–441
5. Wang B, Yang Y, Jiang Z, Zhao J, Mao Y, Liu J et al (2019) Clinicopathological characteristics, diagnosis, and prognosis of pregnancy-associated breast cancer. Thorac Cancer 10(5):1060–1068
6. Gropper AB, Calvillo KZ, Dominici L, Troyan S, Rhei E, Economy KE et al (2014) Sentinel lymph node biopsy in pregnant women with breast cancer. Ann Surg Oncol 21(8):2506–2511
7. Albo D, Wayne JD, Hunt KK, Rahlfs TF, Singletary SE, Ames FC et al (2001) Anaphylactic reactions to isosulfan blue dye during sentinel lymph node biopsy for breast cancer. Am J Surg 182(4):393–398
8. Crivellaro M, Senna G, Dama A, Bonadonna P, Passalacqua G (2003) Anaphylaxis due to patent blue dye during lymphography, with negative skin prick test. J Investig Allergol Clin Immunol 13(1):71–72
9. McEnerney JK, McEnerney LN (1983) Unfavorable neonatal outcome after intraamniotic injection of methylene blue. Obstet Gynecol 61(3 Suppl):35S–37S
10. Pruthi S, Haakenson C, Brost BC, Bryant K, Reid JM, Singh R, Netzel B, Boughey JC, Degnim AC (2011) Pharmacokinetics of methylene blue dye for lymphatic mapping in breast cancer—implications for use in pregnancy. Am J Surg 201(1):70–75
11. Khera SY, Kiluk JV, Hasson DM, Meade TL, Meyers MP, Dupont EL et al (2008) Pregnancy-associated breast cancer patients can safely undergo lymphatic mapping. Breast J 14(3):250–254
12. Peccatori FA, Azim HA Jr, Orecchia R, Hoekstra HJ, Pavlidis N, Kesic V et al (2013) ESMO Guidelines Working Group. Cancer, pregnancy and fertility: ESMO Clinical Practice Guidelines for diagnosis, treatment and follow-up. Ann Oncol 24(suppl_6):vi160–vi170
13. Keleher A, Wendt R III, Delpassand E, Stachowiak AM, Kuerer HM (2004) The safety of lymphatic mapping in pregnant breast cancer patients using Tc-99m sulfur colloid. Breast J 10(6):492–495
14. Pandit-Taskar N, Dauer LT, Montgomery L, Germain JS, Zanzonico PB, Divgi CR (2006) Organ and fetal absorbed dose estimates from 99mTc-sulfur colloid lymphoscintigraphy and sentinel node localization in breast cancer patients. J Nucl Med 47(7):1202–1208
15. Gentilini O, Cremonesi M, Trifiro G, Ferrari M, Baio SM, Caracciolo M et al (2004) Safety of sentinel node biopsy in pregnant patients with breast cancer. Ann Oncol 15(9):1348–1351

16. Gentilini O, Cremonesi M, Toesca A, Colombo N, Peccatori F, Sironi R et al (2010) Sentinel lymph node biopsy in pregnant patients with breast cancer. Eur J Nucl Med Mol Imaging 37(1):78–83
17. Cardonick E, Dougherty R, Grana G, Gilmandyar D, Ghaffar S, Usmani A (2010) Breast cancer during pregnancy: maternal and fetal outcomes. Cancer J 16(1):76–82
18. Han SN, Amant F, Cardonick EH, Loibl S, Peccatori FA, Gheysens O et al (2018) Axillary staging for breast cancer during pregnancy: feasibility and safety of sentinel lymph node biopsy. Breast Cancer Res Treat 168(2):551–557
19. Lohsiriwat V, Peccatori FA, Martella S, Azim HA Jr, Sarno MA, Galimberti V et al (2013) Immediate breast reconstruction with expander in pregnant breast cancer patients. Breast 22(5):657–660
20. Caragacianu DL, Mayer EL, Chun YS, Caterson S, Bellon JR, Wong JS et al (2016) Immediate breast reconstruction following mastectomy in pregnant women with breast cancer. J Surg Oncol 114(2):140–143

13 Local Complications of Breast Surgery during Pregnancy and Lactation

Sadaf Alipour

Abstract

During pregnancy and lactation, breast vascularity increases and edema occurs in the breast. As a consequence, rate of complications of breast biopsy and surgery like bleeding, infection, delayed healing and wound dehiscence is expected to be higher. Milk fistula is a rare event that may complicate surgery or needle biopsy of the breast in a breastfeeding woman, or in late stages of pregnancy. Suppression of lactation has been proposed in the literature as both a preventive and a therapeutic step. However, the advantages of nursing for both mother and child are numerous, and the author do not propose it as a preventive measure nor as a must in treatment of milk fistula. Prevention and management of milk fistula are discussed in this chapter.

Keywords

Adverse effect · Breast surgery · Breastfeeding · Complication · Core needle biopsy · Milk fistula · Pregnancy

13.1 Overview

Like any other intervention on body parts, procedures performed on the breast can result in some complications. Most of these are not related to the site of the operation and are similar to procedures performed in other body areas, and some are mostly site-specific.

Sampling techniques as fine needle aspiration, core needle biopsy, and vacuum assisted biopsy have fewer adverse outcomes, while complications occur more frequently consequent to surgeries; the more extended or complex the operation, the more common and severe the complications. Some of these negative consequences may follow any type of invasive intervention, whereas certain are specific to procedures of the breast.

S. Alipour (✉)
Breast Disease Research Center (BDRC), Tehran University of Medical Sciences, Tehran, Iran

Department of Surgery, Arash Women's Hospital, Tehran University of Medical Sciences, Tehran, Iran
e-mail: salipour@tums.ac.ir

S. Alipour, R. Omranipour (eds.), *Diseases of the Breast during Pregnancy and Lactation*, Advances in Experimental Medicine and Biology 1252,
https://doi.org/10.1007/978-3-030-41596-9_13

13.1.1 Local Complications of Needle Biopsy of Breast Lesions

Hemorrhage, bruising and hematoma of the biopsy site are the most common complications of needle biopsy of the breast. After performing the biopsy, manual compression should be applied for 5–10 min on the site; this normally results in no further bleeding [1, 2]. If hemorrhage continues, it is defined as excessive bleeding. The consequence can be either bruising or a hematoma. Bruising, sometimes extensive, may spread over a large part or all of the breast, and even contiguous structures; this picture is exaggerated 24–48 h after the biopsy. An incidence of around 13% has been reported for post-biopsy bruising (of any size) in the breast, while figures are higher and about 22% for patients under anticoagulant medications. Rarely, hematoma ensues secondary to post-biopsy in-breast hemorrhage. The incidence has been mentioned as 1% [2] and is interestingly no higher in coagulation disorders. Treatment of excessive bleeding generally only consists of manual compression till bleeding stops; this may sometimes take 20–30 min. Very seldom, control of hemorrhage warrants surgery or angiographic embolization [1, 2]. The ecchymosis and swelling generally disappears spontaneously after several days to weeks; warm compresses can accelerate the process. Arteriovenous fistula and pseudoaneurysm are also rare consequences in excessive bleeding and need more sophisticated proceedings for proper diagnosis and treatment [1, 2].

The second most common complication of needle biopsy of the breast is infection, which includes mastitis and abscess [1]. The incidence of infection secondary to needle biopsy of the breast is very low and can be kept to a minimum by observing aseptic technique [3, 4]. Both the clinical picture and management options are alike mammary infectious conditions that occur regardless of any intervention on the breast (see also Chap. 7) [1].

Other complications that can very occasionally occur after needle biopsy of the breast include bloody nipple discharge, which happens secondary to bleeding in a duct traumatized by the needle; and pneumothorax due to intrusion of the needle into the chest cavity [3].

Tumor seeding has been mentioned as a potential adverse consequence of needle biopsy in malignant lesions of the breast, and theoretically consists of dislodgement of malignant cells from the tumor throughout the needle tract. The incidence is unknown but must be very low if any; the probability of tumor recurrence due to growth of implanted cells in the biopsy tract can be minimized by excision of the tract at time of cancer surgery [1].

13.1.2 Local Complications of Surgery of the Breast

As in most other surgeries, operation on the breast can lead to several local complications. Namely, these consist of surgical site infection, hemorrhage, hematoma, seroma, delayed healing, wound dehiscence, bloody nipple discharge, fat necrosis, flap necrosis, and cosmetic deformity. Various factors can affect the incidence of each complication; some of them are extent and type of surgery, the underlying conditions of the patient, and the surgeon's experience. Management is as performed for other surgeries, except for the two latter which sometimes need complex plastic procedures.

13.2 Concerns in Pregnancy and Lactation

During the gestational phase, breast vascularity increases tremendously and edema occurs in skin and parenchyma (see also Chap. 1). As a result, rate of complications is expected to be higher. Figures have not been reproduced, though. Expectedly, bleeding is slightly more frequent due to vascular growth, and infections are more common due to the milk which acts as a culture media for microbial overgrowth; especially in PABC undergoing surgery after neoadjuvant chemotherapy [3, 5]. Following the same rational, delayed healing and wound dehiscence should be anticipated at a higher rate.

Dominici et al. [6] compared complications of breast conserving surgery (BCS) and mastectomy in 67 patients with pregnancy-associated breast cancer (PABC). They detected five cases of infection, four of which were mastitis in mastectomy cases, and one axillary abscess in a case of BCS. One hematoma had also occurred in one patient in the BCS group. They concluded that the rate of complications was comparable in the two techniques.

13.2.1 Milk Fistula

Milk fistula is a rare event which consists of milk discharge from a breaching on the breast or axillary skin, coming from a tract between a milk duct and skin surface. It may very rarely occur spontaneously, but is more commonly seen as a complication of surgery or needle biopsy of the breast in a breastfeeding woman, or in late stages of pregnancy.

Literature about this event is scarce and mainly consists of sporadic reported cases. In a comprehensive search of the literature, Alipour et al [7] found 27 cases of milk fistulae complicating procedures on the breast. They had occurred secondary to needle or excisional biopsy of breast lesions, or after aspiration or drainage of breast abscesses. Most had been managed by cessation of lactation, but healing while still on breastfeeding had also happened in several cases.

Suppression of lactation has been proposed in the literature as both a preventive and a therapeutic action for milk fistula [1, 5, 8–13]. However, the advantages of nursing for both mother and child are numerous, and the author do not propose it as a preventive measure except for cases of PABC, which would have to withhold lactation anyhow.

Some points should be observed in order to prevent formation of a milk fistula in pregnant or breastfeeding women who are going to have an invasive procedure on the breast. Using fine needles for sampling purposes [14] if it does not disturb the results (see also Chap. 4); taking the specimens under guidance of ultrasound [15]; completely evacuating breast milk before the procedure [9]; entering the breast (by needle or incision) from a location as far as possible from the nipple and from the lesion, preferably leaving a curved track [2]; prioritizing minor interventions to surgery in these patients if diagnostic or treatment goals can be fulfilled with the lesser procedure (like aspiration of abscesses instead of surgical drainage) (see also Chap. 7) [3]; using radial incisions for surgery wherever applicable [16, 17]; and applying pressure on the procedure site at time of breastfeeding. Figure 13.1 is a schematic demonstration of suggested measures

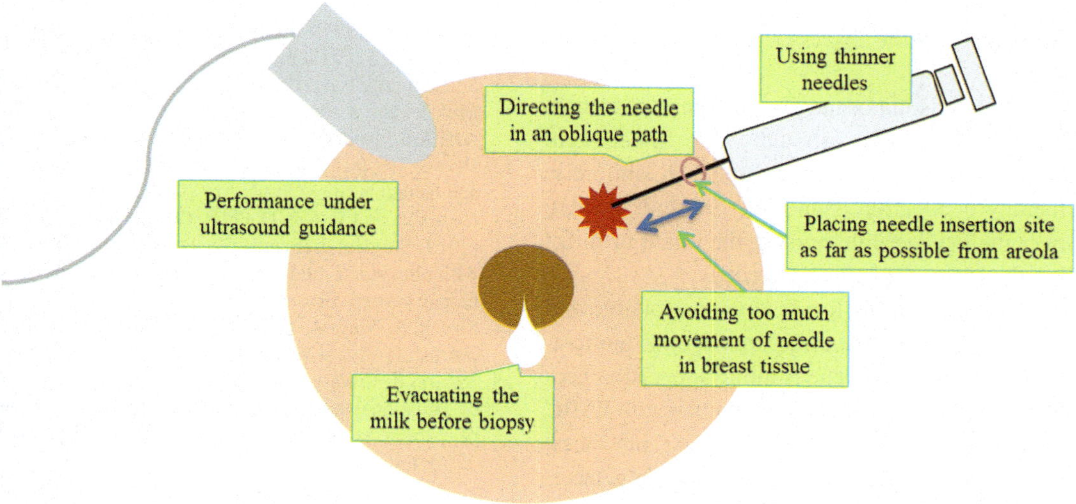

Fig. 13.1 Suggested measures for preventing milk fistula formation after needle sampling in lactating women

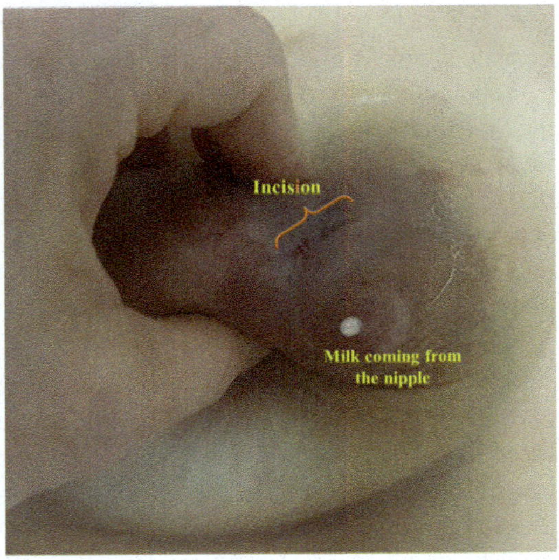

Fig. 13.2 No milk fistula in the healing incision after surgical drainage of a lactating breast abscess. The 40 years-old nursing mother had delivered around 16 months ago and had presented with a very thinned skin over a fluctuating abscess around the areolar margin. Preventive measures had been contemplated during and after surgery and she was able to continue breastfeeding without milk leakage from the wound

for preventing milk fistula formation after needle sampling during lactation. Figure 13.2 shows the healing wound in a breastfeeding mother who had undergone incision and drainage of a breast abscess due to thinning of the skin above the fluctuation (see also Fig. 7.2c); while the relevant preventive measures had been contemplated. She is still nursing her child and has no fistula despite the location of the abscess which was close to the areola.

When a fistula occurs despite these measures, therapeutic steps should be undertaken. Appropriate regular wound care [2, 9], antibiotic therapy if infection is an issue [9], and pressing over the fistula while nursing the infant or milking the breast [18, 19] are needed.

Rarely, when the fistula is uncontrollable and the mother cannot bear the drainage, or intractable infections supervene is milk suppression necessary. This is also the case in women with PABC who would not breastfeed. In these instances, milk can be suppressed by discontinuing lactation, binding and ice-packing the breasts. Medicines can be used; cabergoline and dopamine agonists (bromocriptine) can be prescribed for this purpose [8, 10–13].

References

1. Catani JH, Matsumoto R, Horigome F, Tucunduva T, Costenaro M, Barros N (eds) (2017) A pictorial review of breast biopsy complications. ECR
2. Mahoney MC, Ingram AD (2014) Breast emergencies: types, imaging features, and management. AJR Am J Roentgenol 202(4):W390–W399
3. Lee SS, Hartman HJ, Kuzmiak CM, Crosby KL (2013) The management of breast symptoms in the pregnant and lactating patient. Curr Obstet Gynecol Report 2(1):53–58
4. Ayyappan A, Kulkarni S, Crystal P (2010) Pregnancy-associated breast cancer: spectrum of imaging appearances. Br J Radiol 83(990):529–534
5. Merkel DE (ed) (1996) Pregnancy and breast cancer. Seminars in surgical oncology. Wiley Online Library
6. Dominici LS, Kuerer HM, Babiera G, Hahn KM, Perkins G, Middleton L et al (2010) Wound complications from surgery in pregnancy-associated breast cancer (PABC). Breast Dis 31(1):1–5
7. Alipour S, Dinas Konstantinos (2020) A systematic review of milk fistula in nursing mothers; modifying the perspective toward maintenance of breastfeeding. Clinical lactation Ahead of printing
8. Scott-Conner CE, Schorr SJ (1995) The diagnosis and management of breast problems during pregnancy and lactation. Am J Surg 170(4):401–405
9. Sorosky JI, Scott-Conner CE (1998) Breast disease complicating pregnancy. Obstet Gynecol Clin North Am 25(2):353–363
10. Petrek JA (1994) Breast cancer during pregnancy. Cancer 74(S1):518–527
11. Schackmuth EM, Harlow CL, Norton LW (1993) Milk fistula: a complication after core breast biopsy. AJR Am J Roentgenol 161(5):961–962
12. Hoover HC Jr (1990) Breast cancer during pregnancy and lactation. Surg Clin N Am 70(5):1151–1163
13. Chen C, Luo LB, Gao D, Qu R, Guo YM, Huo JL et al (2019) Surgical drainage of lactational breast abscess with ultrasound-guided Encor vacuum-assisted breast biopsy system. Breast J 25(5):889–897
14. Sharini S, Norie A, Haida H (2017) Chronic lactiferous fistula: A case report. Int Med J Malaysia 6(1)
15. Kataria K, Srivastava A, Dhar A (2013) Management of lactational mastitis and breast abscesses: review of current knowledge and practice. Indian J Surg 75(6):430–435

16. Cantlie HB (1988) Treatment of acute puerperal mastitis and breast abscess. Can Fam Physician 34:2221
17. Surana K, Sagrule D, Lanjewar S (2019) Primary closure with suction drain of acute breast abscess after incision, drainage, and curettage. Indian J Surg:1–5
18. Friedman P, Sanders L (2003) Milk fistula: a conservative approach to treatment after core breast biopsy. J Womens Imag 5(1):40–42
19. Sohail S (2008) Lactiferous fistula in the axillary breast. Editorial Board 2(2):80–82

Aspects of Anesthesia for Breast Surgery during Pregnancy

14

Amirhossein Eskandari and Sadaf Alipour

Abstract

Non-obstetric surgery is needed in 0.75–2% of pregnant women, and safety of anesthesia for mother and child are key points at this time. Some breast diseases need to be approached in a short time interval, and surgery must be performed during pregnancy. In these cases, the technique of anesthesia regarding local, regional or general anesthesia and type of anesthetic medicine are selected based on the extent of the procedure, gestational age, and condition of the mother and child. The ideal timing for any surgery during pregnancy is in the second trimester because the risk of fetal adverse effects as well as preterm labor are lower. However, surgery of breast cancer during pregnancy is performed in any trimester as guided by treatment guidelines and is not deferred based on anesthesia preferences. Various types of anesthesia for breast surgery during pregnancy, preoperative and postoperative considerations are discussed in this chapter.

Keywords

Breast cancer · General anesthesia · Local anesthesia · Regional anesthesia · Pregnancy

14.1 Overview

Non-obstetric surgery is needed in 0.75–2% of pregnant women, most frequently performed for the treatment of appendix or gallbladder inflammation, trauma, and cancer occurring in pregnancy [1]. When this situation materializes, safety of anesthesia and absence of potential harms for mother and child are key points. Accordingly, necessity of the procedure for the mother and fetus [2] and potential risks in case it is postponed should be carefully considered. No emergent surgery should be put off because of pregnancy. However, speaking about breast lesions, they comprise very few, if any, emergent cases and do not match in this category. Nonetheless, many breast diseases need to be

A. Eskandari
Deputy of Education, Ministry of Health, Tehran, Iran

S. Alipour (✉)
Breast Disease Research Center (BDRC), Tehran University of Medical Sciences, Tehran, Iran

Department of Surgery, Arash Women's Hospital, Tehran University of Medical Sciences, Tehran, Iran
e-mail: salipour@tums.ac.ir

S. Alipour, R. Omranipour (eds.), *Diseases of the Breast during Pregnancy and Lactation*, Advances in Experimental Medicine and Biology 1252,
https://doi.org/10.1007/978-3-030-41596-9_14

approached in a short time interval, and surgical management cannot be deferred for several months until termination of pregnancy. These include pregnancy-associated breast cancer (PABC), lesions that convey a high probability of malignancy, and abscesses. In these instances, the technique of anesthesia and type of anesthetics are selected based on the extent of the procedure, gestational age, and condition of the mother and child [3]. The ideal timing for any surgery during pregnancy is in the second trimester because most of the fetal organogenesis has taken place in the previous trimester, and the risk of preterm labor is less than the next trimester [2, 4]. Nonetheless, this can seldom be considered when managing PABC, and the surgery should usually be performed at the time of diagnosis or in an appropriate interval after the end of neoadjuvant treatments (see also Chap. 12).

14.2 Anesthesia for Breast Surgery – Anesthetic Techniques

PABC is the most common cause for operating the breast in a pregnant patient. Anesthesia, along with main therapeutic modalities, affects treatment success in breast cancer surgery [5], albeit to a far less extent.

14.2.1 Local Anesthesia

Among breast operations, minor procedures can be carried out under local anesthesia (LA). These include incisional biopsy of large masses or excisional biopsy of the small ones, which are seldom performed in a pregnant woman, and occasionally lumpectomy of malignant tumors. Mastectomy has also been reported to be successfully performed under LA, but not in pregnant women [6–8].

Lidocaine is the most frequently used drug for LA but is occasionally substituted by bupivacaine, if not effective [3]. Adverse effects during pregnancy are not different from those observed in non-pregnant women, and no significant fetal consequence has been reported [9, 10]. Local anesthetic drugs are more efficacious in pregnant than non-pregnant women, probably owing to gestational hormonal effects; therefore, lower doses of these medications should be used in pregnancy [11]. Another use of LA is for pain control in the postoperative setting, where medicine can be delivered through a thin tube inserted in the surgery site at the time of operation [3].

14.2.2 Regional Anesthesia

Regional anesthesia (RA) consists of blocking the function of specific nerves by injecting appropriate drugs around them (lidocaine and, more commonly for breast surgery, bupivacaine [12]), causing numbness in the related innervated area [13]. Attention must be kept toward use of lower doses of local anesthetics in pregnant women, as described in the Local Anesthesia section [11]. Some physicians have investigated going through breast surgery with RA alone or in combination with general anesthesia (GA) [3, 14–16].

Practical methods of RA for breast surgery include various nerve blocks, including thoracic paravertebral, subpectoral, serratus anterior, and epidural block [3, 13, 14, 17]. It should be kept in mind that innervation of the breast by intercostal nerves involves several thoracic levels, most commonly from the second to sixth thoracic intercostals (Fig. 14.1). Therefore, these are complex procedures and need sufficient expertise to be successfully handled, especially in pregnant patients. Epidural block is sometimes used in larger operations such as immediate breast reconstruction and consists of blocking thoracic nerves supplying the breast by infusing medicine directly in the epidural space or via epidural catheters. Inserted catheters can stay in place and be used for analgesic infusion in the postoperative period. This technique has been used both alone [18] or with GA for controlling postoperative

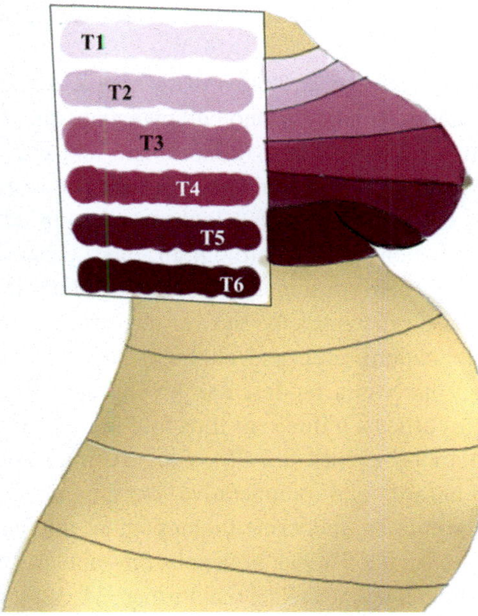

Fig. 14.1 Schematic areas of superficial breast innervation by thoracic intercostal nerves

complications [12]. Considering the position of the breast on the chest wall (see Chap. 1), a paravertebral block involves blocking one to several intercostal nerves near their origin alongside the lateral spinal border in combination with intravenous sedation for performing the surgery, or with GA, to control postoperative pain [13, 14, 19–24]. In the subpectoral block, medicines are injected deep into the pectoralis major muscle. This technique provides good pain relief after surgery, especially for cases including breast reconstruction with subpectoral tissue expanders and prosthesis [17].

RA in combination with intraoperative sedation can be an acceptable alternative to GA for breast surgeries [2], but this has not been studied in pregnancy specifically.

14.2.3 General Anesthesia

Although RA is preferred during gestation when feasible [2, 25, 26], avoiding GA is not always a favorable option while operating on a pregnant woman with PABC, because the procedure could become more complex and the mother and fetus more stressed when awake [27].

14.3 Physiological Changes of Pregnancy – Concerns in Anesthesia

During pregnancy, several physiologic changes occur (see also Chap. 1) in all the body systems, several of which might affect anesthesia for breast surgery.

14.3.1 Respiratory Changes

Mild hyperventilation, decreased pulmonary residual capacity aggravated in the supine position, and chronic respiratory alkalosis occur in gravid mothers. Oxygen consumption increases without affecting PaO_2, but even short-term apnea can quickly cause maternal hypoxemia. In addition, increased vascularization and swelling of upper airways occur, which can lead to difficult intubation and problems in ventilation.

14.3.2 Cardiovascular Changes

In pregnant women, heart rate, stroke volume, blood volume, and cardiac output increase; and hematocrit decreases. Aortocaval compression by the enlarged uterus can reduce cardiac preload and output, causing hypotension. This happens when supine, and is relieved in the lateral position.

14.3.3 Gastrointestinal Changes

Throughout pregnancy, the lower esophageal sphincter pressure decreases, causing a higher risk of gastric content aspiration and warranting more attention focused on airway protection.

14.3.4 Hematological Changes

Gestation produces a hypercoagulable state, that is, a predilection for thromboembolism [2, 4, 25, 28], which needs to be considered through the preoperative, intra-operative, and postoperative phases.

14.4 Breast Surgery-Related Planning before Anesthesia

Issues specific to patients with cancer must be examined by the anesthesiologist in the preoperative assessment (Fig. 14.2). History of recent neoadjuvant treatments for PABC including chemotherapy, endocrine therapy, or, less frequently, radiation therapy should be investigated [3]. Relevant adverse effects of chemotherapy include myelosuppression, which may cause anemia, increased risk of intraoperative bleeding owing to low platelet count, or postoperative infection for leukopenia. Suppression of the bone marrow usually takes around 6 weeks to return to normal [3], and this time interval has been proposed as the adequate delay for anesthesia after chemotherapy, although this timing is subject to controversy. A delay of 21 days or less has recently been shown to be superior in terms of overall survival and recurrence-free survival [29], whereas delays of 4–6 weeks versus 6–8 weeks yielded comparable results elsewhere [30].

Drug-induced cardiomyopathy may also occur secondary to some chemotherapy agents that are used for breast cancer (doxorubicin and cyclophosphamide), revealing the necessity of accurate cardiac assessment including a physical examination, electrocardiogram, and echocardiography before surgery [3] (see also Chap. 15).

Cancer increases the risk for thromboembolic complications [32], and although breast malignancy harbors a modest risk among cancers, it causes at least a three- to four-fold increment in thrombotic events [32–35]. Surgery intensifies the hazard, and preoperative consumption of tamoxifen, an endocrine therapy agent that carries by itself a 5% risk of venous thrombosis [36, 37], would aggravate the risk further [34, 38] (see also Chap. 15). Overall, a ten-fold increment in thromboembolic events has been reported in these settings [39]. It has been suggested that tamoxifen should be discontinued 2–3 weeks before surgery to reduce the risk of perioperative venous thromboembolism [38, 40].

Acid aspiration prophylaxis should be considered in all patients [4]. Another important point to consider is about psychological issues (see also Chap. 28); signs of anxiety or depression that are relatively common in pregnant patients with cancer should be sought and counseled properly to prevent worsening after surgery [31].

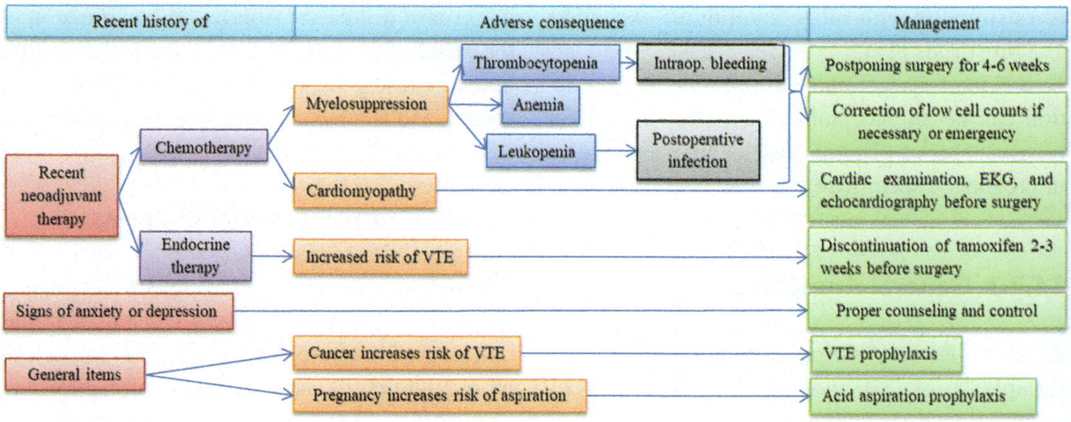

Fig. 14.2 Preoperative assessment of pregnant patients undergoing breast surgery for the anesthesiologist. *Intraop* intraoperative, *VTE* venous thromboembolism

14.5 Management of the Anesthesia

As all the drugs that are used for GA cross the placenta, no ideal technique has been defined; however, the patient and the fetus usually come out of GA without significant adverse consequences [2, 27].

14.5.1 Fetal Concerns

The anesthesiologist must provide safety for the mother and fetus (see also Chap. 21). Central precautionary issues concerning the fetus include avoidance of teratogenic substances, fetal asphyxia, and preterm labor [4].

In animal studies, some negative outcomes for the effects of anesthetics, namely propofol, ketamine, and volatile anesthetics on the fetus have been shown; and nitrous oxide has been shown to induce teratogenic defects in DNA synthesis. However, research on humans has not demonstrated any specific toxicity for the fetus or a birth defect of neonates. Evidence has only suggested slightly increased risks for preterm labor and abortion secondary to GA [1, 3, 4]. Nitrous oxide is better avoided, especially in the first trimester, owing to unproved probable adverse effects for the fetus [4, 25, 26].

Prevention of hypotension and hypoxia is always necessary in any session of GA, but the matter implies more attention to avoid fetal hypoxia and its undesirable perilous consequences [1, 2]. Continuous measurement of fetal heart rate (FHR) is advisable. Any significant decrease of blood pressure needs to be aggressively managed by intravenous fluids and phenylephrine or ephedrine, if necessary [4, 41]. To avoid aortocaval compression, the preferred position for pregnant women throughout anesthesia is the left lateral, though this cannot be offered in most instances of breast surgery, which necessitates specific positioning of the patient depending on the technique of surgery. However, the supine position with head elevation is an acceptable choice both on behalf of the anesthesiologist, for patient safety; and on behalf of the surgeon, for appropriate access to the operation site (see also Chap. 21).

Uterine contractions must be controlled and, as recommended, monitored [1, 25]. Using volatile anesthetics might be useful as a preventive measure in this regard [4, 25], paying attention to the point that pregnancy increases the sensitivity to volatile anesthetic drugs; so that lower dosages are better used during surgery in pregnant patients [4].

14.6 Postoperative Care

Several key points apply to the care of a woman who has undergone breast surgery for cancer while pregnant. Management of postoperative pain is imperative. In cases where an RA catheter has been inserted, infusion of anesthetics via the catheter is an excellent option. However, this is not frequently performed in PABC surgery.

Opioids cross the placenta and sometimes induce decrements in FHR variability, albeit without negative consequences for the neonate; thus, these drugs are regularly used for effective analgesia.

The best method is combination therapy to decrease opioid dosage. Nonsteroidal anti-inflammatory drugs are inappropriate because of fetal adverse effects, but acetaminophen is suitable in combination with opioids for these patients [26, 42, 43].

Monitoring of FHR and uterine contractions has to be undertaken. The patient may not perceive contractions because she is on analgesics, and the medical team must be on alert for the matter. In case of preterm labor, tocolytics should be administered by the gynecologist of the multidisciplinary team [1, 4, 26, 41, 43].

Thromboprophylaxis has not been studied and defined for these cases, and guidelines are not specifically descriptive. Rates of venous thrombosis are higher in patients who have previously received chemotherapy or tamoxifen, have more advanced disease, or have other risk factors for hypercoagulability. Thromboprophylaxis should thus be considered in these patients [41, 43–46].

Psychological support is of utmost importance because these women are facing several stresses simultaneously: pregnancy, cancer, surgery, organ mutilation, and disturbed body image; this deserves special attention and relevant support from the caring multidisciplinary team (see also Chap. 28).

14.7 Anesthesia and Breast Cancer Recurrence

Some studies have paid attention to the probable effect of anesthetics or technique of anesthesia on the prognosis of PABC. Owing to the costs and time that a prospective, well-designed trial would involve, studies are retrospective [47]; some researchers have declared that the combination of propofol and RA can lower the recurrence of PABC [48–52]. This has not been confirmed by other studies [53–56]. Up to the present time, no anesthetic has been known to affect prognosis of the cancer.

References

1. Kuczkowski KM (2004) Nonobstetric surgery during pregnancy: what are the risks of anesthesia? Obstet Gynecol Surv 59(1):52–56
2. Reitman E, Flood P (2011) Anaesthetic considerations for non-obstetric surgery during pregnancy. Br J Anaesth 107(suppl 1):i72–i78
3. Sherwin A, Buggy D (2018) Anaesthesia for breast surgery. BJA Edu 18(11):342–8.0
4. Van De Velde M, De Buck F (2007) Anesthesia for non-obstetric surgery in the pregnant patient. Minerva Anestesiol 73(4):235–240
5. Karvandian K, Zebardast J, Borra NZ (2018) Risk assessment and anesthesia classification in breast cancer surgery. Arch Breast Cancer 5:168–172
6. Carlson GW (2005) Total mastectomy under local anesthesia: the tumescent technique. Breast J 11(2):100–102
7. Kitowski NJ, Landercasper J, Gundrum JD, De Maiffe BM, Chestnut DH, Bottcher ML et al (2010) Local and paravertebral block anesthesia for outpatient elective breast cancer surgery. Arch Surg 145(6):592–594
8. Devereux D (1987) Successful treatment of stages IIIa and IIIb carcinoma of the breast by mastectomy in the elderly high risk patient using local anesthesia. Surg Gynecol Obstet 165(1):38–40
9. Markose G, Graham RM (2017) Anaesthesia: LA in pregnancy. Br Dent J 222(1):3–4
10. Hagai A, Diav-Citrin O, Shechtman S, Ornoy A (2015) Pregnancy outcome after in utero exposure to local anesthetics as part of dental treatment: a prospective comparative cohort study. J Am Dent Assoc 146(8):572–580
11. Fayans EP, Stuart HR, Carsten D, Ly Q, Kim H (2010) Local anesthetic use in the pregnant and postpartum patient. Dental Clinic 54(4):697–713
12. Atanassoff PG, Alon E, Weiss BM (1994) Intercostal nerve block for lumpectomy: superior postoperative pain relief with bupivacaine. J Clin Anesth 6(1):47–51
13. Greengrass R, O'brien F, Lyerly K, Hardman D, Gleason D, D'Ercole F et al (1996) Paravertebral block for breast cancer surgery. Can J Anaesth 43(8):858–861
14. Pusch F, Freitag H, Weinstabl C, Obwegeser R, Huber E, Wildling E (1999) Single-injection paravertebral block compared to general anaesthesia in breast surgery. Acta Anaesthesiol Scand 43(7):770–774
15. Levene JL, Weinstein EJ, Cohen MS, Andreae DA, Chao JY, Johnson M et al (2019) Local anesthetics and regional anesthesia versus conventional analgesia for preventing persistent postoperative pain in adults and children: a Cochrane systematic review and meta-analysis update. J Clin Anesth 55:116–127
16. Tahiri Y, Tran DQ, Bouteaud J, Xu L, Lalonde D, Luc M et al (2011) General anaesthesia versus thoracic paravertebral block for breast surgery: a meta-analysis. J Plast Reconstr Aesthet Surg 64(10):1261–1269
17. Blanco R (2011) The 'pecs block:' a novel technique for providing analgesia after breast surgery. Anaesthesia 66(9):847–848
18. Lynch EP, Welch KJ, Carabuena JM, Eberlein TJ (1995) Thoracic epidural anesthesia improves outcome after breast surgery. Ann Surg 222(5):663–669
19. Cali Cassi L, Biffoli F, Francesconi D, Petrella G, Buonomo O (2017) Anesthesia and analgesia in breast surgery: the benefits of peripheral nerve block. Eur Rev Med Pharmacol Sci 21(6):1341–1345
20. Abdallah FW, Morgan PJ, Cil T, McNaught A, Escallon JM, Semple JL et al (2014) Ultrasound-guided multilevel paravertebral blocks and total intravenous anesthesia improve the quality of recovery after ambulatory breast tumor resection. Anesthesiology 120(3):703–713
21. Coveney E, Weltz CR, Greengrass R, Iglehart JD, Leight GS, Steele SM et al (1998) Use of paravertebral block anesthesia in the surgical management of breast cancer: experience in 156 cases. Ann Surg 227(4):496–501
22. Najarian MM, Johnson JM, Landercasper J, Havlik P (2003) Paravertebral block: an alternative to general anesthesia in breast cancer surgery/discussion. Am Surg 69(3):213–218
23. Klein SM, Bergh A, Steele SM, Georgiade GS, Greengrass RA (2000) Thoracic paravertebral block for breast surgery. Anesth Analg 90(6):1402–1405

24. Terheggen MA, Wille F, Rinkes IHB, Ionescu TI, Knape JT (2002) Paravertebral blockade for minor breast surgery. Anesth Analg 94(2):355–359
25. Kuczkowski KM (2006) The safety of anaesthetics in pregnant women. Expert Opin Drug Saf 5(2):251–264
26. Ravindra G, Madamangalam AS, Seetharamaiah S (2018) Anaesthesia for non-obstetric surgery in obstetric patients. Indian J Anaesth 62(9):710–716
27. Hayashi T, Katayama M, Miyasaka K (2013) General anesthesia for breast cancer surgery during pregnancy: a retrospective review at a breast cancer center in Tokyo between 1999–2011: 11AP5-2. Eur J Anaesthesiol 30:178
28. Rybstein MD, DeSancho MT (2019) Risk factors for and clinical management of venous thromboembolism during pregnancy. Risk 17(7):396–404
29. Omarini C, Guaitoli G, Noventa S, Andreotti A, Gambini A, Palma E et al (2017) Impact of time to surgery after neoadjuvant chemotherapy in operable breast cancer patients. Eur J Surg Oncol 43(4):613–618
30. Sanford RA, Lei X, Barcenas CH, Mittendorf EA, Caudle AS, Valero V et al (2016) Impact of time from completion of neoadjuvant chemotherapy to surgery on survival outcomes in breast cancer patients. Ann Surg Oncol 23(5):1515–1521
31. Pitman A, Suleman S, Hyde N, Hodgkiss A (2018) Depression and anxiety in patients with cancer. BMJ 361:k1415
32. Walker AJ, Card TR, West J, Crooks C, Grainge MJ (2013) Incidence of venous thromboembolism in patients with cancer–a cohort study using linked United Kingdom databases. Eur J Cancer 49(6):1404–1413
33. Horsted F, West J, Grainge MJ (2012) Risk of venous thromboembolism in patients with cancer: a systematic review and meta-analysis. PLoS Med 9(7):e1001275
34. Walker AJ, West J, Card TR, Crooks C, Kirwan CC, Grainge MJ (2016) When are breast cancer patients at highest risk of venous thromboembolism? A cohort study using English health care data. Blood 127(7):849–857
35. Cronin-Fenton DP, Søndergaard F, Pedersen L, Fryzek J, Cetin K, Acquavella J et al (2010) Hospitalisation for venous thromboembolism in cancer patients and the general population: a population-based cohort study in Denmark, 1997–2006. Br J Cancer 103(7):947–953
36. Meier CR, Jick H (1998) Tamoxifen and risk of idiopathic venous thromboembolism. Br J Clin Pharmacol 45(6):608–612
37. Xu X, Chlebowski RT, Shi J, Barac A, Haque R (2019) Aromatase inhibitor and tamoxifen use and the risk of venous thromboembolism in breast cancer survivors. Breast Cancer Res Treat 174(3):785–794
38. Heery M, Corbett P, Zelkowitz R (2018) Precautions for patients taking tamoxifen. J Adv Pract Oncol 9(1):78–83
39. Parikh RP, Odom EB, Yu L, Colditz GA, Myckatyn TM (2017) Complications and thromboembolic events associated with tamoxifen therapy in patients with breast cancer undergoing microvascular breast reconstruction: a systematic review and meta-analysis. Breast Cancer Res Treat 163(1):1–10
40. Hussain T, Kneeshaw PJ (2012) Stopping tamoxifen peri-operatively for VTE risk reduction: a proposed management algorithm. Int J Surg 10(6):313–316
41. Mhuireachtaigh RN, O'Gorman DA (2006) Anesthesia in pregnant patients for nonobstetric surgery. J Clin Anesth 18(1):60–66
42. Lamvu G, Feranec J, Blanton E (2018) Perioperative pain management: an update for obstetrician-gynecologists. Am J Obstet Gynecol 218(2):193–199
43. Stewart MK, Terhune KP (2015) Management of pregnant patients undergoing general surgical procedures. Surg Clin North Am 95(2):429–442
44. Pannucci CJ, Shanks A, Moote MJ, Bahl V, Cederna PS, Naughton NN et al (2012) Identifying patients at high risk for venous thromboembolism requiring treatment after outpatient surgery. Ann Surg 255(6):1093–1099
45. Collins J, Bowles L, MacCallum PK (2016) Prevention and management of venous thromboembolism in pregnancy. Br J Hosp Med 77(12):C194–C200
46. ACOG Practice Bulletin No. 196 (2018) Thromboembolism in pregnancy. Obstet Ggynecol 132(1):e1–e17
47. Heaney A, Buggy DJ (2012) Can anaesthetic and analgesic techniques affect cancer recurrence or metastasis? Br J Anaesth 109(suppl-1):i17–i28
48. Garg R (2017) Regional anaesthesia in breast cancer: benefits beyond pain. Ind J Aanaesth 61(5):369–372
49. Ni Eochagain A, Burns D, Riedel B, Sessler DI, Buggy DJ (2018) The effect of anaesthetic technique during primary breast cancer surgery on neutrophil–lymphocyte ratio, platelet–lymphocyte ratio and return to intended oncological therapy. Anaesthesia 73(5):603–611
50. Exadaktylos AK, Buggy DJ, Moriarty DC, Mascha E, Sessler DI (2006) Can anesthetic technique for primary breast cancer surgery affect recurrence or metastasis? Anesthesiology 105(4):660–664
51. Buckley A, McQuaid S, Johnson P, Buggy DJ (2014) Effect of anaesthetic technique on the natural killer cell anti-tumour activity of serum from women undergoing breast cancer surgery: a pilot study. Br J Anaesth 113(suppl-1):i56–i62
52. Matsumoto M, Flores EM, Kimachi PP, Gouveia FV, Kuroki MA, Barros ACSD et al (2018) Benefits in radical mastectomy protocol: a randomized trial evaluating the use of regional anesthesia. Sci Rep 8(1):7815
53. Kairaluoma P, Mattson J, Heikkilä P, Pere P, Leidenius M (2016) Perioperative paravertebral regional anaesthesia and breast cancer recurrence. Anticancer Res 36(1):415–418

54. Finn DM, Ilfeld BM, Unkart JT, Madison SJ, Suresh PJ, Sandhu NPS et al (2017) Post-mastectomy cancer recurrence with and without a continuous paravertebral block in the immediate postoperative period: a prospective multi-year follow-up pilot study of a randomized, triple-masked, placebo-controlled investigation. J Anesth 31(3):374–379
55. Starnes-Ott K, Goravanchi F, Meininger JC (2015) Anesthetic choices and breast cancer recurrence: a retrospective pilot study of patient, disease, and treatment factors. Crit Care Nurs Q 38(2):200–210
56. Sessler DI, Pei L, Huang Y, Fleischmann E, Marhofer P, Kurz A, Mayers DB, Meyer-Treschan TA, Grady M, Tan EY, Ayad S (2019) Recurrence of breast cancer after regional or general anaesthesia: a randomised controlled trial. Lancet 394(10211):1807–1815

Systemic Treatments in Pregnancy-Associated Breast Cancer

15

Omid S. Tehrani

Abstract

Available data on systemic treatments in pregnancy-associated breast cancer (PABC) is reviewed in this section. These treatments include chemotherapy, endocrine therapy (ET), small molecule inhibitors, monoclonal antibodies against human epidermal growth factor receptor 2 (EGFR-2) also known as HER2; and human epidermal growth factor receptor 3 (EGFR-3), also known as HER3.

In local disease, systemic treatment can be delivered as neoadjuvant (before surgery) or adjuvant (after surgery) treatment. In metastatic disease, systemic therapy is the main modality of treatment.

Approach to PABC is based on available data in the general population, limited only by safety issues for use of medications during gestation and lactation. Therefore, treatments are similar to non-PABC patients while trying to minimize the risk to the fetus. Available data on different chemotherapies, anti-HER2 monoclonal antibodies, ET and small molecule inhibitors are discussed in detail.

Keywords

Anti-HER2 · Chemotherapy · Immunotherapy · Monoclonal antibody · Targeted therapy

15.1 Staging of Pregnancy-Associated Breast Cancer

Every cancer diagnosis needs staging, and breast cancer is no exception. Clinical findings might identify metastatic disease. Distant lymphadenopathy, skin lesions, dry coughs, abnormal liver enzymes or alkaline phosphatase, pathologic fractures, increasing or new pain in the body or skeleton, blurred vision, intractable nausea and vomiting, imbalance, and headaches should alarm the physician to look for metastatic disease. However, there are limitations for imaging techniques during pregnancy. Patients can be evaluated by means of chest radiograph with fetal shielding, ultrasound of the liver, and magnetic resonance imaging (MRI) of the spine without using contrast in order to evaluate metastases. Patients with findings suggestive of distant organ involvement may benefit from further work up including biopsy of the pertinent sites. After staging the disease, patients will be treated accordingly.

O. S. Tehrani (✉)
Department of Hematology and Medical Oncology, Stanford University, Stanford, CA, USA
e-mail: Tehrani@stanford.edu

S. Alipour, R. Omranipour (eds.), *Diseases of the Breast during Pregnancy and Lactation*, Advances in Experimental Medicine and Biology 1252,
https://doi.org/10.1007/978-3-030-41596-9_15

15.2 Loco-Regional Disease: Adjuvant and Neoadjuvant Systemic Treatments during Pregnancy and Lactation

Systemic treatment in early stages or locoregional cases of breast cancer is administered as neoadjuvant or adjuvant forms. PABC is commonly treated with surgery followed by adjuvant treatment (if necessary), although more data is being available on using neoadjuvant therapy.

Most of the discussion here will be focused on adjuvant treatments. Evidence-based practice for adjuvant systemic therapies comes from non-pregnant patients, and therefore recommendations are similar, except for contraindications related to the pregnancy. The goal is to reduce the chance of cancer recurrence while protecting the mother and the embryo or fetus from potential toxicities of the treatments. Decision on the type and duration of treatment relies on the estimated chance of disease recurrence, and the potential of each regimen in reducing that chance. The size and extent of the primary tumor, number of the involved regional lymph nodes, hormone receptor and HER2 overexpression status, as well as genomic tests are among the factors involved in such evaluations. Approximately 60–80% of breast cancers diagnosed in pregnant women are estrogen receptor (ER) negative and about 28–58% of cases have HER2 over-expression [1–6]. It is estimated that more than half of the patients have regional lymph node involvement [6, 7] (see also Chaps. 10 and 11).

15.2.1 Chemotherapy

During systemic therapy, fetal monitoring should be added to usual tests for directing a healthy pregnancy. Present data suggests that exposure to chemotherapy in the first trimester of pregnancy, during which organogenesis occurs, has the greatest risk of fetal malformation [8, 9]. In contrast, it is estimated that the risk of fetal malformation after exposure to chemotherapy in the second and third trimesters is approximately 1.3%, which is not significantly different from the outcome without exposure to chemotherapy [8]. Suitable timing of chemotherapy is discussed later.

The treating physician and the patient must be aware that chemotherapy-induced amenorrhea occurs in about 20–70% of premenopausal breast cancer patients, and recovery depends on the age and the regimen used [10, 11]. This might also happen after delivery in pregnant patients undergoing chemotherapy. After delivery, some level of protection to the ovaries can be achieved with ovarian suppression using gonadotropin releasing hormone agonists; however there is still some risk of chemotherapy-induced ovarian failure [12, 13] (see also Chap. 24). Fertility clinics might help with other options before initiation of the chemotherapy after delivery. Some of those include cryopreservation of ovarian tissue, eggs or oocytes (see also Chap. 25).

There are significant differences in pharmacokinetics of the medications between gestational and non-gestational state (see also Chap. 14). Amniotic fluid can work as a third space fluid, and pregnancy causes significant changes in the blood volume, liver metabolism, renal flow, protein binding, and therefore volume of distribution and clearance of the medications and their active metabolites [14, 15]. With such changes in pharmacokinetics, combined chemotherapy or dose dense regimen can be associated with higher risk of toxicity in this period. As an example, per recommendation of the National Comprehensive Cancer Network (NCCN, 2019), weekly dosing of paclitaxel might be associated with lower risk of toxicity and is preferred over dose dense paclitaxel every 2 weeks. However, a small study reported the outcome of pregnancy in 10 women who received dose-dense chemotherapy every 2 weeks, and 99 women who received conventional chemotherapy, at 3-week intervals. Birth weight, gestational age at delivery, congenital anomalies, and incidence of neutropenia were not significantly different [16]. Dose dense chemotherapy regimen may also necessitate the use of growth factors and transfusions during pregnancy. In each case, potential benefits (decreased relapse rate and increased overall survival) should be weighed against the risks to the mother and fetus' health.

The largest experience in pregnancy has been with anthracycline and alkylating agents [17, 18]. Overall, methotrexate, an antimetabolite, and cyclophosphamide, an alkylating agent, have been reported to be associated with miscarriage and malformations [19–22]. Methotrexate has been associated with fetal methotrexate syndrome, central nervous system, skeletal, gastrointestinal, and cardiac malformations, and even fetal death [23]. Considering the risk of accumulation of methotrexate in the third space, the combination of cyclophosphamide, methotrexate and 5-fluorouracil (CMF) can be replaced with newer regimens. Among chemotherapy agents used in the second and third trimesters, there is more data available on doxorubicin (an anthracycline), cyclophosphamide, and fluorouracil. Several case studies reported no greater risk or specific complications after use of anthracyclines and alkylating agents in the second and third trimesters, either as a single agent or in combination [24–31]. Additionally, a single-institution studied 57 pregnant patients. They received FAC chemotherapy (5-FU 500 mg/m^2 IV days 1 and 4, doxorubicin 50 mg/m^2 by IV infusion over 72 h, and cyclophosphamide 500 mg/m^2 IV day1) safely during the second and third trimesters of pregnancy [7, 18]. There was one child with trisomy-21 (Down syndrome) and two with congenital abnormalities (club foot, congenital bilateral ureteral reflux). These two children are reported to be healthy and progressing well in school [6, 8, 9]. Similarly, single agent anthracyclines have been safely used during pregnancy [32]. In pregnant women, epirubicin has a shorter terminal half-life than doxorubicin because of difference in glucoronidation [33]. Additionally, there is a single institute report of weekly epirubicin (35 mg/m^2) used in the second and third trimesters with no grade III or IV adverse effects and no fetal anomalies [34]. There is also a case report on a patient treated with FEC (5FU, epirubicin and cyclophosphamide) as well as irradiation during conception and the first two trimesters of pregnancy without fetal anomalies [35].

There are some limited data on the use of taxanes during pregnancy [24, 29, 36, 37]. In vitro studies on human primary trophoblasts reported changes in the expression of drug transporters upon exposure to paclitaxel [38]. This finding suggests a different method of transfer of anticancer agents to the fetus in those exposed to paclitaxel. A review of articles published on platinum-based chemotherapy with taxanes (paclitaxel and carboplatin, paclitaxel and cisplatin, and docetaxel and cisplatin, used for ovarian cancer during pregnancy) reported lower average weight of the neonates at the time of delivery (2442.1 g). Exposure to these agents during the second and third trimesters did not increase fetal loss [39].

Overall, when adjuvant chemotherapy is needed, methotrexate containing regimens are not recommended during pregnancy; while combinations comprising anthracyclines and taxanes seem to be safe later in pregnancy. Dose-dense regimens might have higher risks of complications in the gestational period.

15.2.2 Endocrine Therapy

The main adjuvant ET in premenopausal hormone receptor positive, HER2 negative breast cancer is tamoxifen. However, animal studies have reported adverse events in the gravid state, including fetal demise [40, 41]. A review on outcome of pregnancy in 167 cases exposed to tamoxifen reported abnormal fetal development in 12.6% of cases in contrast to 3.9% in unexposed patients. The malformations were non-specific, and the majority of cases had healthy newborns [42]. In addition to the risk of fetal anomalies, animal studies raise a concern about risk of tumor development in the next generation [43, 44]. Such findings support the generally accepted contraindication of tamoxifen in pregnancy.

Among other ETs, aromatase inhibitors are not used in premenopausal patients and fulvestrant is not used in the adjuvant setting.

Genomic tests might help determine the benefit of adding chemotherapy to adjuvant hormonal blockade in hormone receptor positive, HER2 negative breast cancer. Although speculated to be of the same value as the general population, the role of such tests has not been determined in PABC.

15.2.3 Anti-HER2 Therapy

Patients with overexpression of HER2 may benefit from anti-HER2 and anti-HER3 treatments. However, these medications are contraindicated during pregnancy. Outcome of pregnancy has been reported in patients who had unintentional exposure to trastuzumab and/or lapatinib during pregnancy in two trials: the Neoadjuvant Lapatinib and/or Trastuzumab Treatment Optimization (NeoALTTO) and the Adjuvant Lapatinib and/or Trastuzumab Treatment Optimization (ALTTO) trials [45]. Ninety-two patients had a pregnancy, 12 in the exposed group and 80 in the unexposed one. Seven patients (58.3%) in the exposed and 10 (12.5%) in the unexposed group opted for an induced abortion. In the remaining patients in the exposed group, there was a fetus with trisomy 21 (Down syndrome), but no other adverse effects were reported in the pregnancy and delivery period of other patients. As well, disease free survival (DFS) was not reported to be different between the two groups.

There are also case reports of treatment with trastuzumab during pregnancy [46–53]. A case report followed a child exposed to trastuzumab during first trimester, and did not report any complications later [46]. A meta-analysis on 17 case reports with 18 pregnancies and 19 newborns demonstrated that the majority of complications were oligohydramnios or anhydramnios, especially when exposed to trastuzumab in the second or third trimester [54]. Therefore, if possible, trastuzumab should be postponed to the postpartum period.

Ado-trastuzumab is an anti HER2 drug-antibody conjugate. Since the antibody used in this medication is trastuzumab, it has been contraindicated during pregnancy based on trastuzumab data. Food and Drug Administration (FDA) reports delayed kidney development, oligohydramnios, anhydramnios, and even fetal death in monkeys that were exposed during pregnancy at concentrations 2.5–20 times higher than therapeutic concentrations. Similarly, there are case reports on oligohydramnios and anhydramnios in pregnant patients when combined with trastuzumab [55].

Current anti-HER therapies are not used during pregnancy due to their potential complications, but can be considered after delivery.

15.2.4 Timing of Chemotherapy

In general, anti-neoplastic agents should be avoided in the first trimester. As a result, when continuation of pregnancy is considered in the first trimester, chemotherapy is generally postponed to avoid abortion or severe malformations in the fetus. Estimation of the date of conception and the date of the delivery will help with systemic chemotherapy planning. Additionally, chemotherapy is usually avoided within the last 3 weeks before planned delivery, or after week 35 of pregnancy to decrease chance of hematologic complications during child birth (see also Chap. 21).

Postponing chemotherapy till after delivery may be a reasonable option when possible. But if the risk is high and the delay is too long, a decision must be made on the basis of risks versus benefits of postponing treatment. Although the more common approach to the treatment of breast cancer is to do surgical resection before systemic adjuvant therapy, neoadjuvant chemotherapy is a potential option as well. Neoadjuvant treatment is usually used to avoid mastectomy or to postpone surgical resection till after delivery. In comparison, adjuvant treatment is done after surgical resection, which means that the gestational age at the time of chemotherapy is higher. This might be a theoretic advantage as treatment is given farther from embryogenesis.

While information is scarce, there is a study on data of the registry of the German Breast Group (GBG 29/BIG 02-03) and the international Cancer in Pregnancy initiative, on those who received neoadjuvant treatment. They identified 103 patients with PABC, 62 of them received treatment during pregnancy and 41 were treated after delivery. In this group, 26% received non-taxane regimen, 73% had taxane and 24% received trastuzumab containing regimen. They did not report significant differences in disease free survival and overall survival among groups, and compared to 130 non-PABC patients who received neoadjuvant treatment [56].

In general, when possible, current recommendations are to postpone systemic treatments till second or third trimester of pregnancy, or after delivery.

15.3 Metastatic Disease: Systemic Treatments during Pregnancy and Lactation

Similar to treatment of locoregional breast cancer, evidence for the best treatments of metastatic breast cancer is mainly gained from data about non-pregnant women. Thus, recommendations are similar, unless for specific contraindications during pregnancy.

In general, metastatic cancer is considered incurable. The goal of care in this setting is to prolong mother's life, while minimizing the risk to the embryo if possible. Progression of the disease may lead to further spread to vital organs, organ damage and fracture; and may limit therapeutic options. Therefore, postponing treatment can increase morbidity and mortality. The goal of care and the time to initiate the treatment should be discussed with the patient. When delivery of a viable neonate is possible, an obstetrician; and in case of need, a neonatologist should be consulted. Choice of systemic therapy is based on the status of hormone receptors and HER2 expression, involvement of the visceral organs (mainly liver) or central nervous system (CNS), and patient's functional status.

15.3.1 Endocrine Therapy

This is the main treatment in most patients with metastatic hormone receptor positive disease. However tamoxifen, a selective estrogen receptor modulator, is contraindicated due to its potential for causing fetal demise and abnormal fetal development, although the majority of the infants exposed to tamoxifen were born healthy [40–42]. Another concern in using tamoxifen is the chance of developing tumors in the offspring, based on animal studies [44].

Ovarian suppression is another ET used in the non-pregnant population. Although there is limited data on the safety of gonadotropin releasing hormone (GnRH) analogues during pregnancy, these are considered contraindicated due to fetal loss in animal studies. Even so, a report on 6 cases with exposure to leuprolide during pregnancy did not report teratogenicity [57]. However, the efficacy of GnRH analogues and antagonists during pregnancy is also questionable. Beta-hCG levels surge rapidly early in pregnancy and support the gravid corpus luteum, which produces estrogen and progesterone [58]. Therefore, effect of GnRH analogues and antagonists in decreasing estrogen during pregnancy is questionable. Similarly, there is limited information on the effect of oophorectomy on the pregnancy and breast cancer [59].

Aromatase inhibitors are not effective in premenopausal patients and there is limited data on their safety during pregnancy.

Fulvestrant, a selective estrogen receptor modulator degrader, is approved as a single agent or combined with CDK4/6 inhibitors for the treatment of metastatic breast cancer; but is contraindicated during pregnancy (FDA category X). Fetal loss and anomalies occurred in animal studies at much smaller concentrations used in the clinical setting, per FDA accessible data. Therefore, the options in this group of patients will be limited to chemotherapy in second and third trimesters.

15.3.2 Small Molecule Inhibitors and Monoclonal Antibodies

Metastatic hormone receptor positive disease with significant liver involvement (visceral disease) can be treated with chemotherapy or ET [60]; in the general population, CDK4/6 inhibitors are usually added to the ET to increase response rate and duration [61–64]. With exception of abemaciclib, CDK4/6 inhibitors are generally used in combination with ET for treatment of metastatic hormone receptor positive breast cancer [65]. However, it seems that this class of drugs is associated with significant fetal loss and malformations when used during pregnancy. Short of published papers, FDA data shows skeletal variations and reduced body weight in animals treated with palbociclib. Fetal loss and fetal abnormalities have been seen as well in pregnant animals treated with ribociclib. Similarly, abemaciclib caused decreased fetal body weights, and cardiovascular or skeletal malformations. Also, ribociclib and its metabolites were reported in the milk of lactating animals. Therefore, it is recommended to prevent pregnancy when under treatment, and avoid using CDK4/6 inhibitors in pregnant women and during breastfeeding.

PI3K/AKT/mTOR is an important pathway in hormone positive breast cancer. Everolimus and alpelisib are examples of approved drugs working on this pathway. Both medications are used in combination with hormonal blockade. Everolimus has been in the market for a longer time and therefore, there are published data on the outcome of pregnant patients exposed to it. Although congenital malformations have been reported in animal studies, review of several cases treated with everolimus for other diseases showed no congenital malformations in the human fetuses exposed during the entire pregnancy [66–68]. Notably, they reported similar concentration of everolimus in mother and cord blood (66). These data are not enough to prove the safety of the drug during pregnancy and this medication should be avoided in this period. Limited available data provided by FDA on the alpelisib indicates increased risk of fetal demise and malformations. Due to the lack of data in human studies, it is recommended to avoid this medication during pregnancy.

Poly (ADP-ribose) polymerase (PARP) inhibitors are used in metastatic hormone positive breast cancer patients with germline BRCA1 or BRCA2 mutations. Based on FDA data accessible to the public, studies in the pregnant rats showed that olaparib caused teratogenicity and embryo-fetal toxicity at exposures below those used in human patients. Rucaparib exposure has resulted in fetal death in pregnant rats at much lower concentrations than those used in humans. Similarly, talazoparib was reported to cause fetal malformation and death in pregnant rats at lower concentrations than in humans. No such data is published for niraparib. It is recommended that this category of medications should be avoided during pregnancy as well.

In HER2 positive breast cancer, trastuzumab, pertuzumab, fam-trastuzumab deruxtecan and ado-trastuzumab are the frequently used monoclonal antibodies. Among small molecule inhibitors targeting HER2, lapatinib and neratinib are approved. There is not a lot of experience or studies using them in pregnancy. It seems that anti-HER2 monoclonal antibodies increase the chance of fetal kidney damage, oligohydramnios, and adrenal disease [46–53, 69]. Despite reported increased abortions in patients exposed to trastuzumab and/or lapatinib in NeoALTTO and ALTTO trials, there are case reports of trastuzumab used in pregnancy without complications [69]. There is a single case report of a normal newborn despite exposure to lapatinib in the first trimester [70]. As mentioned, a meta-analysis on 17 reported cases indicated that the majority of oligohydramnios or anhydramnios with administration of trastuzumab occurred in those who were exposed in the second or third trimester [54]. Nonetheless, FDA reported that treating animals with lapatinib during organogenesis and lactation led to death of offspring, and neratinib caused fetal death and anomalies. Therefore, FDA recommended avoiding these drugs during pregnancy and lactation.

Atezolizumab is the first checkpoint inhibitor used in triple negative breast cancer. Multiple clinical trials are evaluating the benefit of checkpoint inhibitors in early stage breast cancer. However, there are no trials in PABC. In addition, PD-L1 is shown to be highly expressed in syncytiotrophoblast and to a lower extent in the chorion laeve and the implantation site. These findings suggest potential risks of damage to the placenta with anti-PD-1 or anti-PDL-1 treatments during pregnancy [71].

Considering that methotrexate, trastuzumab, pertuzumab, fam-trastuzumab deruxtecan, ado-trastuzumab, tamoxifen, aromatase inhibitors, LHRH agonists, fulvestrant, CDK4/6 inhibitors, mTOR inhibitor and PIK3 inhibitor should be avoided, other chemotherapeutic agents remain the main option in the second and third trimesters of pregnancy in metastatic PABC.

15.4 Oligometastatic Disease

It seems reasonable to assume that leaving oligometastatic disease untreated may result in widely metastatic disease without a chance of cure. However, there are a variety of approaches to these patients. One may consider local treatments (radiation, microwave ablation, cryotherapy or excision) for single-site metastasis [72–76]. Another approach is to treat these cases similar to widely metastatic disease and meanwhile, monitor the patient for developing additional metastatic foci. If no additional sites of disease are found during systemic treatment, then one may consider local treatments for the oligometastatic sites.

The approach to oligometastatic disease should be based on the potential benefits and risks to the health of the mother and the fetus and availability of the treatments.

15.5 Concerns in Lactation

Among patients receiving chemotherapy during pregnancy, about half were successful in breastfeeding their newborns [7] (see also Chap. 22). Due to the chance of transmission of chemotherapy drugs and small-molecule inhibitors (target therapies) into the milk, breastfeeding is not recommended while receiving these treatments. Examples include cyclophosphamide and methotrexate, which are among drugs that reach relatively high concentrations in breast milk [77].

Guidelines, including that of the Breast Disease Committee of the Society of Obstetricians and Gynecologists of Canada, recommended that women undergoing chemotherapy or tamoxifen treatment should not breastfeed; however, this is weakly supported by the evidence (level of evidence at the time: III-B).

Based on limited available data, paclitaxel appears to be excreted into breast milk in relatively large amounts. In one case, paclitaxel was detectable in the milk for at least 1 week after use, but not at 13 days after a dose of 30 mg/m^2 [77]. A case study on a patient treated with carboplatin and paclitaxel showed concentration of the chemotherapeutic medications in the breast milk samples collected during treatment. Carboplatin had a relative infant dose of 2% and remained measurable after 316 h post-infusion. Paclitaxel had a relative infant dose of 16.7% but was immeasurable by 316 h after infusion [78]. Therefore, with intermittent treatments, breastfeeding might be safe with an appropriate period of breastfeeding abstinence, but the duration of abstinence is not clear.

Endogenous immunoglobulins are found in breast milk, and therefore, it seems reasonable to assume therapeutic monoclonal antibodies, including trastuzumab, fam-trastuzumab deruxtecan and ado-trastuzumab may be present in the milk of nursing mothers as well. Due to the potential for serious adverse reactions in the breastfed infant, the manufacturers recommend that women should not breastfeed during treatment and for 7 months after the last dose of these monoclonal antibodies.

Among targeted therapies, neratinib also has been shown to be present in the milk in animal studies and is recommended to be avoided.

Patients under treatment with chemotherapy, targeted therapy, and ET should avoid breastfeeding till more data confirms the safety of these medications. Duration of abstinence varies by the therapeutic agents and further data is required.

References

1. Aziz S, Pervez S, Khan S, Siddiqui T, Kayani N, Israr M et al (2003) Case control study of novel prognostic markers and disease outcome in pregnancy/

lactation-associated breast carcinoma. Pathol Res Pract 199(1):15–21
2. Bonnier P, Romain S, Dilhuydy JM, Bonichon F, Julien JP, Charpin C et al (1997) Influence of pregnancy on the outcome of breast cancer: a case-control study. Int J Cancer 72(5):720–727
3. Ishida T, Yokoe T, Kasumi F, Sakamoto G, Makita M, Tominaga T et al (1992) Clinicopathologic characteristics and prognosis of breast cancer patients associated with pregnancy and lactation: analysis of case-control study in Japan. Jpn J Cancer Res 83(11):1143–1149
4. Middleton LP, Amin M, Gwyn K, Theriault R, Sahin A (2003) Breast carcinoma in pregnant women: assessment of clinicopathologic and immunohistochemical features. Cancer 98(5):1055–1060
5. Elledge RM, Ciocca DR, Langone G, McGuire WL (1993) Estrogen receptor, progesterone receptor, and HER2/neu protein in breast cancers from pregnant patients. Cancer 71(8):2499–2506
6. Hahn KME, Johnson PH, Gordon N, Kuerer H, Middleton L, Ramirez M et al (2006) Treatment of pregnant breast cancer patients and outcomes of children exposed to chemotherapy in utero. Cancer 107(6):1219–1226
7. Cardonick E, Dougherty R, Grana G, Gilmandyar D, Ghaffar S, Usmani A (2010) Breast cancer during pregnancy: maternal and fetal outcomes. Cancer J 16(1):76–82
8. Doll DC, Ringenberg QS, Yarbro JW (1989) Antineoplastic agents and pregnancy. Semin Oncol 16(5):337–346
9. Ebert U, Löffler H, Kirch W (1997) Cytotoxic therapy and pregnancy. Pharmacol Ther 74(2):207–220
10. Minisini AM, Menis J, Valent F, Andreetta C, Alessi B, Pascoletti G et al (2009) Determinants of recovery from amenorrhea in premenopausal breast cancer patients receiving adjuvant chemotherapy in the taxane era. Anticancer Drugs 20(6):503–507
11. Walshe JM, Denduluri N, Swain SM (2006) Amenorrhea in premenopausal women after adjuvant chemotherapy for breast cancer. J Clin Oncol Off J Am Soc Clin Oncol 24(36):5769–5779
12. Lambertini M, Moore HCF, Leonard RCF, Loibl S, Munster P, Bruzzone M et al (2018) Gonadotropin-releasing hormone agonists during chemotherapy for preservation of ovarian function and fertility in premenopausal patients with early breast cancer: a systematic review and meta-analysis of individual patient-level data. J Clin Oncol Off J Am Soc Clin Oncol 36(19):1981–1990
13. Tao T, Del Valle A (2008) Human oocyte and ovarian tissue cryopreservation and its application. J Assist Reprod Genet 25(7):287–296
14. Redmond GP (1985) Physiological changes during pregnancy and their implications for pharmacological treatment. Clin Invest Med 8(4):317–322
15. Wiebe VJ, Sipila PE (1994) Pharmacology of antineoplastic agents in pregnancy. Crit Rev Oncol Hematol 16(2):75–112
16. Cardonick E, Gilmandyar D, Somer RA (2012) Maternal and neonatal outcomes of dose-dense chemotherapy for breast cancer in pregnancy. Obstet Gynecol 120(6):1267–1272
17. Fuster D, Duch J, Paredes P, Velasco M, Muñoz M, Santamaría G et al (2008) Preoperative staging of large primary breast cancer with [18F]fluorodeoxyglucose positron emission tomography/computed tomography compared with conventional imaging procedures. J Clin Oncol 26(29):4746–4751
18. Groheux D, Moretti J-L, Baillet G, Espie M, Giacchetti S, Hindie E et al (2008) Effect of 18F-FDG PET/CT imaging in patients with clinical Stage II and III breast cancer. Int J Radiat Oncol Biol Phys 71(3):695–704
19. Bawle EV, Conard JV, Weiss L (1998) Adult and two children with fetal methotrexate syndrome. Teratology 57(2):51–55
20. Glantz JC (1994) Reproductive toxicology of alkylating agents. Obstet Gynecol Surv 49(10):709–715
21. Zemlickis D, Lishner M, Degendorfer P, Panzarella T, Sutcliffe SB, Koren G (1992) Fetal outcome after in utero exposure to cancer chemotherapy. Arch Intern Med 152(3):573–576
22. Schapira DV, Chudley AE (1984) Successful pregnancy following continuous treatment with combination chemotherapy before conception and throughout pregnancy. Cancer 54(5):800–803
23. Corona-Rivera JR, Rea-Rosas A, Santana-Ramírez A, Acosta-León J, Hernández-Rocha J, Miguel-Jiménez K (2010) Holoprosencephaly and genitourinary anomalies in fetal methotrexate syndrome. Am J Med Genet 152A(7):1741–1746
24. García-Manero M, Royo MP, Espinos J, Pina L, Alcazar JL, López G (2009) Pregnancy associated breast cancer. Eur J Surg Oncol 35(2):215–218
25. Morris PG, King F, Kennedy MJ (2009) Cytotoxic chemotherapy for pregnancy-associated breast cancer: single institution case series. J Oncol Pharm Pract 15(4):241–247
26. Nieto Y, Santisteban M, Aramendía JM, Fernández-Hidalgo O, García-Manero M, López G (2006) Docetaxel administered during pregnancy for inflammatory breast carcinoma. Clin Breast Cancer 6(6):533–534
27. Lycette JL, Dul CL, Munar M, Belle D, Chui SY, Koop DR et al (2006) Effect of pregnancy on the pharmacokinetics of paclitaxel: a case report. Clin Breast Cancer 7(4):342–344
28. Potluri V, Lewis D, Burton GV (2006) Chemotherapy with taxanes in breast cancer during pregnancy: case report and review of the literature. Clin Breast Cancer 7(2):167–170
29. Gonzalez-Angulo AM, Walters RS, Carpenter RJ, Ross MI, Perkins GH, Gwyn K et al (2004) Paclitaxel chemotherapy in a pregnant patient with bilateral breast cancer. Clin Breast Cancer 5(4):317–319
30. Gadducci A, Cosio S, Fanucchi A, Nardini V, Roncella M, Conte PF et al (2003) Chemotherapy with epirubicin and paclitaxel for breast cancer during pregnancy:

case report and review of the literature. Anticancer Res 23(6D):5225–5229
31. De Santis M, Lucchese A, De Carolis S, Ferrazani S, Caruso A (2000) Metastatic breast cancer in pregnancy: first case of chemotherapy with docetaxel. Eur J Cancer Care (Engl) 9(4):235–237
32. Azim HA, Peccatori FA, Scarfone G, Acaia B, Rossi P, Cascio R et al (2008) Anthracyclines for gestational breast cancer: course and outcome of pregnancy. Ann Oncol 19(8):1511–1512
33. Kushari J, Mukherjea M (1980) Studies on beta-glucuronidase of the developing human placenta. Gynecol Obstet Invest 11(2):119–127
34. Peccatori FA, Azim HA, Scarfone G, Gadducci A, Bonazzi C, Gentilini O et al (2009) Weekly epirubicin in the treatment of gestational breast cancer (GBC). Breast Cancer Res Treat 115(3):591–594
35. Andreadis C, Charalampidou M, Diamantopoulos N, Chouchos N, Mouratidou D (2004) Combined chemotherapy and radiotherapy during conception and first two trimesters of gestation in a woman with metastatic breast cancer. Gynecol Oncol 95(1):252–255
36. Keleher A, Wendt R, Delpassand E, Stachowiak AM, Kuerer HM (2004) The safety of lymphatic mapping in pregnant breast cancer patients using Tc-99m sulfur colloid. Breast J 10(6):492–495
37. Mir O, Berveiller P, Ropert S, Goffinet F, Pons G, Treluyer J-M et al (2008) Emerging therapeutic options for breast cancer chemotherapy during pregnancy. Ann Oncol 19(4):607–613
38. Berveiller P, Mir O, Degrelle SA, Tsatsaris V, Selleret L, Guibourdenche J et al (2019) Chemotherapy in pregnancy: exploratory study of the effects of paclitaxel on the expression of placental drug transporters. Invest New Drugs 37(5):1075–1085
39. Zheng X, Zhu Y, Zhao Y, Feng S, Zheng C (2017) Taxanes in combination with platinum derivatives for the treatment of ovarian cancer during pregnancy: a literature review. Int J Clin Pharmacol Ther 55(9):753–760
40. Iguchi T, Hirokawa M, Takasugi N (1986) Occurrence of genital tract abnormalities and bladder hernia in female mice exposed neonatally to tamoxifen. Toxicology 42(1):1–11
41. Tewari K, Bonebrake RG, Asrat T, Shanberg AM (1997) Ambiguous genitalia in infant exposed to tamoxifen in utero. Lancet 350(9072):183
42. Schuurman TN, Witteveen PO, van der Wall E, Passier JLM, Huitema ADR, Amant F et al (2019) Tamoxifen and pregnancy: an absolute contraindication? Breast Cancer Res Treat 175(1):17–25
43. Cullins SL, Pridjian G, Sutherland CM (1994) Goldenhar's syndrome associated with tamoxifen given to the mother during gestation. JAMA 271(24):1905–1906
44. Diwan BA, Anderson LM, Ward JM (1997) Proliferative lesions of oviduct and uterus in CD-1 mice exposed prenatally to tamoxifen. Carcinogenesis 18(10):2009–2014
45. Lambertini M, Martel S, Campbell C, Guillaume S, Hilbers FS, Schuehly U et al (2019) Pregnancies during and after trastuzumab and/or lapatinib in patients with human epidermal growth factor receptor 2-positive early breast cancer: analysis from the NeoALTTO (BIG 1-06) and ALTTO (BIG 2-06) trials. Cancer 125(2):307–316
46. de Andrade JM, Brito LGO, Moises ECD, Amorim AC, Rapatoni L, Carrara HHA et al (2016) Trastuzumab use during pregnancy: long-term survival after locally advanced breast cancer and long-term infant follow-up. Anticancer Drugs 27(4):369–372
47. Fanale MA, Uyei AR, Theriault RL, Adam K, Thompson RA (2005) Treatment of metastatic breast cancer with trastuzumab and vinorelbine during pregnancy. Clin Breast Cancer 6(4):354–356
48. Pant S, Landon MB, Blumenfeld M, Farrar W, Shapiro CL (2008) Treatment of breast cancer with trastuzumab during pregnancy. J Clin Oncol 26(9):1567–1569
49. Sekar R, Stone PR (2007) Trastuzumab use for metastatic breast cancer in pregnancy. Obstet Gynecol 110(2 Pt 2):507–510
50. Shrim A, Garcia-Bournissen F, Maxwell C, Farine D, Koren G (2007) Favorable pregnancy outcome following Trastuzumab (Herceptin) use during pregnancy – case report and updated literature review. Reprod Toxicol 23(4):611–613
51. Waterston AM, Graham J (2006) Effect of adjuvant trastuzumab on pregnancy. J Clin Oncol Off J Am Soc Clin Oncol 24(2):321–322
52. Watson WJ (2005) Herceptin (trastuzumab) therapy during pregnancy: association with reversible anhydramnios. Obstet Gynecol 105(3):642–643
53. Witzel ID, Müller V, Harps E, Janicke F, Dewit M (2008) Trastuzumab in pregnancy associated with poor fetal outcome. Ann Oncol 19(1):191–192
54. Zagouri F, Sergentanis TN, Chrysikos D, Papadimitriou CA, Dimopoulos M-A, Bartsch R (2013) Trastuzumab administration during pregnancy: a systematic review and meta-analysis. Breast Cancer Res Treat 137(2):349–357
55. Yildirim N, Bahceci A (2018) Use of pertuzumab and trastuzumab during pregnancy. Anticancer Drugs 29(8):810–813
56. Azim HA, Botteri E, Renne G, Dell'orto P, Rotmensz N, Gentilini O et al (2012) The biological features and prognosis of breast cancer diagnosed during pregnancy: a case-control study. Acta Oncol 51(5):653–661
57. Chang SY, Soong YK (1995) Unexpected pregnancies exposed to leuprolide acetate administered after the mid-luteal phase for ovarian stimulation. Hum Reprod 10(1):204–206
58. Hay DL (1988) Placental histology and the production of human choriogonadotrophin and its subunits in pregnancy. Br J Obstet Gynaecol 95(12):1268–1275
59. Villaseca P, Campino C, Oestreicher E, Mayerson D, Serón-Ferré M, Arteaga E (2005) Bilateral oophorec-

tomy in a pregnant woman: hormonal profile from late gestation to postpartum: case report. Hum Reprod Oxf Engl 20(2):397–401
60. Turner NC, Finn RS, Martin M, Im S-A, DeMichele A, Ettl J et al (2018) Clinical considerations of the role of palbociclib in the management of advanced breast cancer patients with and without visceral metastases. Ann Oncol 29(3):669–680
61. Beaver JA, Amiri-Kordestani L, Charlab R, Chen W, Palmby T, Tilley A et al (2015) FDA approval: palbociclib for the treatment of postmenopausal patients with estrogen receptor-positive, HER2-negative metastatic breast cancer. Clin Cancer Res 21(21):4760–4766
62. O'Shaughnessy J, Petrakova K, Sonke GS, Conte P, Arteaga CL, Cameron DA et al (2018) Ribociclib plus letrozole versus letrozole alone in patients with de novo HR+, HER2- advanced breast cancer in the randomized MONALEESA-2 trial. Breast Cancer Res Treat 168(1):127–134
63. Loibl S, Turner NC, Ro J, Cristofanilli M, Iwata H, Im S-A et al (2017) Palbociclib combined with fulvestrant in premenopausal women with advanced breast cancer and prior progression on endocrine therapy: PALOMA-3 results. Oncologist 22(9):1028–1038
64. Sledge GW, Toi M, Neven P, Sohn J, Inoue K, Pivot X et al (2017) MONARCH 2: Abemaciclib in combination with fulvestrant in women with HR+/HER2- advanced breast cancer who had progressed while receiving endocrine therapy. J Clin Oncol 35(25):2875–2884
65. Dickler MN, Tolaney SM, Rugo HS, Cortés J, Diéras V, Patt D et al (2017) MONARCH 1, a Phase II study of Abemaciclib, a CDK4 and CDK6 inhibitor, as a single agent, in patients with refractory HR+/HER2- metastatic breast cancer. Clin Cancer Res Off J Am Assoc Cancer Res 23(17):5218–5224
66. Yamamura M, Kojima T, Koyama M, Sazawa A, Yamada T, Minakami H (2017) Everolimus in pregnancy: case report and literature review. J Obstet Gynaecol Res 43(8):1350–1352
67. Carta P, Zanazzi M, Minetti EE (2015) Unplanned pregnancies in kidney transplanted patients treated with everolimus: three case reports. Transpl Int 28(3):370–372
68. Margoles HR, Gomez-Lobo V, Veis JH, Sherman MJ, Moore J (2014) Successful maternal and fetal outcome in a kidney transplant patient with everolimus exposure throughout pregnancy: a case report. Transplant Proc 46(1):281–283
69. Aktoz F, Yalcin AC, Yüzdemir HS, Akata D, Gültekin M (2018) Treatment of massive liver metastasis of breast cancer during pregnancy: first report of a complete remission with trastuzumab and review of literature. J Matern-Fetal Neonatal Med 11:1–6
70. Kelly H, Graham M, Humes E, Dorflinger LJ, Boggess KA, O'Neil BH et al (2006) Delivery of a healthy baby after first-trimester maternal exposure to lapatinib. Clin Breast Cancer 7(4):339–341
71. Veras E, Kurman RJ, Wang TL, Shih IM (2017) PD-L1 expression in human placentas and gestational trophoblastic diseases. Int J Gynecol Pathol 36(2):146–153
72. Onal C, Guler OC, Yildirim BA (2018) Treatment outcomes of breast cancer liver metastasis treated with stereotactic body radiotherapy. Breast 42:150–156
73. Kwapisz D (2019) Oligometastatic breast cancer. Breast Cancer 26(2):138–146
74. Milano MT, Katz AW, Zhang H, Huggins CF, Aujla KS, Okunieff P (2019) Oligometastatic breast cancer treated with hypofractionated stereotactic radiotherapy: some patients survive longer than a decade. Radiother Oncol 131:45–51
75. Barral M, Auperin A, Hakime A, Cartier V, Tacher V, Otmezguine Y et al (2016) Percutaneous thermal ablation of breast cancer metastases in oligometastatic patients. Cardiovasc Intervent Radiol 39(6):885–893
76. Vargas C, Maiz C, Navarro ME, Oddó D, Sánchez C, Bustos M et al (2019) Surgical treatment in oligometastatic breast cancer. Ecancermedicalscience 13:931
77. Cyclophosphamide (2006) Drugs and lactation database (LactMed). National Library of Medicine (US), Bethesda. Available from: http://www.ncbi.nlm.nih.gov/books/NBK501672/
78. Griffin SJ, Milla M, Baker TE, Liu T, Wang H, Hale TW (2012) Transfer of carboplatin and paclitaxel into breast milk. J Hum Lact 28(4):457–459

Radiotherapy in Pregnancy-Associated Breast Cancer

16

Farnaz Amouzegar Hashemi

Abstract

Breast radiotherapy during pregnancy is a matter of debate as both the efficacy of treatment and the safety of the developing fetus should be considered. Currently there is not enough data to support the safety of in-utero exposure to radiation even with modern radiotherapy techniques. So it is highly recommended that breast radiotherapy is postponed to after delivery, though it might be considered in very selected patients according to risk-benefit assessment.

Keywords

Adjuvant · Breast cancer · Breast radiotherapy · Conservative breast surgery · Pregnancy

F. Amouzegar Hashemi (✉)
Radiation Oncology Research Center, Cancer Institute, Tehran University of Medical Sciences, Tehran, Iran
e-mail: amoozfar@tums.ac.ir

16.1 Overview

Radiotherapy plays a critical role in the management of breast cancer as breast-conserving surgery (BCS) followed by radiotherapy is the most acceptable standard treatment for the majority of early stage breast cancers [1]. Radiotherapy during pregnancy continues to be a matter of debate, and at present, there are only a few reports in the literature regarding this issue. As a general rule, breast cancer during pregnancy should be treated according to the guidelines for non-pregnant women with enough consideration to protect the fetus. Although the modern approach considers the same surgical approach for pregnant women [2], several concerns exist with regard to delaying radiotherapy to after childbirth.

Although mastectomy may have the advantage of eliminating radiation for a group of patients, BCS could still be done considering that radiation can be delayed until after delivery [3]. As radiation dose is typically standard in breast cancer (46–50 Gy to the whole breast in BCS or to the chest wall in advanced mastectomy cases, with an additional 10–16 Gy boost to the tumor site in BCS), the trimester of pregnancy remains the most important factor regarding radiation toxicity to the fetus. As routine non-pregnant patients receive radiotherapy after chemotherapy and 5–6 months after surgery, BCS can be performed during late second trimester and third trimester as

S. Alipour, R. Omranipour (eds.), *Diseases of the Breast during Pregnancy and Lactation*, Advances in Experimental Medicine and Biology 1252,
https://doi.org/10.1007/978-3-030-41596-9_16

the radiation can be easily postponed [4]. The most challenging scenario is breast cancer diagnosed before 16 weeks of gestational age for patients who desire to continue pregnancy and do not wish to proceed with mastectomy. In these situations, delaying radiation until after delivery is very controversial. In a meta-analysis by Chen considering 20 high-quality studies and controlling confounding factors, the author estimates that the risk of local recurrence would increase by 1.0% for every additional month's delay in radiation delivery [5]. Regarding radiation toxicity, with administration of 50 Gy to the chest wall or breast in the first trimester, the fetus receives 0.04–0.15 Gy. As the fetus enlarges and comes out of the pelvic cavity, this dose increases up to 2 Gy because of increased proximity to the radiation field [6, 7]. The threshold of safety dose to the fetus has not been determined, but it definitely differs by gestational age of the embryo or fetus. Health effects including abortion, major malformations, growth retardation, neurological or motor deficiencies, and mental retardation occur at doses above 0.05 Gray of exposure. These are more frequent with exposure during organogenesis, which occurs between the second and seventh weeks of gestation; and up to 16 weeks [8–11]. A dose reduction of 50% to 70% may be achieved by using shields with X-ray energies from 4 to 10 MV [12]. Even with proper shielding, scattered radiation and leakage will be present to some extent. In addition, it is difficult to shield a gravid uterus in advanced pregnancies [13].

Accelerated partial breast irradiation (APBI) with reduced treatment volume is not a suitable option for young women. However, theoretically, it may be attractive as an alternative option for early stages of breast cancer in pregnant patients refusing mastectomy. Galimberti carried out a study on non-pregnant patients with breast cancer to estimate doses to the uterus during electron beam intraoperative radiotherapy (ELIOT). The study performed with thermoluminescence radiation detectors showed a mean dose of 0.09 Gy in the pubic area and 0.17 in the uterus [13]. These show that ELIOT and the shielding apron may be safe for the fetus, but its accuracy needs to be investigated. In December 2011, the first pregnant patient with cancer underwent BCS and ELIOT (21 Gy at 90% isodose) during week 15 of gestation, and the patient underwent whole breast radiotherapy 16 weeks after delivery. It was proposed that ELIOT may be an option to postpone whole breast radiotherapy after delivery for early second-trimester-pregnant women who refused mastectomy [2].

In a study of 129 children whose mothers had cancer during pregnancy (half of which were breast cancer), their development was similar to those in the control group after 22 months [14]. In a subgroup analysis, the same outcomes were reported for 11 children exposed to radiation with gestational age-matched controls. There are case reports of successful radiotherapy of breast cancer during pregnancy and live births and healthy children. However, as it is unclear if there is a safe dose with no increased risk, radiotherapy should be avoided during pregnancy. Even with more advanced radiation technology such as intensity-modulated radiotherapy (IMRT) and image-guided radiotherapy (IGRT), it is still unclear what the long-term effects of radiation exposure to fetal health would be [15]. A complete discussion of the potential risks of postponing radiation therapy until after delivery must be held with the patient in earlier gestational ages, and every treatment decision should be made in a multidisciplinary team in a personalized fashion.

16.2 Breast Radiotherapy and Breastfeeding

If breast cancer is diagnosed during breastfeeding, it is highly recommended that the patient stops further feeding immediately and begins treatment. After finishing cancer treatment, lactation is completely safe from the contralateral breast (the side which is not treated), and there is no evidence to suggest it affects prognosis [16]. Breastfeeding is even possible from the affected side after irradiation in 50% of cases [17]. However the volume of milk production is typically reduced in a significant proportion, depending on various elements like the distance of the previous tumor and incision from the nipple-areola, dosage and type of radiation, social issues and psychological factors [18, 19] (see also Chap. 22).

References

1. Halperin EC, Wazer DE, Perez CA, Brady LW (2019) Perez & Brady's principles and practice of radiation oncology. 7th ed. 4042
2. Leonardi M, Cecconi A, Luraschi R, Rondi E, Cattani F, Lazzari R et al (2017) Electron beam intraoperative radiotherapy (ELIOT) in pregnant women with breast cancer: from in vivo dosimetry to clinical practice. Breast Care 12(6):396–400
3. Botha MH, Rajaram S, Karunaratne K (2018) Cancer in pregnancy. Int Gyn Obstet 143:137–142
4. National Comprehensive Cancer network (NCCN guidelines) (2019) Breast cancer. Available at www.NCCN.org
5. Chen Z, King W, Pearcey R, Kerba M, Mackillop WJ (2008) The relationship between waiting time for radiotherapy and clinical outcomes: a systematic review of the literature. Radiother Oncol 87(1):3–16
6. Petrek JA (1994) Breast cancer during pregnancy. Cancer 74(S1):518–527
7. Antypas C, Sandilos P, Kouvaris J, Balafouta E, Karinou E, Kollaros N et al (1998) Fetal dose evaluation during breast cancer radiotherapy. Int J Radiat Oncol Biol Phys 40(4):995–999
8. Lowe SA (2019) Ionizing radiation for maternal medical indications. Prenat Diagn; ahead of printing
9. Donnelly EH, Smith JM, Farfán EB, Ozcan I (2011 Mar) Prenatal radiation exposure: background material for counseling pregnant patients following exposure to radiation. Disaster Med Public Health Prep 5(1):62–68
10. Mazonakis M, Varveris H, Damilakis J, Theoharopoulos N, Gourtsoyiannis N (2003) Radiation dose to conceptus resulting from tangential breast irradiation. Int J Radiat Oncol Biol Phys 55(2):386–391
11. Needleman S, Powell M (2016) Radiation hazards in pregnancy and methods of prevention. Best Pract Res Clin Obstet Gynaecol 33:108–116
12. Han B, Bednarz B, Xu XG (2009) A study of the shielding used to reduce leakage and scattered radiation to the fetus in a pregnant patient treated with a 6-MV external X-ray beam. Health Phys 97(6):581–589
13. Galimberti V, Ciocca M, Leonardi MC, Zanagnolo V, Paola B, Manuela S et al (2009) Is electron beam intraoperative radiotherapy (ELIOT) safe in pregnant women with early breast cancer? In vivo dosimetry to assess fetal dose. Ann Surg Oncol 16(1):100–105
14. Amant F, Vandenbroucke T, Verheecke M, Fumagalli M, Halaska MJ, Boere I et al (2015) Pediatric outcome after maternal cancer diagnosed during pregnancy. N Engl J Med 373(19):1824–1834
15. Mazzola R, Corradini S, Eidemueller M, Figlia V, Fiorentino A, Giaj-Levra N et al (2019) Modern radiotherapy in cancer treatment during pregnancy. Crit Rev Oncol Hematol 136:13–19
16. Azim HA Jr, Bellettini G, Liptrott SJ, Armeni ME, Dell'Acqua V, Torti F et al (2010) Breastfeeding in breast cancer survivors: pattern, behaviour and effect on breast cancer outcome. Breast 19(6):527–531
17. Leal SC, Stuart SR, Carvalho HD (2013) Breast irradiation and lactation: a review. Expert Rev Anticancer Ther 13(2):159–164
18. Johnson HM, Mitchell KB (2019) Breastfeeding and breast cancer: managing lactation in survivors and women with a new diagnosis. Ann Surg Oncol 26(10):3032–3039
19. Linkeviciute A, Notarangelo M, Buonomo B, Bellettini G, Peccatori FA (2019) Breastfeeding after breast cancer: feasibility, safety, and ethical perspectives. J Hum Lact:0890334419887723

Concerns of Hereditary Breast Cancer in Pregnancy and Lactation

17

Jennifer Chen, Vishnu Prasath, Jennifer Axilbund, and Mehran Habibi

Abstract

Genetic testing should be offered to all women less than 40 years of age who are diagnosed with breast cancer, and patients with PABC are generally among them. However, there is no specific study about these cases, and whether genetic testing should be carried out during or after pregnancy is not known. Generally, testing before delivery should only be performed if positive results change management plans, such as undergoing fetal testing and choosing mastectomy instead of breast conserving surgery.

Keywords

Breast cancer in pregnancy · Genetic testing · Genetic mutations · Prenatal screening · Prognosis · Tumor suppressor genes

J. Chen · V. Prasath · J. Axilbund · M. Habibi (✉)
Department of Surgery, Johns Hopkins University, School of Medicine, Baltimore, MD, USA
e-mail: jennifer.chen@jhmi.edu; vprasat1@jhu.edu; solleje@jhmi.edu; mhabibi2@jhmi.edu

17.1 Overview

Globally, it is agreed that genetic testing should be offered to all women less than 40 years of age who are diagnosed with breast cancer, regardless of family history [1]. Criteria for BRCA testing include patients with (1) breast cancer diagnosed before age 50, (2) bilateral breast cancer, (3) two primary breast cancers, (4) personal or family history of ovarian cancer, (5) two or more primary types of *BRCA1*- or *BRCA2*-related cancers in a single family member, or (6) family history of male breast cancer [2–4]. In addition to those with a positive family history of breast, ovarian, or pancreatic cancer, specific populations to consider include Ashkenazi Jews, who have a tenfold increase in prevalence of *BRCA1* and *BRCA2* mutations compared to the general population [2]. Patients with PABC are generally included in the first criterion and are recommended to undergo genetic testing. However, there is no specific study about these cases, and whether testing should be carried out during pregnancy or after delivery is not known. Testing prior to delivery should generally only be performed if positive results will change management plans, such as undergoing fetal testing and deciding between proceeding with mastectomy versus breast conserving surgery.

S. Alipour, R. Omranipour (eds.), *Diseases of the Breast during Pregnancy and Lactation*, Advances in Experimental Medicine and Biology 1252,
https://doi.org/10.1007/978-3-030-41596-9_17

17.2 Identifying Patients at Risk

No risk factor linked specifically to breast cancer during pregnancy has been identified. However, genetic susceptibility syndromes are typically associated with younger ages at cancer diagnosis than seen in the general breast cancer population. As a result, women with an inherited breast cancer predisposition syndrome are presumably at greater risk for pregnancy-associated breast cancer (PABC). Recent literature suggests that around 5–10% of breast cancers stem from inherited genetic susceptibility, and many of these heritable breast cancers are due to mutations in tumor suppressor genes *BRCA1* and *BRCA2* [2, 5]. Other genes associated with substantially increased risk of young-onset breast cancer include *TP53*, *PTEN*, *STK11* and *CDH1* [6, 7]. For example, patients with a *TP53* mutation have up to 85% absolute risk of breast cancer by age 60 and the cancer often occurs in a woman's 20s or 30s with this particular gene; although inherited *TP53* mutations are much rarer than mutations in *BRCA1* and *BRCA2* [7–9].

17.3 Genetic Testing Logistics

Once inherited predisposition to breast cancer is suspected, many genetic tests are commercially available to potentially identify presence of a gene mutation. Genetic testing for this purpose is non-invasive, as sample types most commonly tested are blood or saliva from the affected individual. While previous sequencing techniques historically evaluated individual genes of interest one at a time, at substantial time and expense to patients and the healthcare system, advances in high throughput sequencing techniques now offer multi-gene panel testing and full genomic testing with greater efficiency and reduced costs. Prior to 2013, *BRCA1* and *BRCA2* genes tests were exclusively available through Myriad Genetics due to patent ownership [2]. However, in 2013 the United States Supreme Court ruled against patenting of human genes, thereby allowing commercial competition. As a result, dozens of companies now offer testing of these two genes, as well as many others genes relevant to breast cancer [2]. It is important to ensure that the laboratory offers full sequencing of the genes, as well as analysis of copy number variants. Additionally, a robust system for variant interpretation is of utmost importance. Other important issues to consider in obtaining genetic testing involve cost, insurance coverage and availability of genetic counseling. While results may allay anxiety among patients, the possibility of identifying variants of uncertain significance (VUS) or other unintended genetic predispositions should be carefully considered before pursuing screening [6]. Additionally, although some protective legislation exists, the potential of breaches in personal data safety and genetic discrimination from the part of employers or health insurers based on positive genetic profiles represent other considerations prior to undergoing genetic testing. Involving a genetic counselor in this process helps ensure that these issues are adequately addressed with the patient.

17.4 Screening Options and Timing

For patients who have BRCA mutations, annual MRI starting at the age of 25 and annual mammography beginning at the age of 30 are recommended for screening; screening recommendations for rarer genetic mutations vary based on estimated prevalence and approximate breast cancer risk portended (Table 17.1) [10–12]. Although routine imaging should be performed in genetically susceptible women, imaging by mammography and MRI with contrast are best postponed during pregnancy. Breast examination, although less sensitive than in non-pregnant women (see also Chap. 2), and ultrasounds can be performed at the time of screening (see also Chap. 3) [13].

Table 17.1 Genes related to increased breast cancer risk, ongoing screening throughout pregnancy

Gene	Estimated prevalence	Reported BC risk	Associated syndrome	NCCN screening recommendations	Suggestions throughout pregnancy
BRCA1 *BRCA2*	1:400 1:40[a]	RR × 10–20 49–57% risk by age 70	HBOC	From 25 y: BSE and biannual CBE, annual mammo and/or MRI	BSE and CBE during pregnancy
TP53	1:20,000	56% risk by 45 y >90% risk by 60 y	Li-Fraumeni syndrome	From 18 y: BSE From 20–25 y, or 5–10 y < earliest family BC: biannual CBE, annual mammo and/or MRI	US for positive findings
PTEN	1:200,000	25–50% lifetime risk	Cowden syndrome	From 18 y: BSE, annual thyroid US From 25 y: biannual CBE From 30–35 y or 5–10 y <earliest family BC: annual mammo or MRI Individualized endometrial cancer screening	Screening mammo/MRI after delivery
STK11	1:25,000–1:280,000	RR × 20 8% risk by 40 y 32% risk by 60 y	Peutz-Jeghers syndrome	From 18 y: monthly BSE From 25 y: biannual CBE, annual mammo	
ATM	1:20,000–1:100,000	RR × 2, less penetrant than other genes	Ataxia-telangiectasia	Screening decisions based on family risk and family history	Diagnostic mammo (with shield) for suspicious US findings
PALB2	1:100,000	RR × 2–6, greater risk at younger age	Fanconi anemia	Screening decisions based on family risk and family history	

BC breast cancer, *BSE* Breast self-examination, *CBE* Clinical breast examination, *mammo* mammography, *HBOC* Hereditary breast and ovarian cancer, *RR* relative risk, *y* year, *US* ultrasound
[a]In Ashkenazi Jewish

17.5 Genetic Risk to the Fetus

Most breast cancer susceptibility genes are inherited in an autosomal dominant pattern, conferring a 50% chance that the fetus also carries the gene mutation. Because of this, parents may request prenatal diagnosis to determine their child's carrier status. The decision to undergo genetic testing of the fetus prior to birth highlights key ethical issues regarding parental rights versus individual autonomy. Most arguments against prenatal diagnosis for cancer predisposition aim to protect the autonomy and medical rights of the unborn child, particularly for adult-onset conditions for which there are no recommended medical interventions before the age of majority. However, prenatal genetic testing may also influence the decision to continue or terminate the current pregnancy, with some patients choosing to abort a carrier fetus [14]. In regard to maternal outcomes, current studies indicate no improvement in disease prognosis with termination of pregnancy. However, in patients with highly aggressive and advanced stage disease, termination may be considered in order to undergo chemotherapy, which is contraindicated in the first trimester (see also Chap. 15).

17.6 Genetic Risk to Family Members

Identification of genetic predispositions in this patient population has implications for other relatives beyond the fetus. The vast majority of these mutations are inherited from a parent, thereby conveying risk to the parent of origin, the patients' siblings, previous offspring, and other extended family members. Additionally, risk to subsequent offspring should also be considered. For future pregnancies, patients with *BRCA1/2* or other known mutations may be able to utilize preimplantation genetic diagnosis (PGD) to select for unaffected embryos through *in vitro* fertilization. Ethical issues regarding parental reproductive rights, rights of embryos, and appropriateness of using PGD for early versus late-onset and full versus incomplete penetrance genetic profiles should be discussed as a part of the genetic counseling process [14].

References

1. Manahan ER, Kuerer HM, Sebastian M, Hughes KS, Boughey JC, Euhus DM et al (2019) Consensus guidelines on genetictesting for hereditary breast cancer from the American Society of Breast Surgeons. Ann Surg Oncol 26(10):3025–3031
2. Lynch JA, Venne V, Berse B (2015) Genetic tests to identify risk for breast cancer. Semin Oncol Nurs 31(2):100–107
3. Moyer VA (2014) Force USPST. Risk assessment, genetic counseling, and genetic testing for BRCA-related cancer in women: U.S. Preventive Services Task Force recommendation statement. Ann Intern Med 160(4):271–281
4. Bevers TB, Helvie M, Bonaccio E, Calhoun KE, Daly MB, Farrar WB et al (2018) Breast cancer screening and diagnosis, version 3.2018, NCCN clinical practice guidelines in oncology. J Natl Compr Cancer Netw 16(11):1362–1389
5. Gage M, Wattendorf D, Henry LR (2012) Translational advances regarding hereditary breast cancer syndromes. J Surg Oncol 105(5):444–451
6. Campeau PM, Foulkes WD, Tischkowitz MD (2008) Hereditary breast cancer: new genetic developments, new therapeutic avenues. Hum Genet 124(1):31–42
7. Lalloo F, Varley J, Ellis D, Moran A, O'Dair L, Pharoah P et al (2003) Prediction of pathogenic mutations in patients with early-onset breast cancer by family history. Lancet 361(9363):1101–1102
8. Gonzalez KD, Noltner KA, Buzin CH, Gu D, Wen-Fong CY, Nguyen VQ et al (2009) Beyond Li Fraumeni Syndrome: clinical characteristics of families with p53 germline mutations. J Clin Oncol 27(8):1250–1256
9. Schon K, Tischkowitz M (2018) Clinical implications of germline mutations in breast cancer: TP53. Breast Cancer Res Treat 167(2):417–423
10. Elezaby M, Lees B, Maturen KE, Barroilhet L, Wisinski KB, Schrager S, Wilke LG, Sadowski E (2019 Apr 30) BRCA mutation carriers: breast and ovarian cancer screening guidelines and imaging considerations. Radiology 291(3):554–569
11. Warner E (2018) Screening BRCA1 and BRCA2 mutation carriers for breast cancer. Cancers 10(12):477
12. Monticciolo DL, Newell MS, Moy L, Niell B, Monsees B, Sickles EA (2018) Breast cancer screening in women at higher-than-average risk: recommendations from the ACR. J Am Coll Radiol 15(3):408–414
13. Carmichael H, Matsen C, Freer P, Kohlmann W, Stein M, Buys SS et al (2017) Breast cancer screening of pregnant and breastfeeding women with BRCA mutations. Breast Cancer Res Treat 162(2):225–230
14. Clancy T (2010) A clinical perspective on ethical arguments around prenatal diagnosis and preimplantation genetic diagnosis for later onset inherited cancer predispositions. Familial Cancer 9(1):9–14

Paget Disease of the Breast in Pregnancy and Lactation

18

Richard Gilmore, Vishnu Prasath, and Mehran Habibi

Abstract

Paget's disease of the breast (PDB) is a rare breast carcinoma believed to arise from an underlying in situ or invasive ductal cancer that migrates through the epidermis causing characteristic skin changes including scaling, redness, and itching of the nipple, areola, and sometimes the surrounding skin. Although Paget's may mimic benign conditions such as contact or allergic eczema and mastitis, it should remain a strong consideration in the differential diagnosis, especially in peripartum women for whom benign conditions such as bacterial mastitis from breastfeeding are common. The workup of Paget's should focus on both making the diagnosis with nipple/skin scrape cytology or punch biopsy as well as evaluating any underlying mass with mammogram, breast ultrasound, and also a core needle biopsy, if required. Treatment focuses on management of the underlying breast cancer as usual. The purpose of this chapter is to describe the presentation of PDB as well as outline an approach to its diagnosis and management, especially in the setting of pregnancy and lactation.

Keywords

Breast neoplasms · Diagnosis · Lactation · Paget's disease · Pregnancy · Treatment

18.1 Overview

Paget's disease of the breast (PDB) is a rare form of breast cancer, which appears in the nipple epithelium, as originally described by Sir James Paget in 1874. Accounting for only 1–3% of new cases of female breast cancer diagnosed annually in the United States, it is far less common than other presentations of breast cancer [1, 2]. Most (>95%) of patients with PDB have underlying ductal breast cancer, either *in situ* or invasive [1, 3]. Malignant cells are believed to migrate through the epidermis where the disease becomes initially apparent in the nipple, followed by the areola, and finally the surrounding skin [1]. The underlying carcinoma may be associated with a palpable mass on exam, but in less than half of cases, the cancer is clinically and radiologically undetectable. Commons symptoms include flaky or scaly skin on the nipple; crusty, hard, or thick skin; a lump in the breast, bloody nipple dis-

R. Gilmore · V. Prasath · M. Habibi (✉)
Department of Surgery, Johns Hopkins University, School of Medicine, Baltimore, MD, USA
e-mail: richard.gilmore@jhmi.edu; vprasat1@jhu.edu; mhabibi2@jhmi.edu

S. Alipour, R. Omranipour (eds.), *Diseases of the Breast during Pregnancy and Lactation*, Advances in Experimental Medicine and Biology 1252,
https://doi.org/10.1007/978-3-030-41596-9_18

charge, itching, or redness. Ulceration and nipple inversion are sometimes present and are usually late findings occurring with more advanced disease [4]. The underlying breast cancer is almost always ductal in subtype, high-grade in nature, and is composed of cells similar to those normally present within the epidermis.

The workup of PDB focuses on both establishing the diagnosis and identifying an underlying breast cancer, if present. Three-dimensional and diagnostic mammograms as well as ultrasound and MRI can be utilized to radiologically identify a mass. Bilateral mammography is helpful for identification of an associated mass and to exclude synchronous cancers or widespread calcifications that might preclude breast conserving surgery (BCS). Breast ultrasound, 3D mammography, and MRI should be used to further evaluate and guide breast core needle biopsy of any palpable mass or mass-like mammographic abnormality [4]. Nipple scrape cytology can accurately diagnose PDB but the diagnosis is usually made by punch biopsy of the nipple or full-thickness wedge biopsy. On histology, intraepithelial adenocarcinoma (Paget) cells are seen; these cells are positive for low molecular weight cytokeratins, a feature useful in distinguishing PDB from squamous carcinoma of the epidermis (Bowen's disease), which has a similar clinical presentation but expresses high molecular weight cytokeratins [5, 6].

Prognosis of PDB is based upon the characteristics of the underlying breast cancer, which also guide treatment. Simple mastectomy has been the historical standard treatment; however the widespread use of BCS for invasive and *in-situ* ductal carcinoma has led to its utilization in PDB as well. Treatment can be divided into those who present with and without a palpable mass or abnormal imaging. If PDB is present in association with a palpable mass or mammographic abnormality, the associated breast cancer tends to be at a more advanced stage than when a mass is absent. Both the nipple-areolar complex (NAC) and the underlying cancer must be excised. Many patients require mastectomy; however, if the NAC and underlying mass can be excised with an acceptable cosmetic outcome and negative margins, BCS followed by whole breast radiation therapy (RT) is an appropriate treatment option. Those who present without a palpable mass or imaging abnormality still often have an underlying associated carcinoma but, in most cases, this tends to be DCIS. Nonetheless, invasive cancer is present in about one-fourth to one-third of cases [1, 7]. Both simple mastectomy and BCS are viable treatment strategies. Additionally, some have described complete resection of the NAC alone followed by whole breast RT as a reasonable alternative with acceptable long-term recurrence and survival outcomes [4, 8]. Management of the axilla in PDB is the same as for any breast cancer, depending on the nature of the underlying carcinoma.

18.2 Concerns in Pregnancy and Lactation

Paget's disease in the lactating breast can often mimic benign eczematous lesions (Fig 18.1). As pregnancy-associated breast cancer (PABC) is reported to have an increased recurrence and mortality rate, potential Paget's cases should be extensively worked up to prevent delays in diagnosis due to physiologic pregnancy-associated breast changes (see also Chap. 1). By promptly detecting PABCs, treatment can be more rapidly initiated. Utmost care can then be taken to safely treat the cancer without affecting the fetus/infant (see also Chaps. 12–16 and 21).

A recent case report details the clinical course of a 32-year old woman who presented in her breastfeeding period, but whose symptoms were assumed to be lactation-related and the disease was not initially identified. She eventually re-presented with ulceration and retraction of the nipple and a 1.5 cm breast lump. Modified radical mastectomy was performed. Final pathology report revealed infiltrating ductal carcinoma, ductal carcinoma *in situ* and Paget's disease of the nipple while eleven dissected lymph nodes were involved. The patient died with metastatic disease one year later despite adequate treatment [9]. This case reminds the need to careful atten-

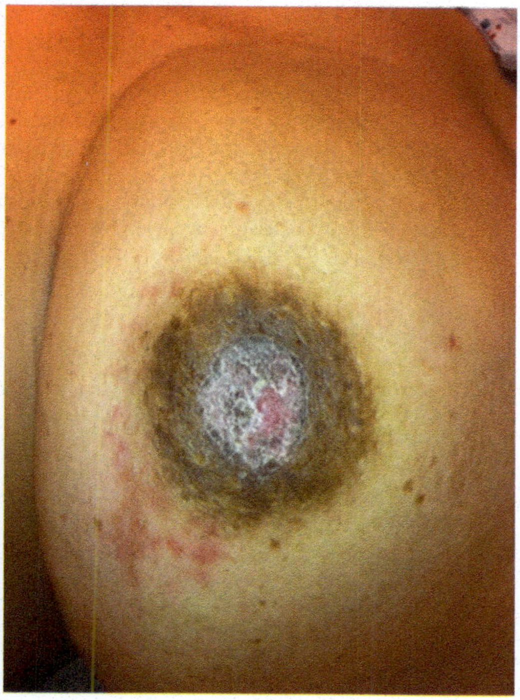

Fig. 18.1 Paget's disease of the breast; scaling of the nipple areolar complex in a 34 years-old pregnant woman, 20 weeks of gestational age

tion to nipple changes during pregnancy and lactation.

Normal physiological changes during pregnancy can make diagnosis of Paget's more challenging. PDB should be considered in the differential diagnosis when encountering a patient with unilateral nipple changes. In addition, eczema of the nipple and areola can occasionally develop in the pregnant or postpartum breastfeeding patient and is associated with pruritus, redness, and pain (see also Chap. 5). Differential diagnosis includes benign conditions such as allergic or contact dermatitis which often respond to elimination of the offending agent. Alternatively, bacterial or fungal mastitis can have a similar presentation with therapies targeted to the most common infectious culprits, generally *staphylococcus aureus* and group *A streptococcus* if bacterial, and candida if fungal. Physicians should suspect PDB if the eczematous changes present unilaterally or if they persist for more than 3 weeks [10]. Up to 20% of patients have signs/symptoms of PDB for longer than a year before seeking medical attention. Several factors contribute to a sometimes profound delay in diagnosis. Patients and providers may conclude that the eczema-like appearance and pruritus of the nipple are due to local inflammation and empirically recommend treatment with topical steroids, resulting in transient or sustained improvement in symptoms and/or appearance of the nipple – so-called healed PDB, which reinforces the misconception that the clinical abnormality is insignificant. Additionally, physicians are often reluctant to perform a punch biopsy of the nipple expecting the problem to be minor and benign, or having a false assumption that a nipple biopsy will result in disfigurement or impair future lactation.

Timely diagnosis of PDB requires a low threshold for obtaining a diagnostic workup and additional biopsies. Any unilateral nipple abnormality that is more severe than expected and any physiologic changes of gestation in a pregnant or breastfeeding woman should be sufficiently worked up until malignancy is ruled out. It is important to remember that treatment options for PDB are safe, even during or immediately following pregnancy. Surgery is possible at any time period and chemotherapy can be used following the completion of the first trimester. Radiation therapy is not feasible during pregnancy but can be used following delivery. The type of surgery in regard to breast conserving survey or mastectomy follows the same guidelines as PABC (see also Chap. 12).

References

1. Jimenez RE, Hieken TJ, Peters MS, Visscher DW (2018) Paget disease of the breast. In: The breast. Elsevier, pp 169–176
2. Nance FC, DeLoach DH, Welsh RA, Becker WF (1970) Paget's disease of the breast. Ann Surg 171(6):864–874
3. Nakai K, Horimoto Y, Semba R, Arakawa A, Saito M (2019) Pathological and radiological assessments of Paget's disease. Ann Breast Surg 3
4. Karakas C (2011) Paget's disease of the breast. J carcinog 10(1):31
5. Aslan F, Demirkesen C, Çağatay P, Tüzüner N (2006) Expression of cytokeratin subtypes in intraepidermal malignancies: a guide for differentiation. J Cutan Pathol 33(8):531–538

6. Lopes Filho LL, Lopes IM, Lopes LR, Enokihara MM, Michalany AO, Matsunaga N (2015) Mammary and extramammary Paget's disease. An Bras Dermatol 90(2):225–231
7. Ikeda DM, Helvie MA, Frank TS, Chapel KL, Andersson IT (1993) Paget disease of the nipple: radiologic-pathologic correlation. Radiology 189(1):89–94
8. Solin LJ, Kurtz J, Fourquet A, Amalric R, Recht A, Bornstein BA et al (1996) Fifteen-year results of breast-conserving surgery and definitive breast irradiation for the treatment of ductal carcinoma in situ of the breast. J Clin Oncol 14(3):754–763
9. Nargotra N, Mendiratta SL (2018) Paget disease of breast (nipple) in a lactating woman – a diagnostic dilemma. Indian Obstet Gynaecol 8(1):20–24
10. Barankin B, Gross MS (2004) Nipple and areolar eczema in the breastfeeding woman. J Cutan Med Surg 8(2):126–130

Phyllodes Tumor of the Breast in Pregnancy and Lactation

19

Sadaf Alipour and Amirhossein Eskandari

Abstract

Phyllodes tumor constitutes around 1% of all and 2.5% of fibroepithelial breast lumps. Three types including benign, borderline, and malignant tumors have been described. The benign variant is the most common, is close to fibroadenoma, but is usually larger and recurs more frequently. The rare malignant type is aggressive. Standard treatment consists of lumpectomy with appropriate margins for benign phyllodes tumor, while the borderline and malignant variants must be treated by wide resection or mastectomy. Phyllodes tumor is a rare tumor in pregnancy and lactation, and the effect of gestational alterations in hormone levels on this tumor have not been discussed in the literature, except for several case reports. In summary and alluding to our recent literature review, large size, fast growth, bilaterality, and probably malignancy are more commonly expected in gestational phyllodes tumors.

Keywords

Benign · Borderline · Breastfeeding · Breast cancer · Malignant · Phyllodes tumor · Pregnancy

19.1 Overview

The first description of phyllodes tumors was provided by Johannes Muller in 1838. Owing to its leaf-like appearance on microscopic examination, Muller named it cystosarcoma phyllodes after phyllon, the Greek word for leaf. The term most widely used today is phyllodes tumor (PT), as designated in the World Health Organization's classification of tumors [1, 2]. PT accounts for up to 1% of all breast tumors and 2.5% of fibroepithelial lesions, and its incidence is no more than 2.1 per million population [2–5].

On the basis of quantitative microscopic features, three types of PT have been defined: benign, borderline, and malignant PT. The benign variant is most frequently encountered and is close to fibroadenoma both in its histologic and clinical presentation, except for a typically larger

S. Alipour (✉)
Breast Disease Research Center (BDRC), Tehran University of Medical Sciences, Tehran, Iran

Department of Surgery, Arash Women's Hospital, Tehran University of Medical Sciences, Tehran, Iran
e-mail: salipour@tums.ac.ir

A. Eskandari
Deputy of Education, Ministry of Health, Tehran, Iran

S. Alipour, R. Omranipour (eds.), *Diseases of the Breast during Pregnancy and Lactation*, Advances in Experimental Medicine and Biology 1252,
https://doi.org/10.1007/978-3-030-41596-9_19

size at presentation and relatively high rate of recurrence. The rare malignant type may follow a very hostile course and recur as extensive local or distant disease [2, 5–8].

Women can be affected at any age, but the average age at presentation is about 40–45 years. The most common clinical finding of PT is a breast lump, which is generally a circumscribed firm mobile mass with rapid growth. Although the average size of PT is around 4 cm to 5 cm, large sizes up to 20 cm or more are not unexpected. Recent imaging technology has led to the detection of much smaller tumors. The occurrence of PT in both breasts is an exceptional event that is known to occur in 0.3–3.5% of cases, whether in a synchronous or metachronous setting [2–4, 9].

An ultrasound (US) scan usually shows a circumscribed, lobulated mass with cystic spaces and heterogenic echogenicity [10, 11].

Standard treatment of benign PT entails excision of the mass to 1 cm clear margins. Borderline and malignant pathology necessitates wider margins of resection. Recent evidence suggests that thinner margins might also be appropriate. Mastectomy is carried out for large tumors, most commonly borderline or malignant types, and recurrences. As this tumor usually does not spread via lymphatics, no axillary intervention is required, except for clinically suspicious lymphadenopathies [5, 8, 12, 13].

Other forms of therapy including chemotherapy, radiotherapy, endocrine therapy, or immune therapy are not recognized as standard treatment in PT but are administered based on individual characteristics of patients and tumors. Chemotherapy can be considered for malignant cases, and effective palliation has been documented in metastatic PT [5, 8, 9, 14].

19.2 Concerns in Pregnancy and Lactation

PT is a rare tumor, and its presentation in pregnancy or breastfeeding and the effect of gestational alterations in hormone levels on preexisting PT have not been discussed in the literature, except for several case reports. All three types of phyllodes tumor, benign (Figs. 19.1 and 19.2), borderline and malignant have been reported in pregnancy and lactation.

Through a comprehensive review of the existing literature , Alipour et al [51] collected data of 37 patients who had presented in the gestational period and had been reported between 1954 and 2018. These patients harbored 43 PTs; Table 19.1 shows the tumor and patient characteristics, and the surgical approach to each tumor. The mean age was 31 (21–43) years, younger than the typical age for PT, which is easily attributed to the reproductive age. Unlike PT occurring in non-pregnant women, bilaterality was not an infrequent event; and the rate of malignancy was unexpectedly high, constituting around 60% of all tumors. As well, the size of lesions at the time of seeking medical attention was much larger than a typical PT, with a mean of around 14 (1.5–40) cm.

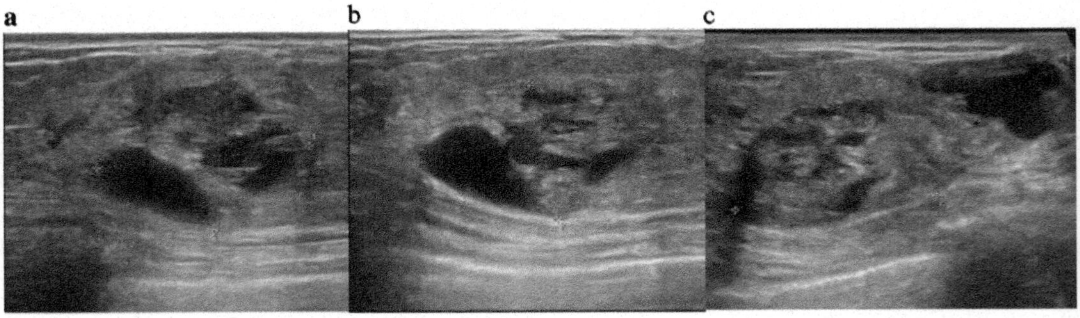

Fig. 19.1 (a–c) Ultrasound image of a 38*15 mm mixed echo tumor in a 33 year-old lactating woman. Histologic exam showed a benign phyllodes tumor. (Courtesy of Dr. Haixian Zhang)

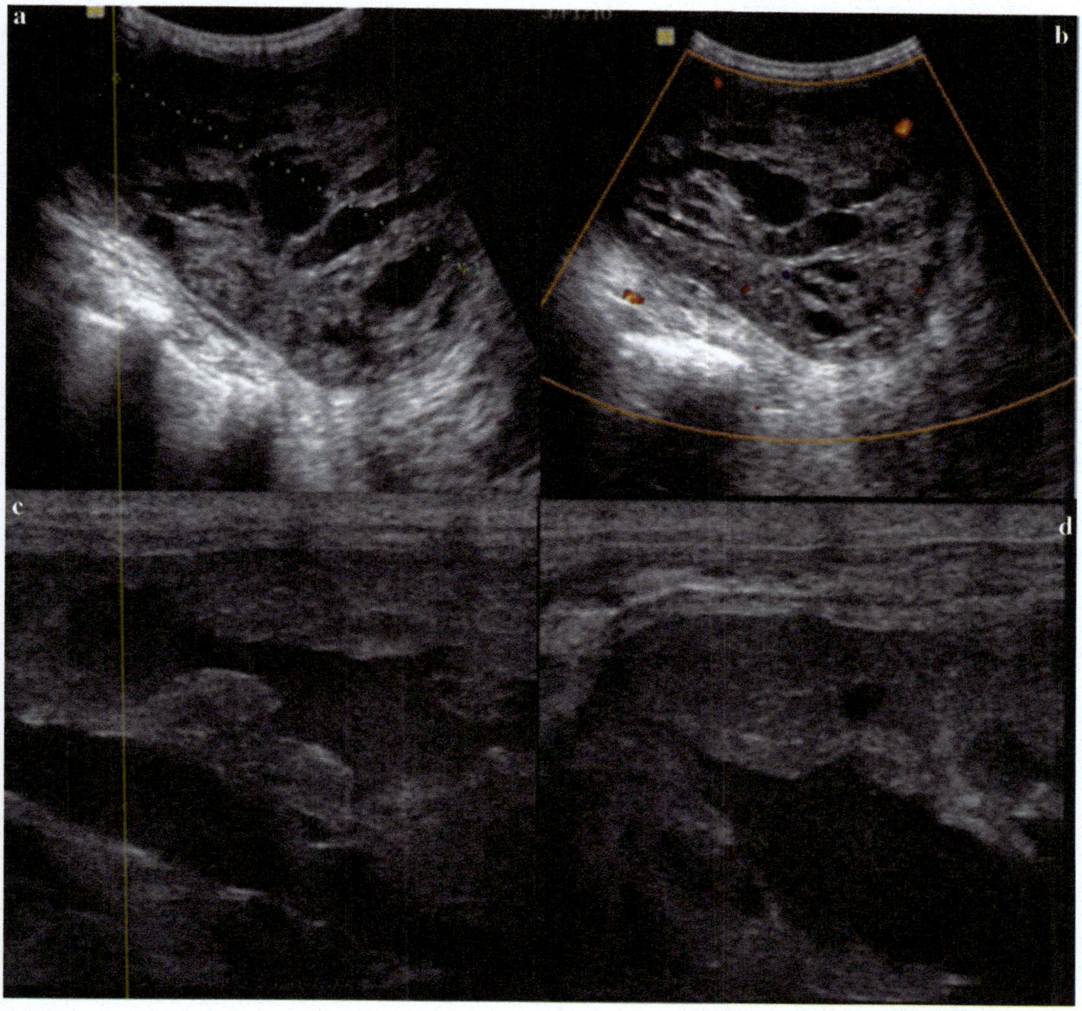

Fig. 19.2 (**a–d**) Ultrasound image of a 7*8*7 cm mixed echo tumor in a 24 year old lactating woman. Histologic exam showed interstitial hemorrhage and inflammatory cell infiltration in a benign phyllodes tumor. (Courtesy of Dr. Haixian Zhang)

The study showed that benign and borderline tumors had been treated by lumpectomy or mastectomy, with a slight predilection for the latter. Results regarding reported follow-up showed good control of all tumors for both types of surgery. In malignant PT, rates of lumpectomy and mastectomy were approximately the same. Follow-up of the patients was not reported in many cases, but as far as reported, recurrence had occurred in six patients. In total, two of these had undergone mastectomy, whereas four had initially been treated using lumpectomy and had undergone mastectomy at recurrence. These figures were too low to be interpreted, but it was observed that recurrence had been reported more frequently in malignant cases that had undergone lumpectomy as opposed to those who had undergone mastectomy as first therapeutic intervention.

In summary and alluding to this recent systematic review, large size, fast growth, and bilaterality may be more common in gestational than non-gestational PT. These features may account for the dependence of PT on sex hormones. Furthermore, considering the high rate of malignancy in these tumors, the probability of malignant transformation of benign PT under the

Table 19.1 Patient and tumor characteristics, surgical approach, and follow up in all tumors

		Gestation period		Laterality		Pathologic type			First surgery	
First author, year	Patient age	Py	PP	Uni	Bil	Ben	Bor	Mal	Lum	Mas
Andreola, 2012 [15]	29	✓			✓		✓		✓[a]	✓[b]
Aranda, 2005 [16]	32	✓		✓		✓				✓
Ariel, 1961 [17]	31		✓	✓				✓		✓
Bal, 2012 [18]	32		✓	✓				✓		✓
Blaker, 2010 [19]	27	✓		✓				✓	✓	
Cha, 2017 [20]	30		✓	✓		✓			✓	
Chai, 2015 [21]	25	✓		✓		✓				✓
De Carvalho, 1999 [22]	28	✓			✓	✓				✓
Furuya, 2012 [23]	32	✓		✓			✓			✓
Gentile, 2016 [24]	22	✓		✓			✓			✓
Hernanz, 2017 [25]	35	✓			✓			✓	✓[b, c]	✓[a]
Kallam, 2017 [26]	32	✓		✓		✓				✓
Kelten, 2016 [27]	37		✓	✓				✓		✓
Lee, 2018 [28]	21	[d]	[d]	✓				✓	✓	
Lester, 1954 [29]	26	✓		✓				✓	✓	
Li, 2008 [30]	27	[d]	[d]	✓				✓		✓
Likhitmasku, 2014 [14]	36	✓		✓		✓				✓
Mrad, 2000 [31]	32	[d]	[d]		✓	✓[a]		✓[b]	✓[a]	✓[b]
Murthy, 2016 [32]	25		✓	✓		✓			✓	
Narla, 2018 [33]	28	[d]	[d]	✓				✓	✓[e]	
Nejc, 2007 [34]	28		✓	✓				✓	✓	
Pacchiarotti, 2011 [35][a]	41	✓		✓				✓	✓	
Pandit, 1985 [36]	32		✓		✓			✓	✓[a]	✓[b]
Pasta, 2012 [37][a]	43	✓		✓				✓		
Pytel, 2009 [38]	25	✓		✓				✓		✓
Ray, 2011 [39]	24	✓		✓				✓		✓
Reich, 1958 [40][b]	23, 33	✓[b]	✓[a]		✓			✓	✓[b, f]	✓[a]
Sharma, 2004 [41]	35	✓		✓		✓				✓
Simpson, 2007 [4]	29	✓		✓				✓		✓
Testori, 2015 [42]	33		✓	✓				✓		✓
Tortoriello, 2017 [43]	37		✓	✓				✓	✓	
Vergine, 2012 [44]	27		✓	✓				✓	✓	
Vintea, 2016 [45]	29	✓		✓		✓			✓	
Ward, 1986 [46]	28		✓	✓				✓	✓[f]	
Way, 1998 [47]	35	✓		✓		✓			✓	
Weledji, 2014 [48]	30		✓	✓			✓		✓[f]	
White, 1954 [49][c]	22	✓		✓				✓	✓	

[a]right breast; [b]left breast; [c]further left mastectomy for potential residue; [d]missing data; left side, for potential residue; [e]further mastectomy for residue; [f]further mastectomy for recurrence; *Ben* benign, *Bil* bilateral, *Bor* borderline; *Lum* lumpectomy, *Mal* malignant, *Mas* mastectomy, *PP* postpartum, *Py* pregnancy, *Uni* unilateral. a: pregnancy after hormonal stimulation due to infertility, b: had metachronous bilateral phyllodes tumor, c: adenofibrosarcoma in intracanalicular fibroadenoma as described by Hill & Stout, who had used this term as equivalent to cystosarcoma phyllodes [50]

stimulation of female sex steroid hormones can be suggested; we propose further research on this aspect.

References

1. Tan PH, Simpson JF, Tse G, Hanby AM, Lee A (2012). Fibroepithelial tumors. In: WHO classifica-

tion of tumours of the breast: International Agency for Research on Cancer (IARC), pp 141–148
2. Brogi E (2015) Fibroepithelial neoplasms. In: Rosen's breast pathology. Wolters Kluwer, pp 213–170
3. Chinyama CN (2013). Benign breast diseases: radiology-pathology-risk assessment. Springer Science & Business Media, pp 114–118
4. Simpson SA, Redstone J, Aziz MS, Bernik SF (2007) Large malignant phyllodes tumor with rapid growth during pregnancy: images of a case. Breast J 13(6):620–621
5. Calhoun KE, Allison KH, Kim JN, Rahbar R, Anderson BO (2014) Phyllodes tumors. In: Diseases of the breast. Wolters Kluwer, pp 826–837
6. Spitaleri G, Toesca A, Botteri E, Bottiglieri L, Rotmensz N, Boselli S et al (2013) Breast phyllodes tumor: a review of literature and a single center retrospective series analysis. Crit Rev Oncol Hematol 88(2):427–436
7. Lu Y, Chen Y, Zhu L, Cartwright P, Song E, Jacobs L et al (2019) Local recurrence of benign, borderline, and malignant Phyllodes tumors of the breast: a systematic review and meta-analysis. Ann Surg Oncol 26(5):1263–1275
8. Tan BY, Acs G, Apple SK, Badve S, Bleiweiss IJ, Brogi E et al (2016) Phyllodes tumours of the breast: a consensus review. Histopathology 68(1):5–21
9. Hayati F, Lian HH, Azizan N, Ali AA, Abidin ZAZ, Suhaili MA (2017) Approaches to phyllodes tumour of the breast: a review article. Int Surg J 4(3):841–845
10. Hamid SA, Rahmat K, Ramli MT, Fadzli F, Jamaris S, See MH et al (2018) Radiopathological characteristics and outcomes of phyllodes tumor of the breast in Malaysian women. Medicine 97(31):e11412
11. Ruvalcaba-Limón E, Bautista-Piña V, Villegas-Carlos F, Espejo-Fonseca R, Moguel-Molina N, Tenorio-Torres JA et al (2017) Phyllodes tumor of the breast-not all are self-detected. J Xiangya Med 2(6)
12. Guillot E, Couturaud B, Reyal F, Curnier A, Ravinet J, Laé M et al (2011) Management of phyllodes breast tumors. Breast J 17(2):129–137
13. Yildirim M, Carti EB, Ucar AD, Erkan N, Vardar E, Yakan S et al (2016) Phyllodes tumor of the breast: clinical presentation, management and follow-up. Acta Med Austriaca 32:157
14. Likhitmaskul T, Asanprakit W, Charoenthammaraksa S, Lohsiriwat V, Supaporn S, Vassanasiri W et al (2015) Giant benign phyllodes tumor with lactating changes in pregnancy: a case report. Gland Surg 4(4):339–343
15. Andreola J, Damin A, Andreola J, Cruz J, Varella M, Canal L et al (2012) Giant borderline phyllodes breast tumor in pregnancy. J Senolog Int Soc 1(3)
16. Aranda C, Sotelo M, Torres A, Zárate M (2005) Phyllodes tumor and pregnancy. A report of a case. Ginecol Obstetricia Mex 73(07):387–392
17. Ariel L (1961) Skeletal metastases in Cystosarcoma Phylloides: a case report and review. Arch Surg 82(2):275–280
18. Bal A, Güngör B, Polat AK, Şimşek T (2012) Recurrent phyllodes tumor of the breast with malignant transformation during pregnancy. Eur J Breast Health 8:45–47
19. Blaker KM, Sahoo S, Schweichler MR, Chagpar AB (2010) Malignant phylloides tumor in pregnancy. Am Surg 76(3):302–305
20. Cha Y, Kim H-W, Kim HS, Won TW (2017) Spontaneous infarction of Phyllodes tumor of the breast in a postpartum woman: a case report. Korean J Radiol 77(5):327–332
21. Chai SC, Umayaal S, Saad AZM (2015) Successful pregnancy "during" pedicled transverse rectus abdominis musculocutaneous flap for breast reconstruction with normal vaginal delivery. Indian J Plast Surg 48:81
22. de Carvalho Jr AW, da Silva SMM, Almeida PBL, da Silva OQ, Schumaltz LEP, de Sousa JA et al (1999) Gravidez e tumor filodes bilateral: Uma Associação Rara. RBGO 21(2):109
23. Furuya S, Miura K, Aikawa T, Nakagomi H, Mri M, Oyama T et al (2012) A case of phyllodes tumor of the breast grown rapidly during the last period of pregnancy. J Jpn Soc Clin Surg 73(4):780–785
24. Gentile LF, Gaillard WF, Wallace JA, Spiguel LR, Alizadeh L, Lentz A et al (2016) A case of a giant borderline phyllodes tumor early in pregnancy treated with mastectomy and immediate breast reconstruction. Breast J 22(6):683–687
25. Hernanz F, González-Noriega M, Arozamena B, Solano J, García J (2018) Bilateral synchronous breast malignant phyllodes in a pregnant woman. Breast J 24(3):412–413
26. Kallam AR, Kanumury V, Korumilli RM, Gudeli V, Polavarapu H (2017) Massive benign phyllodes tumour of breast complicating pregnancy. J Clin Diagn Res 11(5):PD08–PD09
27. Kelten C, Boyaci C, Leblebici C, Behzatoglu K, Trabulus DC, Sari S et al (2016) Malignant phyllodes tumor including aneurysmal bone cyst-like areas in pregnancy-a case report and review of the literature. Breast Care 11(4):291–294
28. Lee MV, Shaw HL, Chi T, Brazeal HA, Holley SO, Appleton CM (2018) Palpable breast abnormalities in women under age 40. Breast J 24(5):798–805
29. Lester J, Stout AP (1954) Cystosarcoma phyllodes. Cancer 7(2):335–353
30. Li X, Yang Y, Wang J, Ma B, Jin Y, Li R (2008) Surgical treatment of giant recurrent breast phyllodes tumor. J Huazhong Univ Sci Technolog Med Sci 28(6):688–692
31. Mrad K, Driss M, Maalej M, Romdhane KB (2000) Bilateral cystosarcoma phyllodes of the breast: a case report of malignant form with contralateral benign form. Ann Diagn Pathol 4(6):370–372
32. Murthy SS, Raju K, Nair HG (2016) Phyllodes tumor in a lactating breast. Clin Med Insights Pathol 9:13–17
33. Narla SL, Stephen P, Kurian A, Annapurneswari S (2018) Well-differentiated liposarcoma of the breast arising in a background of malignant phyllodes tumor

in a pregnant woman: a rare case report and review of literature. Indian J Pathol Microbiol 61(4):577–579
34. Nejc D, Pasz-Walczak G, Piekarski J, Pluta P, Bilski A, Sek P et al (2008) Astonishingly rapid growth of malignant cystosarcoma phyllodes tumor in a pregnant woman—a case report. Int J Gynecol Cancer 18(4):856–859
35. Pacchiarotti A, Frati P, Caserta D, Pacchiarotti A, Frega A, Moscarini M (2011) First case of transformation for breast fibroadenoma to high-grade malignant cystosarcoma in an in vitro fertilization patient. Fertil Steril 96(5):1126–1127
36. Pandit A, Vora I, Shenoy S, Gurjar A (1985) Bilateral cystosarcoma phylloides with osteogenic sarcomatous stroma (a case report with review of literature). J Postgrad Med 31(4):215–216
37. Pasta V, Amabile MI, Bizzarri M, Monti M (2012) Breast sarcoma in a pregnant patient. Ann Ital Chir 28:2012
38. Pytel J, Dedecjus M, Naze M, Stróżyk G, Brzezinski J (2009) Malignant breast phyllodes tumor in pregnancy – a case report. Prz Menopauzalny 13(6):331
39. Ray S, Basak S, Das S, Pal M, Konar H (2011) Malignant phylloides tumor of breast in a pregnant woman with coincidental nulliparous vaginal prolapse. Iran J Med Sci 36(4):315–317
40. Reich T, Solomon C (1958) Bilateral cystosarcoma phyllodes, malignant variant, with a 14-year follow up: a case report. Ann Surg 147(1):39–43
41. Sharma JB, Wadhwa L, Malhotra M, Arora R, Singh S (2004) A case of huge enlargement of cystosarcoma phylloides of breast in pregnancy. Eur J Obstet Gynecol Reprod Biol 115(2):237–239
42. Testori A, Meroni S, Errico V, Travaglini R, Voulaz E, Alloisio M (2015) Huge malignant phyllodes breast tumor: a real entity in a new era of early breast cancer. World J Surgical Oncol 13(1):81
43. Tortoriello MG, Cerra R, Di Bonito M, Botti G, Cordaro FG, Caputo E et al (2017) A giant phyllodes tumor of the breast: a case report in pregnancy. Ann Clin Case Rep 2:1311
44. Vergine M, Pasta V, Redler A, Santucci E, Vasselli I, Ballesio L et al (2012) Cystosarcoma phylloides of the breast: a rare diagnosis. Ann Ital Chir 83(6):547–549
45. Vintea A, Sima R, Burcoş T, Bănceanu G, Toader O (2016) The management of a rare breast tumor in pregnancy. J Surg Sci 3(4):187–191
46. Ward RM, Evans HL (1986) Cystosarcoma phyllodes. A clinicopathologic study of 26 cases. Cancer 58(10):2282–2289
47. Way JC, Culham BA (1998) Phyllodes tumour in pregnancy: a case report. Can J Surg 41(5):407
48. Weledji EP, Enow-Orock G, Ngowe MN, Aminde L (2014) Breast-conserving surgery is contraindicated for recurrent giant multifocal phyllodes tumours of breast. World J Surg Oncol 12(1):213
49. White TT (1954) Carcinoma of the breast and pregnancy: analysis of 920 cases collected from the literature and 22 new cases. Ann Surg 139(1):9–18
50. Hill RP, Stout AP (1942) Sarcoma of the breast. Arch Surg 44(4):723–759
51. Alipour S, Eskandari A, Johar FM, Furuya SH (2020) Phyllodes tumor of the breast during pregnancy and lactation: a systematic review. Arch Iran Med Ahead of Printing

Inflammatory Breast Cancer in Pregnancy and Lactation

20

Samantha Linhares, Tamrah Alrammah, Hattan A. Alghamdi, and Mecker G. Möller

Abstract

Inflammatory breast cancer (IBC) represents only 1% to 5% of all breast malignancies and is an extremely aggressive subtype. At time of diagnosis, up to 85% of patients will present with regional nodal metastases and up to 30 % will have metastasis to distant organs. There is limited medical literature describing treatment guidelines for IBC during gestation. The best diagnostic tools are core needle and full-thickness skin punch biopsies to assess presence of dermal lymphatic invasion. Breast Ultrasound is preferred to mammogram, but mammography could still be done with proper fetal shielding. Ultrasound and Magnetic resonance imaging are used for staging. Pregnant patients should be managed with special attention to the health of the fetus by a multidisciplinary team. Treatment based on current guidelines consist of a sequence of systemic chemotherapy followed by mastectomy with axillary dissection (modified radical mastectomy), and even if good clinical nodal response to neoadjuvant therapy is obtained, sentinel node biopsy is not recommended. Radiation therapy is to be given once the baby has been delivered. Chemotherapy is not recommended in the first trimester, and anti-estrogen hormonal therapy, as well as targeted Her2-neu therapies are contraindicated during the length of the pregnancy. There is no evidence that early termination improves the outcome. However, given the poor prognosis of IBC, patients should be fully counseled on the risks and benefits of continuing or terminating an early pregnancy.

Keywords

Axillary dissection · Inflammatory breast cancer · Neoadjuvant chemotherapy · Mastectomy · Pregnancy

S. Linhares
University of Miami, Miller School of Medicine, Miami, FL, USA
e-mail: slinhares@med.miami.edu

T. Alrammah · H. A. Alghamdi · M. G. Möller (✉)
Department of Surgery, Division of Surgical Oncology, University of Miami, Miller School of Medicine, Miami, FL, USA
e-mail: tamrahalrommah@jhsmiami.org; hattan.alghamdi@jhsmiami.org; mmoller@med.miami.edu

20.1 Overview

Inflammatory breast cancer is an uncommon and extremely aggressive subtype of breast cancer, representing only approximately 1% to 5% of all breast malignancies [1, 2]. The first clinical description and the term of "inflammatory breast

S. Alipour, R. Omranipour (eds.), *Diseases of the Breast during Pregnancy and Lactation*, Advances in Experimental Medicine and Biology 1252,
https://doi.org/10.1007/978-3-030-41596-9_20

cancer (IBC)" was coined by Lee and Tannenbaum in 1924 [2]. It typically presents with concurrent rapid onset of erythematous skin changes with unilateral breast enlargement within 3–6 months with a classic description of a peau d'orange (skin of an orange) appearance [3, 4]. It may be associated with a breast mass or may rather present with breast engorgement and diffuse induration without a mass. The name itself is misleading because IBC does not present with histological features that are typical of an inflammatory process [5]. Based on the 7th edition of the AJCC Cancer Staging Manual, when a woman is diagnosed with IBC, she already is considered advanced stage IIIB, IIIC, or IV depending on the extent of nodal involvement and whether or not distant metastases are present. The primary tumor is classified as T4d by definition [6]. At diagnosis, up to 85% of patients will present with metastasis to regional lymph nodes and up to 30% of patients have distant metastasis [4]. Consequently, this advanced diagnosis at the onset translates to IBC patients having much lower median survival times, and local recurrence rates as high as 50% compared with other common breast cancers [7]. One study examining 398 patients with IBC treated between 1974 and 2005 found that the median overall survival time was 4.2 years [8]. Even though there is ample literature about breast cancer during pregnancy, there is limited work describing IBC during gestation.

The incidence of pregnancy-associated breast cancer (PABC) ranges between 15–35 per 100,000 deliveries [9, 10] (see also Chap. 9). These patients have a similar survival and prognosis rate when matched for stage of disease at diagnosis with non-pregnant patients. A study examining 111 patients with breast cancer diagnosed during pregnancy matched to 253 patients on age, clinical T stage, hormone receptor, HER2 and treatment modality used found that there was no significant difference in the survival outcome rates between the two cohorts [11]. Breast cancer during pregnancy is a challenging situation that requires consideration of the welfare of both mother and fetus. The presentation of breast cancer in pregnant women is similar to non-pregnant females, including a palpable mass or breast thickening. The diagnosis is usually delayed during pregnancy attributed to not suspecting malignancy by neither the patient nor the physician; due to the breast changes that occur normally during gestation [12] (see also Chap. 11). Therefore, women diagnosed with breast cancer during pregnancy often present with an advanced tumor stage and axillary lymph node involvement [13]. A study done by Bonnier et al. [14] comparing 154 pregnancy-associated breast cancer patients with 308 non-pregnancy-associated breast cancer patients revealed that the rates of inflammatory breast cancer, large tumors and negative receptor status breast cancer was higher in the pregnancy-associated group. Histologically most tumors are poorly differentiated ER/PR negative and 20–30% HER2 positive [15] (see also Chap. 10).

20.2 Concerns in Pregnancy

The diagnosis poses a clinical and treatment challenge, as physical examination of the breast during pregnancy is difficult due to increased breast density and firmness (see also Chaps. 1 and 2) as well as the considerations that must be taken from the materno-fetal treatment implications standpoint (see also Chaps. 12 and 21) [13]. The rapid growth of the breast in IBC could be mistaken for the natural growth during pregnancy. The common differential diagnosis of IBC is mastitis, which is common during pregnancy and is a reason for delayed diagnosis [16] (see also Chap. 7). Inflammatory breast cancer can also be mistaken for another type of locally advanced breast cancer (Fig. 20.1). Moreover, clinicians may not be experienced on the diagnosis of IBC and miss it in favor of a dermatologic disease, which postpones the correct diagnosis [17].

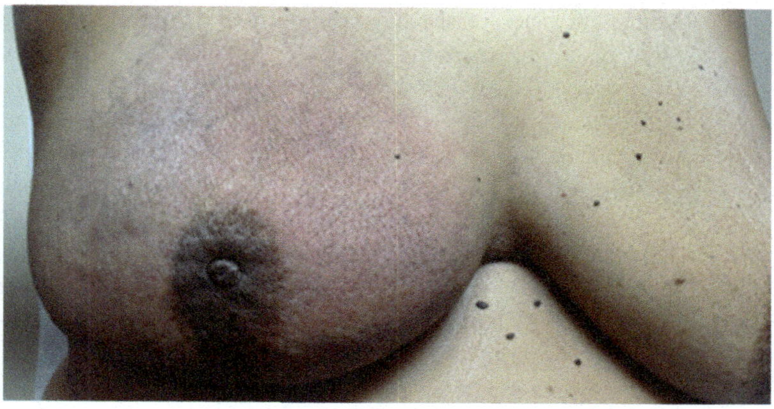

Fig. 20.1 Inflammatory breast cancer with peau d'orange, skin erythema, nipple retraction and breast induration in a 28 years old African American woman in 2nd trimester of pregnancy

There are no specific treatment guidelines for IBC in pregnant patients, and so the best recommendation from the National Comprehensive Cancer Network (NCCN) is that pregnant patients should be treated as non-pregnant cases with special attention to the health of the fetus [18]. The main concern in this regard is limitation for radiation and timing of starting chemotherapy in the second trimester. In this chapter, we will review the recommended workup and treatment options for IBC with attention to each trimester of pregnancy and treatment specific to receptor subtypes as dictated by the recommended standard breast cancer guidelines.

20.3 Current Treatment Guidelines for Inflammatory Breast Cancer

In general, treatment for pregnant patients with breast cancer should adhere as closely as possible with guidelines recommended for non-pregnant patients. The National Comprehensive Cancer Network (NCCN) guidelines and the international IBC expert guidelines recommend an aggressive tri-modality approach, in a sequence of systemic therapy followed by mastectomy with axillary dissection and radiation therapy if there is a good response to the systemic therapy [3, 18]. Response to neoadjuvant therapy to downstage the tumor is necessary before mastectomy is performed [19]. If patients do not respond to neoadjuvant therapy, then additional chemotherapy and radiation options should be performed. If the patient responds to these secondary options, then a mastectomy and adjuvant treatment with radiation should follow [18]. IBC in pregnant women should be managed by a multidisciplinary team including a medical oncologist, a surgical oncologist, a radiation oncologist, an obstetrician and a neonatologist. It is recommended not to perform a skin-sparing

Table 20.1 Summary of diagnostic work up and treatment options for inflammatory breast cancer during pregnancy

Imaging	
Diagnostic	Bilateral diagnostic mammogram with shielding
	Breast US for LN assessment
	Core biopsy (Tru-cut +/− US guided) Skin punch biopsy (recommended)
Staging	Chest X-ray with fetal shielding
	Liver US
	Thoracic & lumbar spine MRI without contrast
Systematic therapy	
Chemotherapy	No in 1st trimester Yes in 2nd and 3rd trimester
Endocrine	No
HER2-Targeted	No
Surgery	Yes Mastectomy + Axillary dissection Fetal monitoring
Radiation therapy	No

HER2 Human Epidermal Growth Factor Receptor 2, *LN* lymph node, *MRI* Magnetic Resonance Imaging, *US* Ultrasound

mastectomy, immediate breast reconstruction or sentinel node mapping, which may be appropriate in other types of locally-invasive breast cancer [3, 20]. Table 20.1 summarizes the workup and treatment of IBC during pregnancy.

20.4 Diagnostic Workup

A pregnant patient who is suspected of having a breast mass in any trimester while pregnant should undergo a careful history taking, physical exam and imaging. A physical inspection of the breast and regional lymph nodes is crucial to help determine severity, although physiologic alteration that occur secondary to hormonal changes of pregnancy make examination difficult (see also Chap. 2). Moreover, inflammatory or infectious problems of the breasts are common during pregnancy since breast milk represents a lactose rich culture medium for bacteria [21] (see also Chap. 7). After a thorough physical examination, the patient should undergo diagnostic imaging by ultrasound to determine the extent of the disease, examine the regional lymph nodes and guide the biopsy (Fig. 20.2) [22]. It is not recommended to perform an MRI with contrast during pregnancy because gadolinium-based contrast has been linked to an increased risk of rheumatological, inflammatory or infiltrative skin conditions in the child, as well as stillbirth and neonatal death in a cohort of 397 women exposed to gadolinium in the first trimester of pregnancy versus 1,418,451 women who did not undergo an MRI while pregnant [23]. Also, animal models have shown that gadolinium can cross the placental barrier and cause fetal abnormalities [24, 25] (see also Chap. 3).

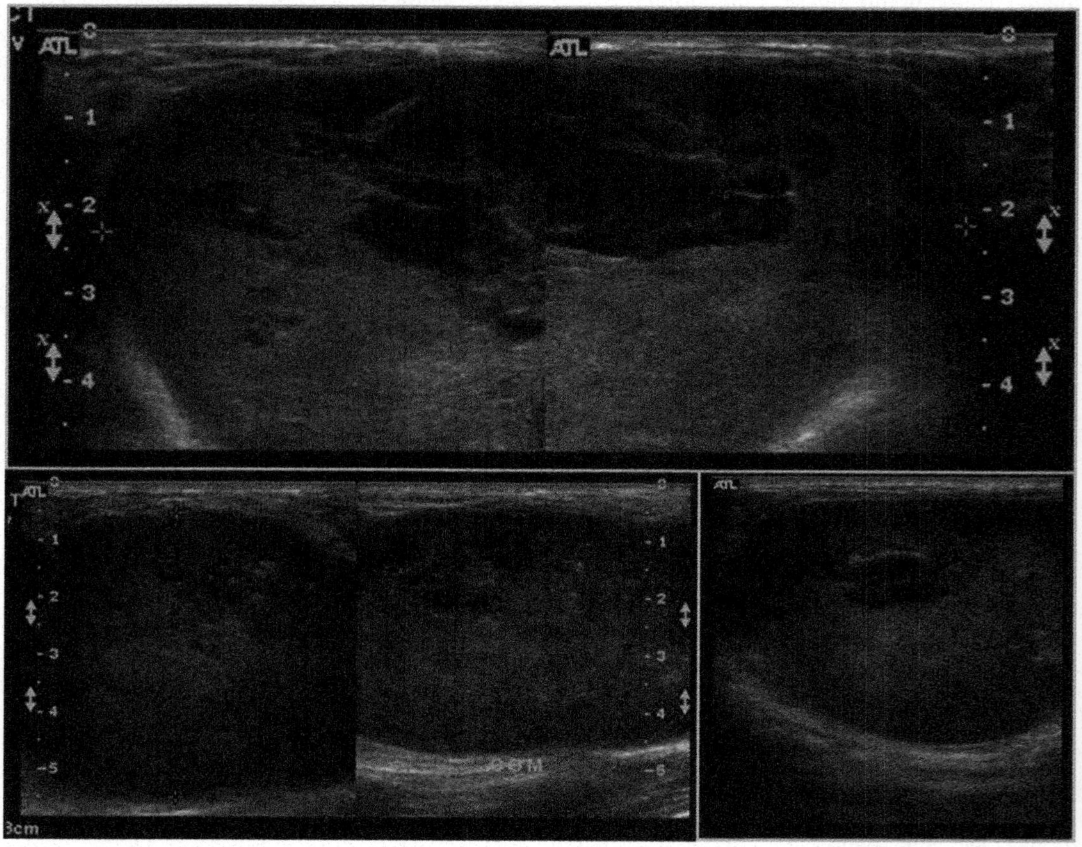

Fig. 20.2 Ultrasound in a 28 years-old African American female in 2nd trimester of pregnancy with inflammatory breast cancer showed an 8 cm oval hypoechoic mass in the left breast at 6 o'clock posterior depth with multiple cystic spaces

According to an international consensus on the clinical management of IBC the preferred type of biopsy is core needle biopsy with two full-thickness skin punch biopsies due to the fact that the defining feature of this disease is dermal lymphatic invasion by tumor cells [19].

For staging of the tumor, the protocol is the same regardless of the trimester of pregnancy; the difference is in the choice of radiological staging trying to minimize fetal exposure to radiation. Since IBC is minimum T4d at initial diagnosis, it is recommended to perform a chest x-ray with shielding, US of the liver and MRI of thoracic and lumbar spine without contrast [18]. These are summarized in Table 20.1.

20.5 Treatment Options

20.5.1 Neoadjuvant Chemotherapy

The standard systemic chemotherapy treatment used today is anthracycline- and taxane- based neoadjuvant therapy [3, 19]. There has not been any large clinical trial data to determine the optimal chemotherapy regimen for patients with IBC; thus, the treatment regimen is based on that used for locally advanced breast cancer. The incorporation of taxanes into the treatment has been shown to improve treatment outcomes. A minimum of six cycles should be administered over a period of 4–6 months before surgery is considered, in order to evaluate the response [19]. It is important to note that any type of chemotherapy should not be administered after week 35 of pregnancy or within 3 weeks of a planned delivery in order to avoid hematologic complications [18].

It is contraindicated to administer anthracycline-based treatment during the first trimester due to the risk of fetal malformations, especially during the first 14 weeks due to organogenesis. A review by Shachar et al. [26] compiled reports of neural tube defects, cleft lip, cleft palate, cardiac defects and even fetal death. Rates of fetal malformations were between 3–5% in the second and third trimesters. Anthracycline-based chemotherapy during the second and third trimesters may increase the risk of preterm labor and low birth weight. One study done by Hahn et al. [27] reported that out of 57 patients, most delivered at 34 weeks or above, and gestational age was 37 weeks in 37%, while 28% of the newborns had difficulty in breathing. However, they did not have any stillbirths, miscarriages, or perinatal deaths.

Taxane-based chemotherapy is also contraindicated during the first trimester for the same reasons as anthracycline-based treatment since there is limited data of use. There also exists limited data during the second and third trimesters, and so it is recommended not to use taxanes unless absolutely necessary [26] (see also Chap. 15).

20.5.2 Anti-HER2 Therapy

The occurrence of HER2 receptor positive IBC ranges between 26–30% [28, 29]. Non-pregnant patients with IBC that over-express HER2 should receive trastuzumab along with the standard chemotherapy regimen for at least one year. This is based on data from a randomized, controlled phase 3 trial by Gianni et al. [30]. They compared the survival rates of a cohort of patients with locally advanced HER2 positive tumors who received chemotherapy plus trastuzumab versus a second cohort of patients with locally advanced HER2 positive tumors who only received standard chemotherapy. The 5-year event-free survival rate in the first and the second group was 58% and 43%, respectively.

However, trastuzumab is contraindicated in pregnancy and in breast-feeding mothers. FDA released a statement in 2010 classifying trastuzumab as a Category D drug, detailing the risks of oligohydramnios potentially leading to pulmonary hypoplasia, skeletal abnormalities and even fetal death. A review done by Zagouri et al. [31] investigatung 17 studies found that overall 61.1% of pregnant patients taking trastuzumab experienced oligohydramnios/anhydramnios and 73.3% of patients in the second and third trimester experienced oligohydramnios/anhydramnios. Pertuzumab is a monoclonal antibody that blocks the formation of HER2:HER3 heterodimers and is often administered with trastuzumab due to higher complete response rates with dual administration. A phase 2 randomized control trial found that patients on trastuzumab and pertuzumab versus just trastuzumab had 5-year survival rates of 86% and 81%, respectively [32]. However, there is not enough data detailing use and outcomes within a pregnant population and so its use is recommended to be avoided. In special high-risk situations, the use of trastuzumab should be thoroughly discussed between the patient and physician to weigh the risks and benefits.

20.5.3 Endocrine Therapy

The occurrence of hormone-receptor positive IBC ranges between 33–56% [28, 29]. The standard recommended endocrine therapy for hormone-receptor positive tumors in IBC is tamoxifen or an aromatase inhibitor for a minimum of 5 years [19]. However these therapies are contraindicated in pregnancy. A review performed by Schuurman et al. [33] investigated the effect of tamoxifen use during pregnancy. They detected a 12.6% and 3.9% risk of fetal abnormality in tamoxifen users and non-users, respectively. Tamoxifen has been classified by the FDA as a Category D drug due to associations with congenital malformations, miscarriage, and fetal death [34, 35]. There is also limited data on the safety of tamoxifen use during breastfeeding and so it is generally recommended to avoid breastfeeding during its use.

20.5.4 Supportive Therapy

Many patients who undergo chemotherapy experience dizziness, nausea and vomiting; for a pregnant patient, these symptoms may be exaggerated due to hormonal changes associated with the pregnancy itself. There are several options available to women without reports of risks to the fetus including promethazine, 5-HT_3 receptor antagonists, neurokinin 1 antagonists, and droperidol combined with diphenhydramine (an H1 receptor antagonist) or dexamethasone. It is recommended that first-trimester use of dexamethasone be avoided due to reports of cleft palate, and long-term use be avoided due to reported cases of attention-deficit disorder [36].

20.5.5 Surgical Intervention

Surgery can be safely performed in all trimesters of pregnancy with minimal risk to the fetus [37] (see also Chap. 12). It is recommended to wait after the 12th week of gestation when the risk of spontaneous abortion may be lower. Mastectomy with axillary dissection is the recommended surgery option for IBC patients who respond to neoadjuvant chemotherapy. This is preferred over breast-sparing techniques due to the fact that the involved breast may have residual disease if it is not fully removed, and mastectomy reduces the risk of locoregional recurrence [19] (Fig. 20.3).

Overall, due to the contraindications of sentinel node biopsy in IBC, this procedure is not recommended. One of the major concerns when performing surgery during pregnancy is the effects of anesthesia on both the mother and fetus. There are no anesthetics which are considered teratogenic, but it is recommended to avoid their use during the first trimester if possible to allow organogenesis to complete [26, 38]. After 24–26 weeks of gestation, intraoperative fetal heartbeat can be monitored, which allows optimization of hemodynamics and temperature to protect both the fetus and mother [38]. It is known that surgery during the third trimester increases the risk for preterm labor due to the stress of the surgery [26, 38] (see also Chaps. 14 and 21).

20.5.6 Radiation Therapy

Radiation is strongly contraindicated during all trimesters of pregnancy due to the high chance of fetal malformation, miscarriage, or even death [26] (see also Chap. 16). Radiation therapy should not be administered before surgery in IBC due to previous reports of high complication rates [3]. Post-mastectomy radiation should include

Fig. 20.3 MRI of breast after completion of modified radical mastectomy, post-delivery radiation and reconstruction: BIRAD 1, normal bilateral breast MRI

the chest wall, axillary (level III), supraclavicular and internal mammary lymph nodes [39]. It is important for radiation in IBC to cover all of the affected areas, which can be quite expansive due to distal lymph node involvement. There is a high likelihood that radiation therapy will cross over the midline to provide adequate coverage or risk recurrence [3]. The typical radiation doses used in IBC are 50.4 or 50 Gy in 1.8 or 2 Gy fractions to locoregional sites followed by a 10 Gy boost to the chest wall, totaling to a dose of 60 Gy. The high likelihood of locoregional recurrence means that patients should have radiation therapy that encompasses the supraclavicular regions and internal mammary lymph nodes as well. For a subset of patients who are <45 years of age, respond poorly to chemotherapy, and have close margins, some studies suggest that escalation of post-mastectomy radiation dose to 66 Gy might be beneficial. In a study of 192 patients who underwent neoadjuvant chemotherapy, surgery and adjuvant radiation, those ≤45 years of age had a significant improvement in their 5-year locoregional control of 86% with 66 Gy versus 65% with 60 Gy [40]. It is imperative in IBC patients to have dose escalation to prevent local recurrence [41]. However, delaying radiotherapy may lead to an increased risk of recurrence. It is recommended that patients discuss the potential benefits and risks of treatment with their physician and interdisciplinary team based on individual case-specific details.

20.6 Concerns in Lactation

Per the NCCN guidelines, active breastfeeding during chemotherapy and endocrine therapy treatments is not recommended due to risks of harm to the fetus [18] (see also Chap. 22). Many of the chemotherapy drugs mentioned above to treat IBC are excreted in the breast milk, which is a risk to the newborn. Thus, it is recommended to stop breastfeeding in order to treat such an aggressive cancer as if the patient were not pregnant. Due to the necessity of mastectomy for adequate treatment, patients will need to use formula to nurse the newborn.

A rare complication of a needle biopsy or surgical intervention while lactating is the formation of a milk fistula [42] (see also Chap. 13). This consist of a connection between a milk duct of the lactating breast and the skin surface [43]. If an intervention is necessary during lactation, it is recommended to discuss the risk of milk fistula with the patient prior to the planned procedure.

20.7 Elective Termination of Pregnancy in Inflammatory Breast Cancer

Because the mother's life span may be limited, and there is a risk of fetal damage with treatment during the first trimester, issues regarding continuation of an early pregnancy should be discussed with the patient and her family; with special attention to religious beliefs or a highly desired pregnancy after infertility treatment. There is no clinical evidence that early termination of pregnancy improves the outcome of PABC [44]. However, given the poor prognosis of IBC even without pregnancy, patients should be fully counseled on the risks and benefits of terminating an early pregnancy.

References

1. Gooch JC, Schnabel F (2019) Inflammatory breast cancer. In: Clinical algorithms in general surgery. Springer, Cham, pp 105–108
2. Walshe JM, Swain SM (2006) Clinical aspects of inflammatory breast cancer. Breast Dis 22(1):35–44
3. Menta A, Fouad TM, Lucci A, Le-Petross H, Stauder MC, Woodward WA et al (2018) Inflammatory breast cancer: what to know about this unique, aggressive breast cancer. Surg Clin North Am 98(4):787–800
4. Van Uden DJ, van Laarhoven HW, Westenberg AH, de Wilt JH, Blanken-Peeters CF (2015) Inflammatory breast cancer: an overview. Crit Rev Oncol Hematol 93(2):116–126
5. Fouad TM, Kogawa T, Reuben JM, Ueno NT (2014) The role of inflammation in inflammatory breast cancer. In: Inflammation and cancer. Springer, Basel, pp 53–73
6. Fouad TM, Barrera AM, Reuben JM, Lucci A, Woodward WA, Stauder MC, Lim B, DeSnyder SM, Arun B, Gildy B, Valero V (2017) Inflammatory breast cancer: a proposed conceptual shift in the UICC–AJCC TNM staging system. Lancet Oncol 18(4):e228–e232
7. Robertson FM, Bondy M, Yang W, Yamauchi H, Wiggins S, Kamrudin S et al (2010) Inflammatory breast cancer: the disease, the biology, the treatment. CA Cancer J Clin 60(6):351–375
8. Gonzalez-Angulo AM, Hennessy BT, Broglio K, Meric-Bernstam F, Cristofanilli M, Giordano SH et al (2007) Trends for inflammatory breast cancer: is survival improving? Oncologist 12(8):904–912
9. Andersson TM, Johansson AL, Hsieh CC, Cnattingius S, Lambe M (2009) Increasing incidence of pregnancy-associated breast cancer in Sweden. Obstet Gynecol 114(3):568–572
10. Smith LH, Danielsen B, Allen ME, Cress R (2003) Cancer associated with obstetric delivery: results of linkage with the California cancer registry. Am J Obstet Gynecol 189(4):1128–1135
11. Ploquin A, Pistilli B, Tresch E, Frenel JS, Lerebours F, Lesur A et al (2018) 5-year overall survival after early breast cancer diagnosed during pregnancy: a retrospective case-control multicentre French study. Eur J Cancer 95:30–37
12. Ishida T, Yokoe T, Kasumi F, Sakamoto G, Makita M, Tominaga T et al (1992) Clinicopathologic characteristics and prognosis of breast cancer patients associated with pregnancy and lactation: analysis of case-control study in Japan. Jpn J Cancer Res 83(11):1143–1149
13. Middleton LP, Amin M, Gwyn K, Theriault R, Sahin A (2003) Breast carcinoma in pregnant women: assessment of clinicopathologic and immunohistochemical features. Cancer 98(5):1055–1060
14. Bonnier P, Romain S, Dilhuydy JM, Bonichon F, Julien JP, Charpin C et al (1997) Influence of pregnancy on the outcome of breast cancer: a case-control study. Societe Francaise de Senologie et de Pathologie Mammaire Study Group. Int J Cancer 72(5):720–727
15. Elledge RM, Ciocca DR, Langone G, McGuire WL (1993) Estrogen receptor, progesterone receptor, and HER-2/neu protein in breast cancers from pregnant patients. Cancer 71(8):2499–2506
16. Peters F, Kießlich A, Pahnke V (2002) Coincidence of nonpuerperal mastitis and noninflammatory breast cancer. Eur J Obstet Gynecol Reprod Biol 105(1):59–63
17. Li J, Xia Y, Wu Q, Zhu S, Chen C, Yang W et al (2017) Outcomes of patients with inflamatory breast cancer by hormone receptor-and HER2-defined molecular subtypes: a population-based study from the SEER program. Oncotarget 8(30):49370–49379
18. National Comprehensive Cancer Network (NCCN guidelines) (2019) Breast cancer. Available at www.NCCN.org
19. Ueno NT, Fernandez JR, Cristofanilli M, Overmoyer B, Rea D, Berdichevski F et al (2018) International consensus on the clinical management of inflammatory breast cancer from the Morgan Welch Inflammatory Breast Cancer Research Program 10th anniversary conference. J Cancer 9(8):1437–1447
20. Stearns V, Ewing CA, Slack R, Penannen MF, Hayes DF, Tsangaris TN (2002) Sentinel lymphadenectomy after neoadjuvant chemotherapy for breast cancer may reliably represent the axilla except for inflammatory breast cancer. Ann Surg Oncol 9(3):235–242
21. Scott-Conner CE, Schorr SJ (1995) The diagnosis and management of breast problems during pregnancy and lactation. Am J Surg 170(4):401–405
22. Nicklas AH, Baker ME (2000) Imaging strategies in the pregnant cancer patient. Semin Oncol 27(6):623–632
23. Ray JG, Vermeulen MJ, Bharatha A, Montanera WJ, Park AL (2016) Association between MRI exposure during pregnancy and fetal and childhood outcomes. JAMA 316(9):952–961
24. Novak Z, Thurmond AS, Ross PL, Jones MK, Thornburg KL, Katzberg RW (1993) Gadolinium-DTPA transplacental transfer and distribution in fetal tissue in rabbits. Investig Radiol 28(9):828–830

25. Webb JA, Thomsen HS, Morcos SK (2005) The use of iodinated and gadolinium contrast media during pregnancy and lactation. Eur Radiol 15(6):1234–1240
26. Shachar SS, Gallagher K, McGuire K, Zagar TM, Faso A, Muss HB et al (2017) Multidisciplinary management of breast cancer during pregnancy. Oncologist 22(3):324–334
27. Hahn KM, Johnson PH, Gordon N, Kuerer H, Middleton L, Ramirez M et al (2006) Treatment of pregnant breast cancer patients and outcomes of children exposed to chemotherapy in utero. Cancer 107(6):1219–1226
28. Copson E, Shaaban AM, Maishman T, Moseley PM, McKenzie H, Bradbury J et al (2018) The presentation, management and outcome of inflamatory breast cancer cases in the UK: data from a multi-centre retrospective review. Breast 42:133–141
29. Li J, Gonzalez-Angulo AM, Allen PK, Yu TK, Woodward WA, Ueno NT et al (2011) Triple-negative subtype predicts poor overall survival and high locoregional relapse in inflammatory breast cancer. Oncologist 16(12):1675–1683
30. Gianni L, Eiermann W, Semiglazov V, Lluch A, Tjulandin S, Zambetti M et al (2014) Neoadjuvant and adjuvant trastuzumab in patients with HER2-positive locally advanced breast cancer (NOAH): follow-up of a randomised controlled superiority trial with a parallel HER2-negative cohort. Lancet Oncol 15(6):640–647
31. Zagouri F, Sergentanis TN, Chrysikos D, Papadimitriou CA, Dimopoulos MA, Bartsch R (2013) Trastuzumab administration during pregnancy: a systematic review and meta-analysis. Breast Cancer Res Treat 137(2):349–357
32. Gianni L, Pienkowski T, Im YH, Tseng LM, Liu MC, Lluch A et al (2016) 5-year analysis of neoadjuvant pertuzumab and trastuzumab in patients with locally advanced, inflammatory, or early-stage HER2-positive breast cancer (NeoSphere): a multicentre, open-label, phase 2 randomised trial. Lancet Oncol 17(6):791–800
33. Schuurman TN, Witteveen PO, van der Wall E, Passier JLM, Huitema ADR, Amant F et al (2019) Tamoxifen and pregnancy: an absolute contraindication? Breast Cancer Res Treat 175(1):17–25
34. Cullins SL, Pridjian G, Sutherland CM (1994) Goldenhar's syndrome associated with tamoxifen given to the mother during gestation. JAMA 271(24):1905–1906
35. Tewari K, Bonebrake RG, Asrat T, Shanberg AM (1997) Ambiguous genitalia in infant exposed to tamoxifen in utero. Lancet 350(9072):183
36. Loibl S, Schmidt A, Gentilini O, Kaufman B, Kuhl C, Denkert C et al (2015) Breast cancer diagnosed during pregnancy: adapting recent advances in breast cancer care for pregnant patients. JAMA Oncol 1(8):1145–1153
37. Amant F, Deckers S, Van Calsteren K, Loibl S, Halaska M, Brepoels L et al (2010) Breast cancer in pregnancy: recommendations of an international consensus meeting. Eur J Cancer 46(18):3158–3168
38. Kuczkowski KM (2004) Nonobstetric surgery during pregnancy: what are the risks of anesthesia? Obstet Gynecol Surv 59(1):52–56
39. Walker GV, Niikura N, Yang W, Rohren E, Valero V, Woodward WA et al (2012) Pretreatment staging positron emission tomography/computed tomography in patients with inflamatory breast cancer influences radiation treatment field designs. Int J Radiat Oncol Biol Phys 83(5):1381–1386
40. Bristol IJ, Woodward WA, Strom EA, Cristofanilli M, Domain D, Singletary SE et al (2008) Locoregional treatment outcomes after multimodality management of inflammatory breast cancer. Int J Radiat Oncol Biol Phys 72(2):474–484
41. Woodward WA (2014) Postmastectomy radiation therapy for inflammatory breast cancer: is more better? Int J Radiat Oncol Biol Phys 89(5):1004–1005
42. Schackmuth E, Harlow C, Norton L (1993) Milk fistula: a complication after core breast biopsy. AJR Am J Roentgenol 161(5):961–962
43. Larson KE, Valente SA (2016) Milk fistula: diagnosis, prevention, and treatment. Breast J 22(1):111–112
44. Nugent P, O'Connell TX (1985) Breast cancer and pregnancy. Arch Surg 120(11):1221–1224

Prenatal Care during and after Breast Cancer Treatment

21

Mina Mhallem Gziri and Khadija Bouhna

Abstract

Cancer associated with pregnancy is defined by diagnosis during pregnancy, lactation, or the first year after delivery. The decision about type of treatment depends on the cancer stage and gestational age. Termination of pregnancy does not seem to modify the maternal prognosis for breast cancers. Interdisciplinary meetings and discussions are needed to evaluate and balance the maternal and fetal risks. In this chapter, we discuss about how to prevent or treat maternal and fetal complications of surgery and chemotherapy in pregnancy-associated breast cancer.

Keywords

Breast Cancer · Chemotherapy · Fetal complications · Fetal monitoring · Maternal hypoxia · Prenatal care · Surgery

M. Mhallem Gziri (✉) · K. Bouhna
Cliniques Universitaires Saint-Luc, Obstétrique, Brussels, Belgium
e-mail: mina.mhallem@uclouvain.be; khadija.bouhna@uclouvain.be

21.1 Overview

Cancer associated with pregnancy is defined by diagnosis during pregnancy or during the first year after delivery. Pregnant patients with cancer can be treated with surgery, chemotherapy, radiotherapy, or a combination thereof. The decision about type of treatment depends on the cancer stage and gestational age. Termination of pregnancy does not seem to modify the maternal prognosis for breast cancers [1, 2]. Interdisciplinary meetings and discussions are needed to evaluate and balance the maternal and fetal risks [3]. (see also Chaps. 3, 4).

21.2 Prevention and Treatment of Maternal Complications

21.2.1 During and after Breast Cancer Surgery

The multidisciplinary team must consider the physiological, anatomical, and pharmacological maternal adaptations to ensure maternal and fetal pre-, intra-, and postoperative care [4]. Pregnancy is associated with increases in cardiac output, plasma volume, oxygen consumption, glomerular filtration, and coagulation state as well as decreased gastric motility, anemia, leukocytosis, and aortocaval compression. Therefore, main-

S. Alipour, R. Omranipour (eds.), *Diseases of the Breast during Pregnancy and Lactation*, Advances in Experimental Medicine and Biology 1252,
https://doi.org/10.1007/978-3-030-41596-9_21

taining the left lateral tilt position, adequate anesthesia, and prophylactic thrombosis prophylaxis are recommended [5, 6]. Surgery can be performed under local or general anesthesia. Anesthetic drugs may cross the placenta depending on the gestational age and the dose. During the first two weeks of pregnancy, anesthesia can be associated with an all or nothing phenomenon, and then between the second and eighth week, it can be associated with structural abnormalities. Opioids, volatile agents, muscle relaxants, and local anesthesia are known to be safe during pregnancy, but nitrous oxide alters the DNA synthesis and should be avoided [6, 7]. Non-emergent surgery should be postponed after the first trimester of pregnancy to avoid miscarriages and malformations [8]. One of the major serious complications during anesthesia is maternal hypoxia causing reduced utero-placental perfusion, fetal hypoxia, asphyxia, and fetal death. Therefore, the anesthesiologist must strictly maintain maternal oxygenation, blood pressure, and uterine tonus. Fetal heart rate monitoring is not routinely recommended [9] (see also Chap. 14).

Among relevant studies, the largest series was published by Mazze et al. [10] and included 5405 surgeries during 720,000 pregnancies in Sweden between 1973 and 1981. No increase of malformation and stillbirth was observed; nevertheless, the incidence of low-birth-weight infants and live infants dying within the first 7 days increased. Cohen-Kerem et al. [11] reviewed 12,452 patients who underwent surgery during pregnancy, and no increase of miscarriages and malformations was observed. Patients with peritonitis had an increased risk of fetal loss. Van Calsteren et al. [12] reviewed 215 patients with a cancer diagnosis during pregnancy. Treatment started during pregnancy in 122 cases (56.7%), and surgery alone and/or other treatments were also reported in 80 cases. In total, seven patients treated by surgery had complications (8.75%): preterm contractions, sepsis, intrauterine growth restriction (IUGR), and preterm labor. Table 21.1 reports obstetrical complications secondary to breast cancer treatment.

Table 21.1 Obstetrical complications in 80 patients undergoing surgery for treatment of breast cancer during pregnancy. Adapted from Van Calsteren et al. [12]

Type of treatment	Number of patients (%)	Complications
Surgery	49 (40.2)	Preterm contractions, sepsis, IUGR
Surgery and chemotherapy	25 (20.5)	Preterm contractions, preterm labor, sepsis
Surgery and radiotherapy	3 (2.5)	No
Surgery and chemotherapy and radiotherapy	3 (2.5)	No

21.2.2 During and after Chemotherapy

Chemotherapy is the second possible treatment during pregnancy and must be avoided during the first trimester. The chemotherapeutic agents can induce an all or nothing phenomenon during the implantation days and then induce malformations between days 10 and 56 of pregnancy (organogenesis). Chemotherapy should be started after the 14th week of pregnancy to preserve fetal development; it can be administered until the week 35 [13, 14].

Cardonick reported 157 neonates exposed to chemotherapy after the first trimester. No increase of malformations, prematurity and IUGR was observed but a significant difference in the birth weight was reported [15]. Surprisingly, Aviles described 54 patients exposed to chemotherapy for hematological cancers during the first trimester without an increase in malformations and chromosomal abnormalities [16]. These patients refused abortion or had a higher risk of death and/or complications in case of abortion. The authors propose that renal clearance, hepatic function, and chemotherapy metabolism are different during the first trimester of pregnancy, but they do not have specific and clear explanations. In total, 4 fetuses were lost, but autopsies were performed in only two cases where no congenital malformations were observed [17]. Nevertheless, chemotherapy is proposed to be withheld until 14 weeks of gestational age because of a higher teratogenic risk [18]. This risk rises with first trimester exposure to chemotherapy. But when chemotherapy is administered after the first trimester, there are no more and/or other malformations when compared with the background population [19].

Chemotherapeutic agents act by killing the dividing cells and consequently induce maternal complications such as myelosuppression, gastrointestinal distress, alopecia, bleeding, and fatigue. They can also damage specific organs such as the heart, brain, liver, kidneys, ovaries, and the inner ear. The frequently used types of chemotherapy include alkylating agents, antimetabolites, alkaloids, topoisomerase inhibitors, and cytotoxic antibiotics [13] (see also Chap. 15).

21.3 Prevention and Treatment of Fetal Complications

21.3.1 During and after Breast Cancer Surgery

The major complications of surgery are linked to maternal hypotension and hypoxia. Prematurity and IUGR have been described after surgical treatment of malignant tumors [2]. Fetal heart rate is commonly monitored but not yet recommended during surgery between the second and third trimesters of pregnancy [5, 9]. However, fetal ultrasound and monitoring are recommended before and after surgery (see also Chap. 14).

21.3.2 During and after Chemotherapy

Prematurity, IUGR, fetal malformations, cardiotoxicity, and death are reported in the literature [13, 15, 20, 21]. The European breast cancer registry compared children exposed and not exposed to chemotherapy for maternal breast cancer. Chemotherapy during pregnancy was associated with a higher risk of low birth weight and adverse fetal outcomes, but these complications were more likely related to prematurity than to chemotherapy use during pregnancy [22] (see also Chap. 15). Table 21.2 summarizes the largest studies including chemotherapy during pregnancy. These studies show no increase of malformations after chemotherapy. This phenomenon can be explained by the use of antineoplastic agents after 14 weeks of pregnancy and also by the placenta function. The transplacental passage determines the concentrations and impacts of the compounds on the fetus. This was tested in placenta perfusion models and in vivo experiences (mouse and baboons). These fetal concentrations depend on multiple factors such as maternal pharmacokinetics, placental blood flow, and the physicochemical drug properties [23]. Different ways of transfer are possible such as facilitated or

Table 21.2 Fetal impact of chemotherapy during pregnancy

	Systemic case review [13]	North American registry [15]	European multicentre study [12]	European breast cancers registry [21]
Total number included	321	231	215	447
Total number exposed to chemotherapy	321	157	62	197
Fetal demise	5.10%	6.4%	2.3%	1%
Termination of pregnancy	N/A	5.6%	14%	12%
Preterm delivery	5.10%	5.8%	54.2%	51%
Intrauterine growth restriction	7.10%	7.7%	14.9%	9%
Malformations	3.40%	3.8%	6.5%	3.6%

Table 21.3 Rate of transplacental passage of chemotherapy drugs

Chemotherapeutic agent	Perfusion model	Pregnant animal model
Anthracyclines[a]	2.21–4.73% (22;23)	1–10.8% (24;25)
Paclitaxel	NA	0.8–2.4% (26)
Vinblastine	NA	3–34% (24;25)
4-hydroxyphosphamide	NA	18.8–31.4% (24)
Cytarabine	NA	34.1–78.6% (25)
Carboplatin	NA	43.3–155.9% (25;26)

Adapted from Amant et al. [14]
[a]epirubicin and doxorubicin

passive diffusion and active transport. The role of placental transporters such as P-glycoproteins for transfer of antineoplastic products has not been yet evaluated [14]. Table 21.3 reviews the current transplacental passage results.

Chemotherapy, and especially anthracyclines, is currently used during pregnancy; but specific effects on the maternal and fetal heart are relatively poorly studied. Fetal myocardium differs from the adult's; it is more sensitive and more vulnerable because of several cellular factors. Animal models have demonstrated that fetal myocardium is much smaller, contains only one nucleus, has fewer not well-organized myofibrils, and is more vascular. The fetal heart grows by hyperplasia and not by hypertrophy [24].

Only case reports, case series, and reviews are published in the literature about effects of anthracyclines on the fetus. In a series of 160 fetuses exposed to anthracyclines, two cases of myocardial dysfunction were identified [22]. One resulted in fetal demise in the third trimester, whereas in the second case, fetal cardiac function normalized during follow-up. In total, three cases of fetal cardiac complications were reported after idarubicin exposure. One fetus was born after idarubicin exposure at 28 weeks and had several problems related to prematurity including respiratory distress syndrome, necrotizing enterocolitis, and cerebral hemorrhage but was diagnosed with severe biventricular dysfunction that normalized after three days of life [25]. A second fetus died 2 days after administration of idarubicin but no information is available regarding the cause of death [26]. A third fetus developed reversible cardiac dysfunction at 24 weeks following administration of fludarabine, idarubicin, and gemtuzumab. The fetus delivered at 33 weeks, and follow-up at six months demonstrated complete normalization of cardiac function [27]. Acute reversible and irreversible fetal cardiac dysfunction is therefore possible after in utero chemotherapy exposure. However, not all fetuses develop cardiac dysfunction as shown by Meyer-Wittkopf who serially monitored fetal cardiac function before and after 4 cycles of doxorubicin in a single fetus. The systolic and diastolic function of the fetus remained normal during serial fetal follow-up [28]. The long-term outcome is poorly described, and the chronic cardiotoxicity also remains a question mark. Avilés et al. [29] conducted the first large study that described the cardiac outcome of 81 cases aged between 9 and 29 years, exposed to anthracyclines in utero. The results are reassuring but are only based on left ventricular end-diastolic and end-systolic dimensions (fractional shortening). Van Calsteren et al. [30] observed mildly reduced ventricular wall thickness and reduced left ventricular mass index in a pilot group of 10 children exposed to chemotherapy in utero (see also Chaps. 15 and 23).

References

1. Loibl S, Schmidt A, Gentilini O, Kaufman B, Kuhl C, Denkert C et al (2015) Breast cancer diagnosed during pregnancy: adapting recent advances in breast cancer care for pregnant patients. JAMA Oncol 1(8):1145–1153
2. Amant F, Han SN, Gziri MM, Vandenbroucke T, Verheecke M, Van Calsteren K (2015) Management of cancer in pregnancy. Best Pract Res Clin Obstet Gynaecol 29(5):741–753
3. Lishner M, Avivi I, Apperley JF, Dierickx D, Evens AM, Fumagalli M et al (2016) Hematologic malignancies in pregnancy: management guidelines from

an international consensus meeting. J Clin Oncol 34(5):501–508
4. Evans SR, Sarani B, Bhanot P, Feldman E (2012) Surgery in pregnancy. Curr Probl Surg 49(6):333–388
5. Reitman E, Flood P (2011) Anaesthetic considerations for non-obstetric surgery during pregnancy. Br J Anaesth 107((Suppl-1)):i72–i78
6. Van De Velde M, De Buck F (2007) Anesthesia for non-obstetric surgery in the pregnant patient. Minerva Anestesiol 73(4):235–240
7. Rowland AS, Baird DD, Shore DL, Weinberg CR, Savitz DA, Wilcox AJ (1995) Nitrous oxide and spontaneous abortion in female dental assistants. Am J Epidemiol 141(6):531–538
8. Moran BJ, Yano H, Al Zahir N, Farquharson M (2007) Conflicting priorities in surgical intervention for cancer in pregnancy. Lancet Oncol 8(6):536–544
9. Amant F, Loibl S, Neven P, Van Calsteren K (2012) Breast cancer in pregnancy. Lancet 379(9815):570–579
10. Mazze RI, Kallén B (1989) Reproductive outcome after anesthesia and operation during pregnancy: a registry study of 5405 cases. Am J Obstet Gynecol 161(5):1178–1185
11. Cohen-Kerem R, Railton C, Oren D, Lishner M, Koren G (2005) Pregnancy outcome following non-obstetric surgical intervention. Am J Surg 190(3):467–473
12. Van Calsteren K, Heyns L, Smet FD, Van Eycken L, Gziri MM, Van Gemert W, Halaska M, Vergote I, Ottevanger N, Amant F (2010) Cancer during pregnancy: an analysis of 215 patients emphasizing the obstetrical and the neonatal outcomes. J Clin Oncol 28(4):683–689
13. Cardonick E, Iacobucci A (2004) Use of chemotherapy during human pregnancy. Lancet Oncol 5(5):283–291
14. Amant F, Han SN, Gziri MM, Dekrem J, Van Calsteren K (2012) Chemotherapy during pregnancy. Curr Opin Oncol 24(5):580–586
15. Cardonick E, Usmani A, Ghaffar S (2010) Perinatal outcomes of a pregnancy complicated by cancer, including neonatal follow-up after in utero exposure to chemotherapy: results of an international registry. Am J Clin Oncol 33(3):221–228
16. Avilés A, Neri N, Nambo MJ (2012) Hematological malignancies and pregnancy: treat or no treat during first trimester. Int J Cancer 131(11):2678–2683
17. Avilés A, Neri N, Nambo MJ (2013) Author's reply: chemotherapy during first trimester of pregnancy. Int J Cancer 132(7):1729
18. Han SN, Gziri MM, Calsteren KV, Amant F (2013) Is chemotherapy during the first trimester of pregnancy really safe? Int J Cancer 132(7):1728
19. Doll DC, Ringenberg QS, Yarbro JW (1989) Antineoplastic agents and pregnancy. Semin Oncol 16(5):337–346
20. Cardonick E, Dougherty R, Grana G, Gilmandyar D, Ghaffar S, Usmani A (2010) Breast cancer during pregnancy: maternal and fetal outcomes. Cancer J 16(1):76–82
21. Loibl S, Han SN, von Minckwitz G, Bontenbal M, Ring A, Giermek J et al (2012) Treatment of breast cancer during pregnancy: an observational study. Lancet Oncol 13(9):887–896
22. Germann N, Goffinet F, Goldwasser F (2004) Anthracyclines during pregnancy: embryo-fetal outcome in 160 patients. Ann Oncol 15(1):146–150
23. Gedeon C, Behravan J, Koren G, Piquette-Miller M (2006) Transport of glyburide by placental ABC transporters: implications in fetal drug exposure. Placenta 27(11-12):1096–1102
24. Rudolph AM (2007) Myocardial growth before and after birth: clinical implications. Acta Paediatr 89(2):129–133
25. Achtari C, Hohlfeld P (2000) Cardiotoxic transplacental effect of idarubicin administered during the second trimester of pregnancy. Am J Obstet Gynecol 183(2):511–512
26. Reynoso EE, Huerta F (1994) Acute leukemia and pregnancy--fatal fetal outcome after exposure to idarubicin during the second trimester. Acta Oncol 33(6):709–710
27. Baumgärtner AK, Oberhoffer R, Jacobs VR, Ostermayer E, Menzel H, Voigt M et al (2009) Reversible foetal cerebral ventriculomegaly and cardiomyopathy under chemotherapy for maternal AML. Onkologie 32(1–2):40–43
28. Meyer-Wittkopf M, Barth H, Emons G, Schmidt S (2001) Fetal cardiac effects of doxorubicin therapy for carcinoma of the breast during pregnancy: case report and review of the literature. Ultrasound Obstet Gynecol 18(1):62–66
29. Avilés A, Neri N, Nambo MJ (2006) Long-term evaluation of cardiac function in children who received anthracyclines during pregnancy. Ann Oncol 17(2):286–288
30. Van Calsteren K, Berteloot P, Hanssens M, Vergote I, Amant F, Ganame J et al (2006) In utero exposure to chemotherapy: effect on cardiac and neurologic outcome. J Clin Oncol 24(12):e16–e17

Lactation during and after Breast Cancer

22

Fedro A. Peccatori, Bruna Migliavacca Zucchetti, Barbara Buonomo, Giulia Bellettini, Giovanni Codacci-Pisanelli, and Micaela Notarangelo

Abstract

Breastfeeding is an important aspect of mother-newborn relationship and is of great benefit for the baby. Unfortunately, many drugs taken by the mother may pass into her milk and exert an effect on the newborn. Very limited data is available and a cautionary approach is warranted especially when the woman receives anticancer treatment including chemotherapy, hormonal treatment and the recently introduced target agents as well as monoclonal antibodies. In all these conditions breastfeeding should be put on hold.

More and more often physicians are faced with women that are pregnant years after the diagnosis of cancer: this has long been considered dangerous for the mother, but data show that prognosis is definitely not worse. If the woman is no longer being actively treated, breastfeeding is advisable every time it is possible, even if patients that received breast radiation may be unable to produce a sufficient amount of milk on that side.

Keywords

Breastfeeding · Breast cancer · Drug transfer · Lactation · Weaning

F. A. Peccatori (✉)
Gynecologic Oncology Program, European Institute of Oncology IRCCS, Milan, Italy
e-mail: fedro.peccatori@ieo.it

B. Migliavacca Zucchetti
European Institute of Oncology IRCCS, Milan, Italy

Medical Oncology Department, Hospital Sirio-Libanes, Sao Paulo, Brazil

B. Buonomo
Gynecologic Oncology Program, European Institute of Oncology IRCCS, Milan, Italy

Gynecology and Obstetrics Unit, Department of Neuroscience, Reproductive Sciences and Dentistry, School of Medicine, University of Naples Federico II, Naples, Italy
e-mail: Barbara.buonomo@ieo.it

G. Bellettini
Pediatrician, Milan, Italy

G. Codacci-Pisanelli
Department of Medical and Surgical Sciences and Biotechnology, Sapienza University of Rome, Rome, Italy
e-mail: Giovanni.codacci-pisanelli@uniroma1.it

M. Notarangelo
Private Lactation Consultant Practice, Lerici, La Spezia, Italy

S. Alipour, R. Omranipour (eds.), *Diseases of the Breast during Pregnancy and Lactation*, Advances in Experimental Medicine and Biology 1252,
https://doi.org/10.1007/978-3-030-41596-9_22

22.1 Overview

The occurrence of breast cancer during breastfeeding is referred to as "pregnancy-associated breast cancer," as the definition includes cancers that appear not only during pregnancy but also within 12 months after delivery [1]. Breast cancer diagnosed during lactation has a worse prognosis than cancer diagnosed during pregnancy. This is partly, but not only, because of a delay in diagnosis and treatment. Physicians, and the patients themselves, may be reluctant to consider the possibility of a malignancy in such a phase of life and at such a young age. Furthermore, the physiological changes associated with lactation may mask the appearance of a nodule and delay its evaluation (see also Chaps. 1 and 2). It must be noted that diagnostic tools, both imaging and biopsies, are safe and effective during breastfeeding, provided the operator has enough experience and the breast is well drained before the examination (see also Chaps. 3, 4 and 13). Prognosis, however, remains worse even when the tumor stage is the same: this may in part be due to a higher prevalence of biological markers of tumor aggressiveness (higher percentage of HER2-positive or triple-negative cancers) or local release of inflammatory cytokines during mammary gland involution that may favor metastatic spread [2] (see also Chap. 10).

Table 22.1 Breastfeeding tips

Tips for a good breastfeeding start	Tips for gently discontinuing breastfeeding
Start breastfeeding or expressing milk as soon as possible after birth (within 1 h)	Gradual discontinuation of breastfeeding can prevent breast inflammation
Breastfeed on cue (at least 8–12 times/24 h)	The breast must remain full, so that FIL can trigger involution, but not hard (gentle manual expression), so that less inflammation occurs
Latch should be deep (150° at the angle of the mouth), chin to the breast, head slightly extended	
Help milk transfer by massaging and compressing the breast during the feed or the use of a breast pump	At advanced stages of lactation suppression of prolactin through medication alone does not prevent inflammation
Seek help if breastfeeding is painful or the infant is sleepy, if voids or stooling is scant, if feedings are excessively long or frequent	Gentle lymph drainage massage can help prevent inflammation
	Fans and ice packs can ease discomfort and reduce inflammation

FIL feedback inhibitor of lactation, a polypeptide in milk

22.2 Transfer of Chemotherapy Drugs into Human Milk

Several agents taken up by the mother may pass into her milk; they are then absorbed by the baby and may exert an effect that may be difficult to predict. This may also happen with anticancer agents. Cyclophosphamide, doxorubicin, cisplatin, and their metabolites have all been detected in the milk of nursing mothers who were under chemotherapy. In the case of a mother receiving cyclophosphamide, it was also possible to determine the effect of the drug on blood cell counts of the newborn. Unfortunately, the active drugs and their metabolites persist in the milk for a long time after administration: in a patient treated for lymphoma, cyclophosphamide and doxorubicin were still measurable for 21 days after drug injection. Considering the acute and delayed toxicity of anticancer drugs, it is reasonable to advise women to avoid breastfeeding during treatment with traditional chemotherapy [3]. This also applies to women receiving anti-hormones (tamoxifen or aromatase inhibitors) or target agents (eg, lapatinib, imatinib, or monoclonal antibodies) (see also Chaps. 15 and 21). To prevent inflammation during suppression of lactation, it is advisable to gently express some milk to decrease intramammary pressure while prolactin is being suppressed by the administration of cabergoline (Table 22.1).

22.3 Pregnancy and Breastfeeding after Breast Cancer

Many physicians and patients remain concerned about the safety of pregnancy in breast cancer survivors, especially in women previously diagnosed with estrogen receptor (ER)–positive diseases in whom pregnancy could be regarded as potentially detrimental owing to endocrine stimulation [4, 5]. Several studies were conducted to address this question. Some studies have suggested that pregnancy is associated with a better prognosis. However, these studies may be subjected to a selection bias that has been described as the "healthy mother effect" [6]. This refers to the fact that although the pregnancy-exposed group was matched with controls of a similar age and stage, women who became pregnant still represent a group that is on average healthier and free of relapse. Lambertini et al. [7] reported about 333 patients with pregnancy after breast cancer who were matched (1:3) to 874 non-pregnant patients of similar characteristics, adjusting for a guaranteed time bias. At a median follow-up of 7.2 years after pregnancy, no difference in disease-free survival was observed between pregnant and non-pregnant patients with ER-positive (HR, 0.94; 95% CI, 0.70-1.26; $P = .68$) or ER-negative disease (HR, 0.75; 95% CI, 0.53-1.06; $P = .10$). No overall survival (OS) difference was observed in ER-positive patients (HR, 0.84; 95% CI, 0.60-1.18; $P = .32$); ER-negative patients in the pregnant cohort had better OS (HR, 0.57; 95% CI, 0.36-0.90; $P = .01$). Abortion, time to pregnancy, breastfeeding, and type of adjuvant therapy had no impact on patients' outcomes. This study provides reassuring evidence on the long-term safety of pregnancy in breast cancer survivors, including those with ER-positive disease [7]. No clear evidence about the timing of pregnancy after breast cancer is available. Waiting for 2 years following completion of breast cancer therapy in patients with ER-negative tumors is a reasonable option. For those with ER-positive breast cancer, 5 years of endocrine therapy (ET) should be completed before attempting to become pregnant [8]. The ongoing POSITIVE study addresses the possibility, following 18–30 months of ET, to temporarily suspend ET for up to 2 years to allow conception and pregnancy, following which ET can be completed (5–10 years) [9] (see also Chap. 23).

Limited available evidence suggests that breastfeeding is feasible and safe after breast cancer. Azim et al. [10] performed a survey among patients with breast cancer who completed their pregnancy following breast cancer management to examine their lactation behavior and its effect on breast cancer outcomes. Out of 32 women identified, 20 were reachable and accepted to take the questionnaire. At a median follow-up of 48 months following delivery, all 20 women were alive with two relapses, one in the group of the 10 lactating women and one in the group of the 10 non-lactating women. Thus, in this study, breastfeeding did not seem to have any detrimental effect on breast cancer outcomes in survivors who succeeded to complete their pregnancies [10]. Apart from safety considerations, addressing how to manage breastfeeding with unilateral milk production or reduced milk production from the irradiated breast is an important issue. As more young women have breast-conserving surgery (BCS) and subsequent radiotherapy, the long-term effects of surgery and ionizing radiation on the mammary gland have become increasingly relevant. The proximity of the incision to the areola and nipple, the location of the tumor, the dose and type of radiotherapy are all contributing factors to lactation success of the treated breast. Several small studies have shown that around 80% of patients treated with BCS and radiotherapy experience diminished breast enlargement and engorgement during pregnancy and around 50% have limited postnatal milk production from the ipsilateral breast [11, 12]. However, patients and physicians should be informed that milk produced by one breast is sufficient for the nutritional need of the newborn. Breastfeeding education is of utmost importance: early initiation of breastfeeding (within 1 hour from parturition) and frequent feeding in the first days, as recommended by the Baby Friendly Hospital Initiative, increase milk production and improves breastfeeding outcomes at later stages [13]. Mothers should be shown how to position the infant to achieve a deep and wide latch (150° at the angle of the mouth) and be encouraged to offer the breast as many times as requested by their infants, using an effective electric pump if the baby does not empty the breast completely. A laid-back position of the mother and side-lying holds may also result in easier latching. Frequent

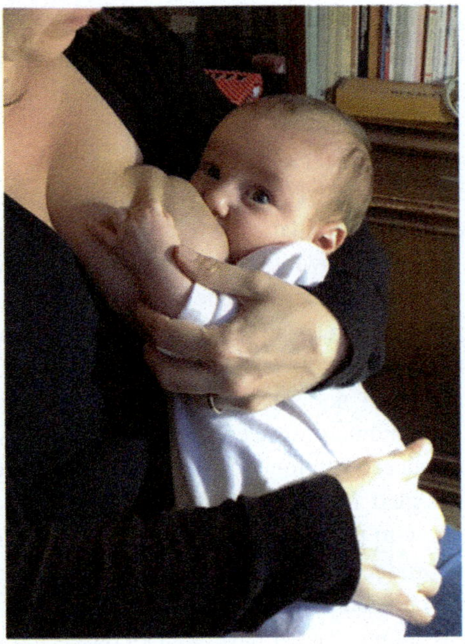

Fig. 22.1 Breastfeeding is good!

changes in the positioning of the baby and gentle massage and compression by stimulating the milk ejection reflex and increasing intramammary pressure [14] improve breast drainage in all quadrants, thus reducing the risk of engorgement and increasing milk production (Fig. 22.1).

If pain occurs, the mother should improve the baby's latch-on, trying to cover the entire nipple-areola complex with the baby's mouth and seek specialist professional advice before abrasions develop or worsen (Table 22.1).

As stated in the Society of Obstetricians and Gynecologists of Canada clinical practice guidelines [15]: "There is no evidence that breastfeeding increases the risk of recurrence of breast cancer or of the development of a second breast cancer, nor that it carries any risk for the child. Women previously treated for breast cancer that do not show any evidence of residual tumor should be encouraged to breastfeed their children. Unilateral breastfeeding should be encouraged and supported in breast cancer patients because it is frequently enough for baby's growth. Great importance should be given to breastfeeding counseling and to supporting patients, since misinformation is the main cause for avoiding breastfeeding."

References

1. Azim HA Jr, Santoro L, Russell-Edu W, Pentheroudakis G, Pavlidis N, Peccatori FA (2012) Prognosis of pregnancy-associated breast cancer: a meta-analysis of 30 studies. Cancer Treat Rev 38(7):834–842
2. Faupel-Badger JM, Arcaro KF, Balkam JJ, Eliassen AH, Hassiotou F, Lebrilla CB et al (2012) Postpartum remodeling, lactation, and breast cancer risk: summary of a National Cancer Institute–sponsored workshop. J Natl Cancer Inst 105(3):166–174
3. Pistilli B, Bellettini G, Giovannetti E, Codacci-Pisanelli G, Azim HA Jr, Benedetti G et al (2013) Chemotherapy, targeted agents, antiemetics and growth-factors in human milk: how should we counsel cancer patients about breastfeeding? Cancer Treat Rev 39(3):207–211
4. Biglia N, Torrisi R, D'Alonzo M, Codacci Pisanelli G, Rota S, Peccatori FA (2015) Attitudes on fertility issues in breast cancer patients: an Italian survey. Gynecol Endocrinol 31(6):458–464
5. Lambertini M, Di Maio M, Pagani O, Demeestere I, Del Mastro L, Loibl S et al (2017) A survey on physicians' knowledge, practice and attitudes on fertility and pregnancy issues in young breast cancer patients. Breast 32:S85–S86
6. Sankila R, Heinävaara S, Hakulinen T (1994) Survival of breast cancer patients after subsequent term pregnancy:"healthy mother effect". Am J Obstet Gynecol 170(3):818–823
7. Lambertini M, Kroman N, Ameye L, Cordoba O, Pinto A, Benedetti G et al (2018) Long-term safety of pregnancy following breast cancer according to estrogen receptor status. J Natl Cancer Inst 110(4):426–429
8. Azim HA, Peccatori FA, De Azambuja E, Piccart MJ (2011) Motherhood after breast cancer: searching for la dolce vita. Expert Rev Anticancer Ther 11(2):287–298
9. Pagani O, Ruggeri M, Manunta S, Saunders C, Peccatori F, Cardoso F et al (2015) Pregnancy after breast cancer: are young patients willing to participate in clinical studies? Breast 24(3):201–207
10. Azim HA Jr, Bellettini G, Liptrott SJ, Armeni ME, Dell'Acqua V, Torti F et al (2010) Breastfeeding in breast cancer survivors: pattern, behaviour and effect on breast cancer outcome. Breast 19(6):527–531
11. Tralins AH (1995) Lactation after conservative breast surgery combined with radiation therapy. Am J Clin Oncol 18(1):40–43

12. Moran MS, Colasanto JM, Haffty BG, Wilson LD, Lund MW, Higgins SA (2005) Effects of breast-conserving therapy on lactation after pregnancy. Cancer J 11(5):399–403
13. Procaccini D, Curley AL, Goldman M (2018) Baby-friendly practices minimize newborn infants weight loss. Breastfeed Med 13(3):189–194
14. Brimdyr K, Blair A, Cadwell K, Turner-Maffei C (2003) The relationship between positioning, the breastfeeding dynamic, the latching process and pain in breastfeeding mothers with sore nipples. Breastfeed Rev 11(2):5–10
15. Helewa M, Levesque P, Provencher D, Lea RH, Rosolowich V, Shapiro HM (2002) Breast cancer, pregnancy, and breastfeeding. J Obstet Gynaecol Can 24(2):164–180

Pregnancy in Breast Cancer Survivors

23

Vesna Bjelic-Radisic, Mohsen Esfandbod, and Sadaf Alipour

Abstract

Safety of pregnancy occurring after breast cancer treatment has been studied largely, but it is still debatable. These studies have generally showed that overall and disease-free survival in breast cancer survivors with subsequent pregnancy is not less than those without future pregnancy. Also, breast cancer survivors treated with chemotherapy, radiation therapy, or both had no increased risk of congenital anomalies, single gene disorders, or chromosomal syndromes in their offspring. However, it appears that the incidence of preterm labor, low birth weight, and fetal anomalies is higher in these cases.

These issues as well as safe time interval from breast cancer treatment to pregnancy, safe contraceptive method after breast cancer, counseling about pregnancy in survivors, and how to follow up the patient for breast cancer recurrence during pregnancy are discussed in this chapter.

Keywords

Breast cancer · Chemotherapy · Obstetric complications · Pregnancy · Survivors

23.1 Overview

About 7% of all breast cancers are diagnosed in women younger than the age of 40, with a cumulative risk of 0.4–0.45% by this age [1, 2]. Because of the inherent characteristics of the tumors in young women, frequently a more complex and prolonged treatment including chemotherapy and additional years of ovarian suppression in hormone receptor positive cancer is carried out for these patients [3–5]. The positive impact of chemotherapy on disease-free survival (DFS) and overall survival (OS) in breast cancer patients can influence fertility negatively [6–8]. Actually, potential detrimental effects of systemic treatments on fertility, long-term hor-

V. Bjelic-Radisic
Breast Unit, Helios University Hospital, University Witten Herdecke, Wuppertal, Germany
e-mail: vesna.bjelic-radisic@medunigraz.at

M. Esfandbod
Department of Oncology, Sina Hospital, Tehran University of Medical Sciences, Tehran, Iran
e-mail: sfandbod@tums.ac.ir

S. Alipour (✉)
Breast Disease Research Center (BDRC), Tehran University of Medical Sciences, Tehran, Iran

Department of Surgery, Arash Women's Hospital, Tehran University of Medical Sciences, Tehran, Iran
e-mail: salipour@tums.ac.ir

S. Alipour, R. Omranipour (eds.), *Diseases of the Breast during Pregnancy and Lactation*, Advances in Experimental Medicine and Biology 1252,
https://doi.org/10.1007/978-3-030-41596-9_23

monal therapy, and worry of women and their family that a future pregnancy would increase their risk of BC recurrence are some of the central reasons for a low incidence of pregnancy in breast cancer survivors.

Published data about the incidence of pregnancy after breast cancer are diverse, caused by different definitions of pregnancy; some studies include only cases of full-term pregnancy, while others include pregnancies as a baseline [9, 10]. Overall and according to several reports, the incidence of pregnancy in women after diagnosis of breast cancer ranges between 8–10%. This is approximately 50% less compared to the age-matched groups without breast cancer [6–8, 11]. Accordingly, in other studies, only 8% of breast cancer survivors aged less than 35 years experienced a full-term pregnancy. In addition and as expected, figures were lower in older age groups; and comprised only 3% of women less than 45 years of age [8, 12]. In a 2011 published population-based study from Norway, women with previous breast cancer had a 70% lower chance of becoming subsequently pregnant compared to the age-matched population, even after adjusting for education and previous parity [13].

Although some results are controversial, overall incidence of pregnancy in breast cancer survivors is low; this most likely implies a need toward better education of both patients and physicians.

23.2 Safety of Pregnancy and Childbirth in Breast Cancer Survivors

23.2.1 Effect of Pregnancy on Breast Cancer Prognosis

Whether pregnancy occurring after breast cancer treatment can have detrimental effects on prognosis of the disease- due to the pronounced rise of female sex hormones- has for long been the concern of both patients and physicians. Many studies have been carried out about the issue, mostly retrospective in nature because of the limitations for prospective work in this context. One major bias is the "healthy mother effect", which suggests that survivors who are in better health and those who have had more favorable pathologies opt for subsequent maternity [9, 14, 15]; but this point also has been considered in some of the studies [16, 17].

Overall survival in breast cancer survivors with subsequent pregnancy has been shown to be similar or even better than those without future pregnancy [16–26]; DFS also had no poorer outcome in these women [10, 20, 23, 26, 27]. Similar effects on OS and DFS have been demonstrated for estrogen receptor positive (ER+) tumors [28–30]. Figure 23.1 shows studies that have performed a systematic review of previous works, and the results regarding survival and recurrence in breast cancer survivors who had a subsequent pregnancy, compared to those without further pregnancies.

Some theories have been proposed to explain the protective effects of pregnancy on breast cancer prognosis. Alloimmunization suggests that fetal cells that have common antigens with the mother enter maternal circulation, inducing antibodies that would remove probable metastatic cells. Another hypothesis suggests that extreme levels of estrogen and progesterone, added to the high placental hCG, act like endocrine therapies [31]. Other theories include inducing repair of the DNA and better cellular differentiation, alterations in genes of cell differentiation and death [32], and reduction of breast multipotent stem cells [33].

23.2.2 Obstetrical Outcomes in Breast Cancer Survivors

A recent systematic review has investigated the rate of various obstetrical outcomes in women previously treated for breast cancer. The meta-analysis showed that the incidence of preterm labor, low birth weight, intrauterine fetal and born fetal anomalies was higher in these patients compared to the general population [34].

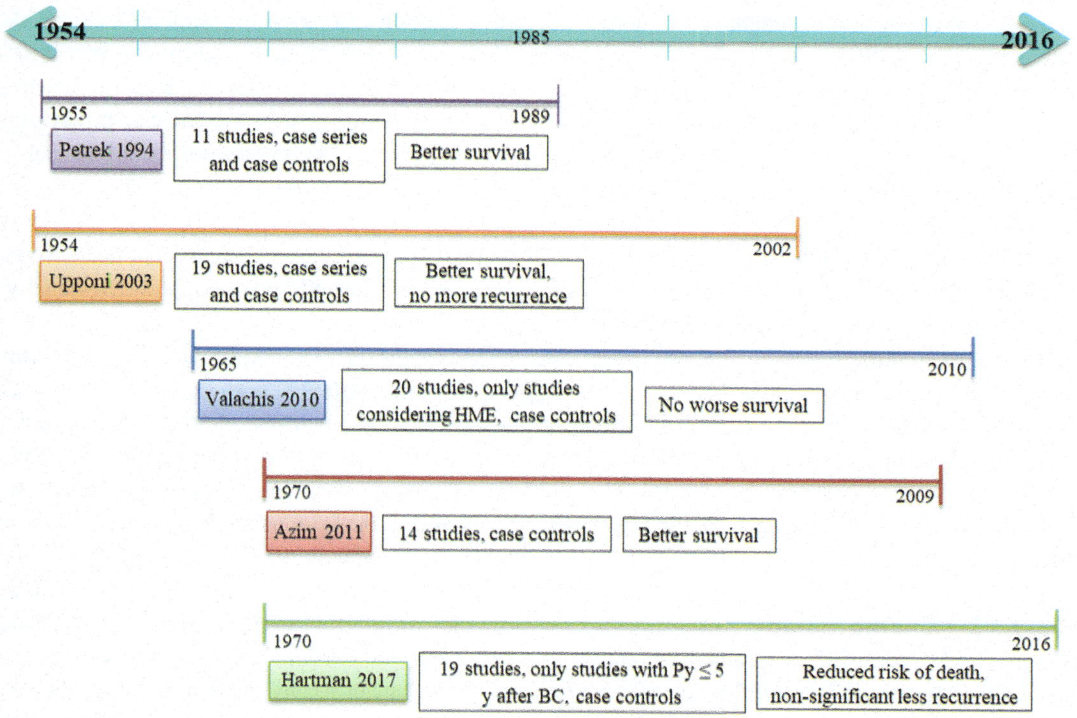

Fig. 23.1 Systematic reviews about pregnancy following breast cancer, and their results regarding survival and recurrence in breast cancer survivors with a subsequent pregnancy compared to those without. *BC* breast cancer, *HME* healthy mother effect, *Py* pregnancy, *y* years

23.2.3 Effect of Adjuvant Treatments on Subsequent Obstetric Complications

One of the strongest predictors of emotional well-being in breast cancer survivors, besides sexual function and appearance, is feeling healthy enough to be a good parent. Parenthood can represent normalcy, happiness, and life fulfilment. Breast Cancer survivors are often fearful that their history of cancer or its treatment will have an adverse impact on offspring conceived after their cancer treatment, such as placing them at risk for malignancy, congenital anomalies, or impaired growth and development (see also Chap. 28). They are also concerned about the risks of cancer recurrence, infertility, miscarriage, and achieving a successful pregnancy outcome.

23.2.3.1 Risk of Congenital and Chromosomal Abnormalities

In studies including several thousands offspring, female breast cancer survivors treated with chemotherapy, radiation therapy, or both had no increased risk of congenital anomalies, single gene disorders, or chromosomal syndromes in their offspring [35–44]. These studies primarily evaluated pregnancies that were conceived years after treatment (see also Chap. 15).

23.2.3.2 Risk of Adverse Pregnancy Outcomes

The risk of miscarriage, preterm birth, fetal growth restriction, and stillbirth in female breast cancer survivors depends, in part, on the type of therapy they received (chemotherapy, radiation therapy, or target therapy) and to non-treatment factors (eg, age at start of pregnancy). There is no strong evidence of an increased risk of adverse pregnancy outcome among female breast cancer survivors who received chemotherapy [45]. Chemotherapy does not appear to damage the uterus, which may account for the generally favorable pregnancy outcome in exposed patients [46]. However, women treated for breast cancer appear to be at increased risk for obstetric complications (eg, oligohydramnios due to trastuzumab). In a birth registry study of over 2.3 million births, women with a history of breast cancer had an increased risk of preterm birth, low birth weight, and small for gestational age, especially if they received chemotherapy or gave birth within 2 years of their breast cancer diagnosis date [47].

23.2.3.3 Chemotherapy

Women who received chemotherapy alone or with other therapies (surgery, radiation therapy, or both) had lower rates of live birth than their female siblings. This appeared to be due primarily to a higher rate of pregnancy termination, as the rates of miscarriage and stillbirth were generally statistically similar for the survivors and their siblings. In addition, the male: female sex ratio of 1.09:1.0 in offspring of survivors was similar to that in the general population and to that in offspring of female siblings of the survivors, suggesting that exposure to mutagenic agents (chemotherapy, radiation therapy) did not increase transmission of lethal X-linked mutations. Lastly, the rate of live birth was not lower for patients treated with any particular drug compared with those not treated with that drug [48] (see also Chap. 15).

23.3 Safe Time Interval from Breast Cancer Treatment to Pregnancy

The safe time interval from breast cancer to pregnancy is unknown, whereas "safe time" should ideally stand for no further risk for relapse or breast cancer-specific death. Very often, patients will be advised to wait at least 2 years after the end of therapy to become pregnant. One possible reason for this recommendation is the assumption that highest recurrence and breast cancer-specific death risks occur during the first 2 years after diagnosis. Nonetheless, this is only partially correct. Although overall, breast cancer patients have a higher risk for recurrence and death in the years 2–4 after their first diagnosis, comparison of risk of recurrence and death trends in women with ER-positive and ER-negative tumors shows that the risks cross at 6–8 years after diagnosis. In addition, the comparison implies that patients with ER-negative breast cancer have a much greater risk for recurrence and death in earlier years after the initial diagnosis, whereas women with ER-positive breast cancer have a more consistent long-term risk of breast cancer death [49, 50].

Retrospective data are difficult to compare because of different definitions of "time". These definitions vary from time from initial diagnosis of breast cancer to any pregnancy, time from the end of breast cancer treatment to any pregnancy, or any of the above till full term pregnancy. Nevertheless, published data implicate that a longer interval between diagnosis and pregnancy seems to be safer. Clark et al. [51] reported a better survival rate for breast cancer patients with pregnancies between 6 and 24 months after diagnosis (78%) compared to only 54% survival in patients with pregnancies in the first 6 months after breast cancer treatment. Similar results were reported in the series of 136 women by Clark and Chua [52], which showed that the 5 year survival rate goes up to 92% in patients with an interval of 2 years to pregnancy compared to 59% with pregnancy within the first 6 months. In a case series of 96 women, Sankila et al. [9] reported a better survival for women with breast cancer and following pregnancies compared to the control group. As well, published results of a multicenter retrospective cohort study comprising 333 patients with pregnancy occurring at any time after an ER positive breast cancer compared to 874 matched controls showed no difference in DFS between patients with and without subsequent pregnancy at a median follow-up of 5 years. Additionally, there were no difference in DFS between patients who became pregnant within 2 years and those who became pregnant later [28].

23.4 Active Counselling and Planning for Pregnancy after Breast Cancer Treatment

As one of the major concerns of young women with breast cancer, discussion addressed to fertility and family planning is an important component of quality oncology care [53, 54].

Published data suggests that less than one half of breast cancer patients have undergone counselling regarding reproductive issues at time of diagnosis and therapy of their breast cancer [55–57]. In contrast to these data, more than 90% have a discussion with the physician regarding side effects of the therapy and the impact on normal daily activity [57]. Counselling about reproductive health is related to the age of the patient, her insurance status and her income; especially very young women are concerned regarding their reproductive health [55–58].

Duffy at al. [57] found out that an active discussion addressing fertility and reproductive health issues was less frequently held in anxious women, as well as in cases who had difficulty in their communication with the medical team. They also demonstrated that health teams are failing to fully inform young women about the risks and benefits of adjuvant treatment regarding fertility issues and early menopause. Young women with breast cancer constitute a vulnerable patient group. The decision for or against explicit therapy has to be made carefully. This includes not only benefits of the therapy regarding recurrence and survival, but also potential risks in regard to fertility, premature menopause, sexual dysfunction and body image. Early consultation with a fertility specialist and active counselling over this topic are integral parts of the care of young patients [59]. In agreement with this advice, ESO-ESMO third international consensus guidelines for breast cancer in young women (BCY3) highly recommended counselling on fertility, sexual health and socio-economic impact as part of the individual treatment planning [60].

Notably, use of GnRH analogue during chemotherapy should be discussed on a case by case basis to preserve ovarian function and possibly fertility. All young women should be counselled about the risks of systemic therapy (both chemotherapy and hormonal therapy) for amenorrhea and premature menopause before starting the therapy (see also Chaps. 24 and 25).

23.5 Contraception after Breast Cancer

Clinical guidelines including BCY3 and World Health Organization (WHO) highly recommend active counselling of young women regarding risk of pregnancy occurring while undergoing systemic chemotherapy, immunotherapy and hormonal therapy; and the possibility of getting pregnant even with amenorrhoea [60, 61].

Despite this recommendation, and although young breast cancer survivors in reproductive age have serious contraceptive concerns, studies show that they may not get fully informed about available and safe contraceptive options. In a qualitative study of 10 women between 18 and 50 years of age with a history of breast cancer, patients reported that their physicians had not focussed on reproductive issues, and had provided only limited information about contraception. Nevertheless, women were anxious about unintended pregnancy and wished to receive the information soon after diagnosis [62]. As well, in a survey among medical oncologists in Switzerland, only 20% reported that they informed young women (under 40 years of age) about the necessity of reliable contraception during the therapy, asked patients about their contraceptive methods while under treatment, and referred patients to a gynaecologist [63].

23.5.1 Oral Contraceptives and Subcutaneous Implants

The use of subcutaneous implants, which are effective contraceptives for 3 years, has not been tested in breast cancer patients and at the time being, cannot be recommended in survivors [5, 64]. Regarding combined oral contraceptives, a meta-analysis of 54 studies showed a relative risk of 1.24 for a diagnosis of breast cancer, and consequently these cannot be recommended in breast cancer survivors [65]. Overall, according to BCY3, hormonal contraception is contraindicated in women in reproductive age with a history of breast cancer; and non-hormonal contraception is the method of choice. Also, according to the Collège National des Gynécologues et Obstétriciens Français (CNGOF) Contraception Guidelines, all hormonal contraceptives are contra-indicated after breast cancer, regardless of time since treatment and biology of breast cancer (hormone receptor status and histological subtype) [66].

23.5.2 Barrier Methods

Aternative methods such as condom, female barrier methods, and male sterilisation could be considered [67, 68]. The WHO Medical Eligibility Criteria for Contraceptive Use also recommends use of condoms or copper IUDs within 5 years after diagnosis of breast cancer [61].

Many breast cancer patients remain sexually active after diagnosis and during treatment stages and a large number use condoms [69, 70]. While barrier and behavioural methods have a high risk of unintended pregnancy with a 1-year failure rate of 15–32% [71, 72], the copper IUD is an effective non-hormonal contraceptive, with a failure rate of 0.3–0.6% [72, 73].

The insertion of Copper IUDs is simple, and it can be used in all women, nulliparous and parous [61, 74–76]. The copper IUD can be used for up to 10 years, and some studies reported successful contraception for up to 12–20 years [77, 78]. Side effects of the copper IUD consist of increased menstrual cramping and vaginal bleeding in some women. Levonorgesterel- IUD was developed for a better control of these symptoms. The level of progestin is low, but levonorgesterel-IUD produces detectable levels of levonorgestrel in the serum [79]. For breast cancer survivors, this might account for hormonal stimulation [61], and include a risk of higher recurrence. Although one study reported no higher recurrence in breast cancer patients using a levonorgestrel IUD, a subgroup of patients who were users at time of diagnosis and continued to use levonorgestrel IUD afterwards had a higher rate of recurrence [80].

23.6 Screening for Breast Cancer Recurrence during Pregnancy

General guiding principles for follow up after treatment of breast cancer have been described for non-pregnant women. These consist of history taking as well as breast and general physical examination [81–83]. The history should consist of usual components, as well as changes in the patient family history, and inquiry about symptoms that can be related to local or metastatic recurrence. Breast, axillary and chest examination have to be performed in addition to general physical examination. Mammography is also regularly done in order to identify ipsilateral local recurrence or contralateral breast cancer [84–86]. Other imaging or routine laboratory tests are not recommended in asymptomatic patients for detection of recurrence [87].

ASCO recommendations for surveillance in breast cancer survivors consist of visits undertaken every 3–6 months up to 1 year, then every 6–12 months up to 2 years; and annual visits thereafter [87]. The effect of shorter or longer intervals of visits has not been studied so far [88]. Mammography is performed on a yearly schedule, except for additional imaging which might be needed in between. For very high risk women, as those with a very strong family history or genetically positive cases, MRI is recommended in the follow up period [89].

Survivors who opt for pregnancy after treatment of the cancer are generally young. Commonly, breast cancer is more aggressive in young women, and is more frequently hormone receptor negative or triple negative [90–92] (see also Chaps. 10 and 11). Nevertheless, the same follow up guidelines apply for young survivors; except that they might more frequently fulfill the indications for MRI, genetic testing and family screening. Guidelines for follow up issues during the pregnancy itself have not been described, and the best plan of action has not been studied. The followings are suggestions based on prevailing knowledge and current practice in our institutions.

History taking and physical exam should be performed at the appropriate time regarding follow up schedules. Breast ultrasound has not still been recognized as a screening modality for cancers, but can be used as a diagnostic tool if needed. Mammography should not be done for follow-up purpose in the prenatal period and it would better be performed at the time of planning for pregnancy and before conception. Therefore, if the pregnancy had not been planned earlier and imaging had not been done, the x-ray study would be postponed until after delivery. It should be emphasized that pregnancy is not an absolute contraindication for mammography, but its use should be limited to suspicious ultrasound or clinical exam findings that are awaiting mammographic assessment (see also Chap. 3).

During breastfeeding, screening can be performed as usual, although false positive imaging findings are more frequent in this period [93].

References

1. Ferlay J, Shin HR, Bray F, Forman D, Mathers C, Parkin DM (2010) Estimates of worldwide burden of cancer in 2008: GLOBOCAN 2008. Int J Cancer 127(12):2893–2917
2. DeSantis C, Ma J, Bryan L, Jemal A (2014) Breast cancer statistics, 2013. CA Cancer J Clin 64(1):52–62
3. Early Breast Cancer Trialists' Collaborative Group (EBCTCG) (2011) Relevance of breast cancer hormone receptors and other factors to the efficacy of adjuvant tamoxifen: patient-level meta-analysis of randomised trials. Lancet 378(9793)a:771–784
4. Davies C, Pan H, Godwin J, Gray R, Arriagada R, Raina V et al (2013) Long-term effects of continuing adjuvant tamoxifen to 10 years versus stopping at 5 years after diagnosis of oestrogen receptor-positive breast cancer: ATLAS, a randomised trial. Lancet 381(9869):805–816
5. Gray RG, Rea D, Handley K, Bowden SJ, Perry P, Earl HM et al (2013) aTTom: Long-term effects of continuing adjuvant tamoxifen to 10 years versus stopping at 5 years in 6,953 women with early breast cancer. J Clin Oncol 31(18-suppl):5
6. Reichman BS, Green KB (1994) Breast cancer in young women: effect of chemotherapy on ovarian function, fertility, and birth defects. J Natl Cancer Inst Monogr 16:125–129
7. Del Mastro L, Catzeddu T, Venturini M (2006) Infertility and pregnancy after breast cancer: current knowledge and future perspectives. Cancer Treat Rev 32(6):417–422
8. Mueller BA, Simon MS, Deapen D, Kamineni A, Malone KE, Daling JR (2003) Childbearing and survival after breast carcinoma in young women. Cancer 98(6):1131–1140
9. Sankila R, Heinävaara S, Hakulinen T (1994) Survival of breast cancer patients after subsequent term pregnancy: "healthy mother effect". Am J Obstet Gynecol 170(3):818–823
10. von Schoultz E, Johansson H, Wilking N, Rutqvist LE (1995) Influence of prior and subsequent pregnancy on breast cancer prognosis. J Clin Oncol 13(2):430–434
11. Pagani O, Partridge A, Korde L, Badve S, Bartlett J, Albain K, Gelber R, Goldhirsch A (2011) Pregnancy after breast cancer: if you wish, ma'am. Breast Cancer Res Treat 129(2):309–317
12. Blakely LJ, Buzdar AU, Lozada JA, Shullaih SA, Hoy E, Smith TL et al (2004) Effects of pregnancy after treatment for breast carcinoma on survival and risk of recurrence. Cancer 100(3):465–469
13. Stensheim H, Cvancarova M, Møller B, Fosså SD (2011) Pregnancy after adolescent and adult cancer: a population-based matched cohort study. Int J Cancer 129(5):1225–1236
14. Rippy EE, Karat IF, Kissin MW (2009) Pregnancy after breast cancer: the importance of active counseling and planning. Breast 18(6):345–350
15. Azim HA, Peccatori FA, De Azambuja E, Piccart MJ (2011) Motherhood after breast cancer: searching for la dolce vita. Expert Rev Anticancer Ther 11(2):287–298
16. Cooper DR, Butterfield J (1970) Pregnancy subsequent to mastectomy for cancer of the breast. Ann Surg 171(3):429–433
17. Valachis A, Tsali L, Pesce LL, Polyzos NP, Dimitriadis C, Tsalis K et al (2010) Safety of pregnancy after primary breast carcinoma in young women: a meta-analysis to overcome bias of healthy mother effect studies. Obstet Gynecol Surv 65(12):786–793
18. Petrek JA (1994) Pregnancy safety after breast cancer. Cancer 74(1-Suppl):528–531
19. Kroman N, Jensen MB, Melbye M, Wohlfahrt J, Mouridsen HT (1997) Should women be advised against pregnancy after breast-cancer treatment? Lancet 350(9074):319–322
20. Kopeika J, Bhaduri M, Kugadas A, Reddy N, Shewbridge A, Mukherji D, Sandri I, Mansi J (2019) Planned and unplanned pregnancies in breast cancer survivors. Breast 46:75–80
21. Ives A, Saunders C, Bulsara M, Semmens J (2007) Pregnancy after breast cancer: population based study. BMJ 334(7586):194

22. Kroman N, Jensen MB, Wohlfahrt J, Ejlertsen B (2008) Pregnancy after treatment of breast cancer—a population-based study on behalf of Danish Breast Cancer Cooperative Group. Acta Oncol 47(4):545–549
23. Kranick JA, Schaefer C, Rowell S, Desai M, Petrek JA, Hiatt RA et al (2010) Is pregnancy after breast cancer safe? Breast J 16(4):404–411
24. Azim HA Jr, Santoro L, Pavlidis N, Gelber S, Kroman N, Azim H et al (2011) Safety of pregnancy following breast cancer diagnosis: a meta-analysis of 14 studies. Eur J Cancer 47(1):74–83
25. Córdoba O, Bellet M, Vidal X, Cortés J, Llurba E, Rubio IT et al (2012) Pregnancy after treatment of breast cancer in young women does not adversely affect the prognosis. Breast 21(3):272–275
26. Hartman EK, Eslick GD (2016) The prognosis of women diagnosed with breast cancer before, during and after pregnancy: a meta-analysis. Breast Cancer Res Treat 160(2):347–360
27. Largillier R, Savignoni A, Gligorov J, Chollet P, Guilhaume MN, Spielmann M et al (2009) Prognostic role of pregnancy occurring before or after treatment of early breast cancer patients aged< 35 years: a GET (N) a Working Group analysis. Cancer 115(22):5155–5165
28. Azim HA Jr, Kroman N, Paesmans M, Gelber S, Rotmensz N, Ameye L et al (2013) Prognostic impact of pregnancy after breast cancer according to estrogen receptor status: a multicenter retrospective study. J Clin Oncol 31(1):73
29. Nye L, Rademaker A, Gradishar WJ (2017) Breast Cancer outcomes after diagnosis of hormone-positive breast cancer and subsequent pregnancy in the tamoxifen era. Clin Breast Cancer 17(4):e185–e189
30. Lambertini M, Kroman N, Ameye L, Cordoba O, Pinto A, Benedetti G et al (2017) Long-term safety of pregnancy following breast cancer according to estrogen receptor status. J Natl Cancer Inst 110(4):426–429
31. de Bree E, Makrigiannakis A, Askoxylakis J, Melissas J, Tsiftsis DD (2010) Pregnancy after breast cancer. A comprehensive review. J Surg Oncol 101(6):534–542
32. Russo J, Moral R, Balogh GA, Mailo D, Russo IH (2005) The protective role of pregnancy in breast cancer. Breast Cancer Res 7(3):131–142
33. Siwko SK, Dong J, Lewis MT, Liu H, Hilsenbeck SG, Li Y (2008) Evidence that an early pregnancy causes a persistent decrease in the number of functional mammary epithelial stem cells—implications for pregnancy-induced protection against breast cancer. Stem Cells 26(12):3205–3209
34. D'Ambrosio V, Vena F, Di Mascio D, Faralli I, Musacchio L, Boccherini C, Brunelli R, Piccioni MG, Panici PB, Giancotti A (2019) Obstetrical outcomes in women with history of breast cancer: a systematic review and meta-analysis. Breast Cancer Res Treat 178:485–492
35. Byrne J, Rasmussen SA, Steinhorn SC, Connelly RR, Myers MH, Lynch CF et al (1998) Genetic disease in offspring of long-term survivors of childhood and adolescent cancer. Am J Hum Genet 62:45–52
36. Dodds L, Marrett LD, Tomkins DJ, Green B, Sherman G (1993) Case-control study of congenital anomalies in children of cancer patients. BMJ 307(6897):164–168
37. Green DM, Fiorello A, Zevon MA, Hall B, Seigelstein N (1997) Birth defects and childhood cancer in offspring of survivors of childhood cancer. Arch Pediatr Adolesc Med 151:379–383
38. Chiarelli AM, Marrett LD, Darlington GA (2000) Pregnancy outcomes in females after treatment for childhood cancer. Epidemiology 11(2):161–166
39. Kenney LB, Nicholson HS, Brasseux C, Mills JL, Robison LL, Zeltzer LK et al (1996) Birth defects in offspring of adult survivors of childhood acute lymphoblastic leukemia. A Childrens Cancer Group/National Institutes of Health Report. Cancer 78(1):169–176
40. Signorello LB, Mulvihill JJ, Green DM, Munro HM, Stovall M, Weathers RE et al (2010) Stillbirth and neonatal death in relation to radiation exposure before conception: a retrospective cohort study. Lancet 376(9741):624–630
41. Winther JF, Olsen JH, Wu H, Shyr Y, Mulvihill JJ, Stovall M et al (2012) Genetic disease in the children of Danish survivors of childhood and adolescent cancer. J Clin Oncol 30:27
42. Signorello LB, Mulvihill JJ, Green DM, Munro HM, Stovall M, Weathers RE et al (2012) Congenital anomalies in the children of cancer survivors: a report from the childhood cancer survivor study. J Clin Oncol 30(2):239
43. Boice JD Jr, Tawn EJ, Winther JF, Donaldson SS, Green DM, Mertens AC et al (2003) Genetic effects of radiotherapy for childhood cancer. Health Phys 85(1):65–80
44. van der Kooi AL, Brewster DH, Wood R, Nowell S, Fischbacher C, van den Heuvel-Eibrink MM et al (2018) Perinatal risks in female cancer survivors: a population-based analysis. PLoS One 13(8):e0202805
45. Mueller BA, Chow EJ, Kamineni A, Daling JR, Fraser A, Wiggins CL et al (2009) Pregnancy outcomes in female childhood and adolescent cancer survivors: a linked cancer-birth registry analysis. Arch Pediatr Adolesc Med 163(10):879–886
46. Critchley HO, Wallace WH (2005) Impact of cancer treatment on uterine function. J Natl Cancer Inst Monogr 2005(34):64–68
47. Black KZ, Nichols HB, Eng E, Rowley DL (2017) Prevalence of preterm, low birthweight, and small for gestational age delivery after breast cancer diagnosis: a population-based study. Breast Cancer Res 19(1):11
48. Green DM, Whitton JA, Stovall M, Mertens AC, Donaldson SS, Ruymann FB et al (2002) Pregnancy outcome of female survivors of childhood cancer: a report from the Childhood Cancer Survivor Study. Am J Obstet Gynecol 187(4):1070–1080
49. Cardoso F, Loibl S, Pagani O, Graziottin A, Panizza P, Martincich L et al (2012) The European Society of Breast Cancer Specialists recommendations for the management of young women with breast cancer. Eur J Cancer 48(18):3355–3377

50. Jatoi I, Chen BE, Anderson WF, Rosenberg PS (2007) Breast cancer mortality trends in the United States according to estrogen receptor status and age at diagnosis. J Clin Oncol 25(13):1683–1690
51. Clark RM, Reid J (1978) Carcinoma of the breast in pregnancy and lactation. Int J Radiat Oncol Biol Phys 4(7–8):693–698
52. Clark RM, Chua T (1989) Breast cancer and pregnancy: the ultimate challenge. Clin Oncol 1(1):11–18
53. Howard-Anderson J, Ganz PA, Bower JE, Stanton AL (2012) Quality of life, fertility concerns, and behavioral health outcomes in younger breast cancer survivors: a systematic review. J Natl Cancer Inst 104(5):386–405
54. Loren AW, Mangu PB, Beck LN, Brennan L, Magdalinski AJ, Partridge AH et al (2013) Fertility preservation for patients with cancer: American Society of Clinical Oncology clinical practice guideline update. J Clin Oncol 31(19):2500
55. Thewes B, Meiser B, Rickard J, Friedlander M (2003) The fertility-and menopause-related information needs of younger women with a diagnosis of breast cancer: a qualitative study. Psycho-Oncology 12(5):500–511
56. Schover LR, Rybicki LA, Martin BA, Bringelsen KA (1999) Having children after cancer: a pilot survey of survivors' attitudes and experiences. Cancer 86(4):697–709
57. Duffy CM, Allen SM, Clark MA (2005) Discussions regarding reproductive health for young women with breast cancer undergoing chemotherapy. J Clin Oncol 23(4):766–773
58. Avis NE, Crawford S, Manuel J (2004) Psychosocial problems among younger women with breast cancer. Psycho-Oncology 13(5):295–308
59. Freedman RA, Partridge AH (2013) Management of breast cancer in very young women. Breast 22:S176–S179
60. Paluch-Shimon S, Pagani O, Partridge AH, Abulkhair O, Cardoso MJ, Dent RA et al (2017) ESO-ESMO 3rd international consensus guidelines for breast cancer in young women (BCY3). Breast 35:203–217
61. World Health Organization (2015) Medical eligibility criteria for contraceptive use, 5th edn. World Health Organization, Geneva
62. Mody SK, Panelli DM, Hulugalle A, Su HI, Gorman JR (2017) Contraception concerns, utilization and counseling needs of women with a history of breast cancer: a qualitative study. Int J Women's Health 9:507–512
63. Güth U, Huang DJ, Bitzer J, Tirri BF, Moffat R (2016) Contraception counseling for young breast cancer patients: a practical needs assessment and a survey among medical oncologists. Breast 30:217–221
64. International Collaborative Post-Marketing Surveillance of Norplant (2001) Post-marketing surveillance of Norplant contraceptive implants: IContraceptive efficacy and reproductive health. Contraception 63(4):167–186
65. Collaborative Group on Hormonal Factors in Breast Cancer (1996) Breast cancer and hormonal contraceptives: collaborative reanalysis of individual data on 53 297 women with breast cancer and 100 239 women without breast cancer from 54 epidemiological studies. Lancet 347(9017):1713–1727
66. Pragout D, Laurence V, Baffet H, Raccah-Tebeka B, Rousset-Jablonski C (2018) Contraception et cancer. RPC Contraception CNGOF. Gynecol Obstet Fertil Senol 46(12):834–844
67. Blanc B, Lazard A, Estrade JP, Agostini A, Gurriet B (2010) Contraceptive methods after gynecological and breast cancer. Bull Acad Natl Med 194(3):521–527
68. Schwarz EB, Hess R, Trussell J (2009) Contraception for cancer survivors. J Gen Intern Med 24(2-Suppl):401–406
69. Cook-Andersen H, Komrokian S, DeMichele A, Su HI (2011) Breast cancer patients have lower rates of contraception use. Fertil Steril 96(3):S201–S202
70. Dominick SA, McLean MR, Whitcomb BW, Gorman JR, Mersereau JE, Bouknight JM et al (2015) Contraceptive practices among female cancer survivors of reproductive age. Obstet Gynecol 126(3):498–507
71. Hatcher R, Trussell J, Stewart F, Nelson A, Cates W, Guest F et al (2004) Contraceptive technology, 18th edn. Ardent Media, New York
72. Trussell J (2011) Contraceptive efficacy. In: Hatcher RA, Trussell J, Nelson AL, Cates W, Kowal D, Policar M (eds) Contraceptive technology, 20th edn. Ardent Media, New York
73. Thonneau PF, Almont T (2008) Contraceptive efficacy of intrauterine devices. Am J Obstet Gynecol 198(3):248–253
74. Mohllajee AP, Curtis KM, Peterson HB (2006) Does insertion and use of an intrauterine device increase the risk of pelvic inflammatory disease among women with sexually transmitted infection? A systematic review. Contraception 73(2):145–153
75. Curtis KM, Jatlaoui TC, Tepper NK, Zapata LB, Horton LG, Jamieson DJ, Whiteman MK (2016) US selected practice recommendations for contraceptive use
76. Van Houdenhoven K, Van Kaam KJ, Van Grootheest AC, Salemans TH, Dunselman GA (2006) Uterine perforation in women using a levonorgestrel-releasing intrauterine system. Contraception 73(3):257–260
77. United Nations Development Programme (1997) Long-term reversible contraception. Twelve years of experience with the TCu380A and TCu220C. Contraception 56(6):341–352
78. Sivin I (2007) Utility and drawbacks of continuous use of a copper T IUD for 20 years. Contraception 75(6-Suppl):S70–S75
79. Lockhat FB, Emembolu JE, Konje JC (2005) Serum and peritoneal fluid levels of levonorgestrel in women with endometriosis who were treated with an intrauterine contraceptive device containing levonorgestrel. Fertil Steril 83(2):398–404
80. Trinh XB, Tjalma WA, Makar AP, Buytaert G, Weyler J, van Dam PA (2008) Use of the levonorgestrel-releasing intrauterine system in breast cancer patients. Fertil Steril 90(1):17–22

81. De Bock GH, Bonnema J, van Der Hage J, Kievit J, Van de Velde CJ (2004) Effectiveness of routine visits and routine tests in detecting isolated locoregional recurrences after treatment for early-stage invasive breast cancer: a meta-analysis and systematic review. J Clin Oncol 22(19):4010–4018
82. Montgomery DA, Krupa K, Cooke TG (2007) Alternative methods of follow up in breast cancer: a systematic review of the literature. Br J Cancer 96(11):1625–1632
83. Lu W, de Bock GH, Schaapveld M, Baas PC, Wiggers T, Jansen L (2011) The value of routine physical examination in the follow up of women with a history of early breast cancer. Eur J Cancer 47(5):676–682
84. Grunfeld E, Noorani H, McGahan L, Paszat L, Coyle D, Van Walraven C et al (2002) Surveillance mammography after treatment of primary breast cancer: a systematic review. Breast 11(3):228–235
85. Lash TL, Fox MP, Buist DS, Wei F, Field TS, Frost FJ et al (2007) Mammography surveillance and mortality in older breast cancer survivors. J Clin Oncol 25(21):3001–3006
86. Schootman M, Jeffe DB, Lian M, Aft R, Gillanders WE (2008) Surveillance mammography and the risk of death among elderly breast cancer patients. Breast Cancer Res Treat 111(3):489–496
87. Runowicz CD, Leach CR, Henry NL, Henry KS, Mackey HT, Cowens-Alvarado RL et al (2016) American cancer society/American society of clinical oncology breast cancer survivorship care guideline. J Clin Oncol 34(6):611–635
88. Montgomery DA, Krupa K, Cooke TG (2007) Alternative methods of follow up in breast cancer: a systematic review of the literature. Br J Cancer 96(11):1625–1632
89. Rojas MP, Telaro E, Russo A, Moschetti I, Coe L, Fossati R et al (2005) Follow-up strategies for women treated for early breast cancer. Cochrane Database Syst Rev 1:CD001768
90. Bauer KR, Brown M, Cress RD, Parise CA, Caggiano V (2007) Descriptive analysis of estrogen receptor (ER)-negative, progesterone receptor (PR)-negative, and HER2-negative invasive breast cancer, the so-called triple-negative phenotype: a population-based study from the California cancer registry. Cancer 109(9):1721–1728
91. Dawood S (2010) Triple-negative breast cancer: epidemiology and management options. Drugs 70(17):2247–2258
92. van Maaren MC, de Munck L, Strobbe LJ, Sonke GS, Westenend PJ, Smidt ML et al (2019) Ten-year recurrence rates for breast cancer subtypes in the Netherlands: a large population-based study. Int J Cancer 144(2):263–272
93. Johnson HM, Lewis TC, Mitchell KB (2020) Breast cancer screening during lactation: ensuring optimal surveillance for breastfeeding women. Obstet Gynecol 135(1):194–198

Impact of Breast Cancer Treatment on Fertility

24

Konstantinos D. Dinas

Abstract

Malignancy may unfortunately present quite early in a woman's life. In the case of breast cancer, rescue of the breast cancer patient's life is the top priority, but after completion of the effective treatment, the question about the ability to accomplish a pregnancy arises. The treatment strategies in breast cancer patients include surgical interventions, chemotherapy, radiotherapy, hormonal therapy and other special types of mainly targeted biologic therapies. Under normal circumstances, surgery for breast cancer does not involve any intervention in the ovaries or the uterus. Thus, even after an extended operation, the anatomic integrity of the gynecological system is guaranteed, and fertility is unaffected.

The chemotherapeutic factors that influence fertility are the drug category used, the total dose given, the patient's age at treatment, the drug combination and finally whether targeted therapy is used or not. Alkylating agents are considered to be the most toxic ones. In young breast cancer patients there is a trend to modify regimens to achieve less gonadotoxicity.

Evidence regarding tamoxifen, the main used endocrine drug, is scarce and controversial on its direct effect on ovarian reserve. There are not enough studies on the impact of aromatase inhibitors on fertility. Also, HER2-directed agents have not yet demonstrated significant ovarian toxicity and there are scarce data on their effect on fertility.

Keywords

Breast cancer treatment · Fertility · Infertility · Pregnancy · Preservation

24.1 Overview

Malignancy may unfortunately present quite early in a woman's life. Fortunately though, this is not the rule and therefore it is less common. Obviously, rescue of the breast cancer patient's life is the top priority and always comes first; but after completion of the effective treatment, the patient returns to normal life and regains previous expectations. Thus, the distressing vague question about ability to accomplish a pregnancy arises, and contributes to anxieties and sense of insecurity (see also Chap. 28).

Although breast cancer is fortunately less common in women younger than 40 years of age,

K. D. Dinas (✉)
Second Department of Obstetrics and Gynaecology, Medical School, Aristoteles University of Thessaloniki, Hippokrateion Hospital, Thessaloniki, Greece
e-mail: dinas@auth.gr

S. Alipour, R. Omranipour (eds.), *Diseases of the Breast during Pregnancy and Lactation*, Advances in Experimental Medicine and Biology 1252,
https://doi.org/10.1007/978-3-030-41596-9_24

involving 7% of its total cases; it is the most common cancer in women, accounting for 23% of all female malignancies [1–3]. Young survivors most often want to conserve and maintain their fertility. In the vast majority of cases, the surgical treatment of breast cancer warranties the integrity of the female genital system.

Unfortunately, surgery is not always the sole modality in treatment of breast cancer, and additional forms of treatment such as chemotherapy, endocrine therapy and radiotherapy may also be necessary. These might seriously affect the possibility of childbearing and pregnancy (see also Chaps. 15 and 16) by acting upon or affecting the neuroendocrine-hormonal axis, or by the exponential reduction of ovarian reserves; or finally by mutations, involving the pregnant woman, the course of pregnancy or the embryo both in utero and after birth. Therefore, although anatomical integrity of the reproductive system is preserved, its function might get compromised.

Oogenesis takes place during fetal life and discontinues before birth. Therefore, the number of available oocytes is not infinite. As a result every woman is born having about 2,000,000 follicles (and hence oocytes) in her ovaries. However, at menarche, only 400,000 will remain in situ and she will have to go for about 40 years with this number of follicles; at menopause only the last 1000 follicles are left. These occur gradually, so at 35 years of age there are still around 150,000 follicles, 80,000 follicles at the age of 41 and 15,000 follicles at the age of 45 [4]. Only a small percentage of the total available number of oocytes will eventually mature and undergo ovulation during reproductive years because the vast majority will be lost through a specific independent mechanism called atresia.

But the issue is not only about the decreasing number of follicles, since their quality is also worsening through the years. Thus, the percentage of poor-quality follicles is increasing as we move from young ages to the age of 50. Broekmans et al. in 2004 have shown that after the age of 41, where the bad quality follicles represent 25% of the total, there is a rapid quality deterioration since the above rate is 60% at the age of 45 and nearly 100% at the age of 50 [5].

24.2 Impact of Treatments on Fertility

The treatment strategies in breast cancer patients include surgical interventions, chemotherapy, radiotherapy, hormonal therapy and other special types of mainly targeted biologic therapies (see also Chaps. 12–16).

24.2.1 Surgical Treatment

Under normal circumstances, surgery for breast cancer does not involve any intervention in the ovaries or the uterus. Thus, even after an extended operation, the anatomic integrity of the gynecological system is guaranteed, and fertility is unaffected. An exception is surgery for breast cancer ovarian metastases, however breast cancers have low risk of ovarian metastasis in women of reproductive age, and thus the possibility of removing the adnexa is low.

24.2.2 Chemotherapy

There are various important factors that influence the effects of chemotherapeutic protocols on fertility. The most important ones are the drug category used, the total dose given, the patient's age at treatment, the drug combination and finally whether targeted therapy is used or not [6, 7].

Modern chemotherapy regimens include administration of more than one agent. Their action regarding the ovaries specifically affects the sensitive cells like oocytes, but also the granulosa and theca cells. Thus the ovarian reserve will be reduced significantly. This is supported by data from various studies, documenting that in breast cancer survivors, chemotherapy administration significantly lowered serum anti-Mullerian hormone (AMH), which can be used as a marker of fertility [8]. Most commonly, post-chemotherapy amenorrhea is induced, which may often be reversible.

The factor *age* is very important, affecting greatly the risk of chemotherapy-induced ovarian failure. In general, the younger is the patient undergoing chemotherapy treatment, the less are the problems to the gonads. Thus, older women who have less primordial follicles in their ovaries seem to have more often a permanent ovarian damage compared with young women. Complete ovarian failure after chemotherapy occurs in more than half of breast cancer patients in their forties, while in patients less than 35 years of age it does not affect more than one third (15–30%) [9]. According to Minton et al. [10], patients treated after the age of 31 years have low fertility expectancy. It has been demonstrated that women treated with chemotherapy for breast cancer have eventually serum AMH level corresponding to the levels of a healthy woman who is 12 years older [11, 12]. Furthermore, high doses of chemotherapeutic medicines, specifically alkylating agents, and combined chemotherapy and radiotherapy can severely impair fertility [13].

Alkylating agents (like cyclophosphamide, ifosfamide, melphalan, and busulfan) are considered to be the most toxic chemotherapy agents in this regard. Their action is not cell-cycle specific and they may damage all the germ cells, including the resting oocytes and primordial follicles. Compared to unexposed patients, breast cancer cases who received cyclophosphamide had a four to nine times increased risk for premature ovarian failure. According to long lasting studies, [6, 14] their toxicity is dose dependent and it is reported that their ovarian detrimental effect has been observed even at very low doses. However, the same authors have concluded that chemotherapy for breast cancer has an average chance of approximately 50% for damaging ovarian tissue, which is an intermediate rate compared to the detrimental effects of chemotherapy for other malignancies in females of the same age group. Lately, modern chemotherapeutic strategies with new drugs, in a dose-dependent manner, are aiming to treat breast cancer effectively while protecting the ovaries and causing as less ovarian damage as possible by affecting only the actively developing follicles and proliferating granulosa cells [15, 16]. Other chemotherapeutics like platinum agents and anthracyclines seem to have moderate risk on damaging the ovarian reserve, whereas taxanes still have questionable effects on fertility [10, 15–17].

Combination of various chemotherapeutics is the routine nowadays, most commonly consisting of alkylating agents, anthracyclines and taxanes. In young breast cancer patients, the trend is to modify these regimens by reducing cyclophosphamide as much as possible, in order to result in less gonadotoxicity [15, 16] (see also Chaps. 15 and 21).

24.2.3 Radiotherapy

Radiotherapy has also detrimental effects on the ovaries. Human oocyte is extremely sensitive to ionizing radiation.

Irradiation can affect fertility either directly, if applied holistically on the body, targeted at the abdomen, the pelvis or the spine; or indirectly as there is always a proportion of escaping radiation that is scattered. This may be the reason for ovarian failure even when ovaries are outside of the radiation field. Nevertheless, the main concern is to keep the ovaries outside the radiation target by precisely focusing the radiation beam, or by proper displacement or coverage of the ovaries [6, 18, 19].

It is documented that direct radiation may cause reduction in the ovarian reserve, which is dose- and age-dependent. According to some authors [6, 19, 20], radiation therapy itself can have detrimental effects in the ovaries, causing oocyte destruction, proportional to the dose. Thus, with doses of 3, 3–5, and 5 Gy, 11%, 60%, and 100% of the follicles are destroyed, respectively. Furthermore, due to the lower oocyte reserve in older patients compared to young women, they are at higher risk of developing infertility after radiotherapy. Thus, although treatments in very young patients usually do not cause major problems to the ovaries, the impact on patients is serious after the age of 30 [21, 22]. Also, if the total amount of radiation is divided to many small doses instead of few high fractions, it would be less toxic [23].

Apart from directly affecting the gametes, radiation may also influence fertility due to effects on other tissues that anatomically contribute to fertility; like the vascular and smooth muscle damages and fibrosis in the uterus, tubes and cervix that occurs secondary to radiotherapy. In cases of cranial irradiation due to relevant metastases, the hypothalamic-pituitary axis is damaged [24].

Finally, the combination of radiotherapy with chemotherapy can severely impair fertility [6, 13].

24.2.4 Endocrine Therapy

Tamoxifen is now used for up to 10 years after the documentation of positive estrogen receptors in breast cancer patients of reproductive age, but it has been noted to increase the rate of amenorrhea and menopause. Nevertheless, there is scarce and controversial evidence on the direct effect of tamoxifen on the ovarian reserve. Some authors found no significant effect of tamoxifen on serum AMH while others have demonstrated lower serum AMH in women taking tamoxifen [25–27]. For now, pregnancy is contraindicated while under tamoxifen.

Aromatase inhibitors (like letrozole) are mainly used in menopausal patients and therefore there are not enough studies on their impact on fertility [28]. Nevertheless, they are now routinely used in young breast cancer patients undergoing IVF treatment before their chemotherapy, as the basic protective medication minimizing the production of circulating estradiol (see also Chap. 15).

24.2.5 Targeted Biologic Therapies

Trastuzumab is a targeted treatment for HER2+ breast cancer. HER2-directed agents have not yet demonstrated significant ovarian toxicity [11]. There are scarce data on the effect of trastuzumab on fertility. Interestingly, in very few studies, the use of trastuzumab with chemotherapy was associated with significantly higher serum AMH levels compared to patients who received only chemotherapy [11]. Although encouraging, these data need further confirmation through larger and powerful studies. The current knowledge is to avoid pregnancy while on trastuzumab. Furthermore, a period of at least 6 months after completing trastuzumab treatment is considered to be safe before any pregnancy occurs [29].

References

1. Siegel RL, Miller KD, Jemal A (2015) Cancer statistics, 2015. Cancer J Clin 65(1):5–29
2. Anders CK, Johnson R, Litton J, Phillips M, Bleyer A (2009) Breast cancer before age 40 years. Semin Oncol 36:237–249
3. Howlader N, Noone AM, Krapcho M eds (2014), SEER Cancer Statistics Review, 1975–2011. National Cancer Institute: Bethesda. Based on November 2013 SEER data submission, posted to the SEER website, April 2014
4. Tarlatzis B, Zepiridis L (2003) Perimenopausal conception. Ann N Y Acad Sci 997(1):93–104
5. Broekmans FJ, Faddy MJ, Scheffer G, te Velde ER (2004) Antral follicle counts are related to age at natural fertility loss and age at menopause. Menopause 11:607–614
6. Meirow D, Nugent D (2001) The effects of radiotherapy and chemotherapy on female reproduction. Hum Reprod Update 7(6):535–543
7. Warne GL, Fairley KF, Hobbs JB, Martin FI (1973) Cyclophosphamide-induced ovarian failure. New Engl J Med 289(22):1159–1162
8. Su HI, Sammel MD, Green J, Velders L, Stankiewicz C, Matro J et al (2010) Antimullerian hormone and inhibin B are hormone measures of ovarian function in late reproductive-aged breast cancer survivors. Cancer 116(3):592–599
9. Goodwin PJ, Ennis M, Pritchard KI, Trudeau M, Hood N (1999) Risk of menopause during the first year after breast cancer diagnosis. J Clin Oncol 17(8):2365–2369
10. Minton SE, Munster PN (2002) Chemotherapy-induced amenorrhea and fertility in women undergoing adjuvant treatment for breast cancer. Cancer Control 9(6):466–472
11. Morarji K, McArdle O, Hui K, Gingras-Hill G, Ahmed S, Greenblatt EM et al (2017) Ovarian function after chemotherapy in young breast cancer survivors. Curr Oncol 24(6):e494–e502
12. Peigne M, Decanter C (2014) Serum AMH level as a marker of acute and long-term effects of chemotherapy on the ovarian follicular content: a systematic review. Reprod Biol Endocrinol 12(1):26

13. Nagarajan R, Robison LL (2005) Pregnancy outcomes in survivors of childhood cancer. J Natl Cancer Inst Monogr 2005(34):72–76
14. Meirow D (1999) Ovarian injury and modern options to preserve fertility in female cancer patients treated with high dose radio-chemotherapy for hemato-oncological neoplasias and other cancers. Leuk Lymphoma 33(1-2):65–76
15. Ronn R, Holzer HE (2013) Oncofertility in Canada: the impact of cancer on fertility. Curr Oncol 20(4):e338–e344
16. Turan V, Oktay K (2014) Sexual and fertility adverse effects associated with chemotherapy treatment in women. Expert Opin Drug Saf 13(6):775–783
17. Crozier JA, Swaika A, Moreno-Aspitia A (2014) Adjuvant chemotherapy in breast cancer: to use or not to use, the anthracyclines. World J Clin Oncol 5(3):529–538
18. Jensen JR, Morbeck DE, Coddigton CC (2011) Fertility preservation. Mayo Clin Proc 86(1):45–49
19. Sonmezer M, Oktay K (2004) Fertility preservation in female patients. Hum Reprod Update 10(3):251–266
20. Pierce LJ, Strawderman M, Narod SA, Oliviotto I, Eisen A, Dawson L et al (2000) Effect of radiotherapy after breast-conserving treatment in women with breast cancer and germline BRCA1/2 mutations. J Clin Oncol 18(19):3360–3369
21. Hickey M, Peate M, Saunders CM, Friedlander M (2009) Breast cancer in young women and its impact on reproductive function. Hum Reprod Update 15(3):323–329
22. Hawkins MM, Smith RA (1989) Pregnancy outcomes in childhood cancer survivors: probable effects of abdominal irradiation. Int J Cancer 43(3):399–402
23. Lee SJ, Schover LR, Partridge AH, Patrizio P, Wallace WH, Hagerty K et al (2006) American Society of Clinical Oncology recommendations on fertility preservation in cancer patients. J Clin Oncol 24(18):2917–2931
24. Critchley HO, Bath LE, Wallace WH (2002) Radiation damage to the uterus-review of the effects of treatment of childhood cancer. Hum Fertil 5(2):61–66
25. Partridge AH, Ruddy KJ, Gelber S, Schapira L, Abusief M, Meyer M et al (2010) Ovarian reserve in women who remain premenopausal after chemotherapy for early stage breast cancer. Fertil Steril 94(2):638–644
26. Torino F, Barnabei A, De Vecchis L, Sini V, Schittulli F, Marchetti P et al (2014) Chemotherapy-induced ovarian toxicity in patients affected by endocrine-responsive early breast cancer. Crit Rev Oncol Hematol 89(1):27–42
27. American College of Obstetricians and Gynecologists (2014) Committee Opinion No. 601: tamoxifen and uterine cancer. Obstet Gynecol 123(6):1394–1397
28. Azim AA, Costantini-Ferrando M, Oktay K (2008) Safety of fertility preservation by ovarian stimulation with letrozole and gonadotropins in patients with breast cancer: a prospective controlled study. J Clin Oncol 26(16):2630–2635
29. Zagouri F, Sergentanis TN, Chrysikos D, Papadimitriou CA, Dimopoulos MA, Bartsch R (2013) Trastuzumab administration during pregnancy: a systematic review and meta-analysis. Breast Cancer Res Treat 137(2):349–357

Fertility Counseling and Preservation for Breast Cancer Patients

25

Konstantinos D. Dinas

Abstract

Fertility preservation includes all the necessary steps that have to be taken in order to implement a woman's aim for preserving and enhancing her future chances of having her own biological offspring.

There are two available choices for patients to maintain their fertility.

The first category includes all the preventive methods following the principle of reduction of gonadotoxicity, in terms of using fewer toxic drugs to the ovaries and similarly milder radiation regimens. It also includes the administration of agents that would keep the ovaries suppressed, to minimize the harmful effects of systemic therapy on the oocytes.

The second includes necessary artificial reproductive technology procedures for collecting and freezing either gametes (oocytes) or zygotes (embryos). This can be accomplished with immature oocyte pickup followed by in vitro maturation (IVM). Alternatively, mature oocytes can be collected after minimal ovarian stimulation with very small doses or no medication at all, or through a natural cycle.

Beyond that, promising is the modern option of ovarian tissue cryopreservation for future transplantation, while the option of oocyte donation is actually an option for childbearing, not for fertility.

Keywords

Breast Cancer · Cryopreservation · Fertility Counseling · Fertility Preservation · In Vitro Maturation · Oocyte Collection · Oocyte donation

25.1 Overview

Fertility preservation and its restoration include all the necessary steps that have to be taken in order to implement a woman's aim for preserving and enhancing the chances of having her own biological offsprings in the future.

There are two available choices for patients to maintain their fertility. The first include preventive methods, and the second consists of using artificial reproductive technology options.

The first category includes all the methods obeying the principle of reduction of gonadotoxicity, in terms of using as less toxic drugs to the ovaries as possible and similarly milder radiation. Furthermore, it includes the administration

K. D. Dinas (✉)
Second Department of Obstetrics and Gynaecology, Medical School, Aristoteles University of Thessaloniki, Hippokrateion Hospital, Thessaloniki, Greece
e-mail: dinas@auth.gr

S. Alipour, R. Omranipour (eds.), *Diseases of the Breast during Pregnancy and Lactation*, Advances in Experimental Medicine and Biology 1252,
https://doi.org/10.1007/978-3-030-41596-9_25

of medication that would keep the ovaries in suppression mode, so that their oocytes will not be harmed by chemo- or radiotherapy.

The second includes necessary procedures for collecting and freezing either gametes (oocytes) or zygotes (embryos). This can be accomplished with immature oocyte pickup from small follicles followed by in vitro maturation (IVM). Alternatively, mature oocytes can be collected after minimal ovarian stimulation with very small doses of medication, use of special medication or after no medication at all through a natural cycle. Beyond that, promising is the modern option of ovarian tissue cryopreservation while the option of oocyte donation could be applied.

25.2 Preventive Method-Reduction of Gonadotoxicity

Ancient Greeks always told that "it is better to prevent than to treat". But is there anything to do in advance in order to avoid the impairment of fertility? The answer is yes. Reduction of gonadotoxicity means that, during and after any treatment protocol, as less oocytes or/and ovarian tissue as possible will be destroyed.

Whether the suppressing action of Gonadotropin-releasing hormone (GnRH) agonists can protect the ovaries from gonadotoxic chemotherapy is still a point of controversy. These are administered every 28 or 84 days in parallel and after chemotherapy. There are a lot of studies that tried to elucidate this interesting issue by including an adequate sample of breast cancer patients. However available data derived from various meta-analyses and systematic reviews [1–7] cannot support the hypothesis that GnRH agonists significantly help to maintain fertility in women with breast cancer. Nonetheless, there is evidence that this may be the case for women with cancer in general, and the American Society of Clinical Oncology (ASCO) and the American Society for Reproductive Medicine (ASRM) opinion is that data on the effectiveness of the method is insufficient and controversial. Thus, GnRH-analogues may be administrated to patients in terms of reducing the chances of chemotherapy-induced ovarian insufficiency [8–10].

25.3 Fertility Conservation Options

Considering that chances for modifying the toxicity of chemotherapy or radiotherapy are limited, the alternative option would be to preserve the ability to reproduce by taking out fertility elements like oocytes, embryos or ovarian tissue from the body and storing them safely; then using them when the disease no longer exists.

These can be achieved by oocyte pickup either after a natural cycle with no medication or following ovarian stimulation and ovulation induction with minimal doses of gonadotropins. Alternatively oocytes could be collected again, without administration of medication from the immature follicles in the ovaries. Then, these oocytes are matured in the laboratory with IVM. Oocyte collection is performed as soon as the follicles have grown adequately.

25.3.1 Oocyte Collection in Natural Cycles

In the recent past, the choice of controlled ovarian stimulation would have been a reckless option. Having in mind the importance of estrogen and progesterone receptor in breast cancer pathogenesis and development, the administration of normal doses of gonadotropins looked indisputably unrealistic, since they resulted in a remarkable increase in estradiol and progesterone serum levels. As a result, oocyte pickup was done without any gonadotropin administration in one or more natural cycles. Each oocyte pickup in natural cycles yields a maximum of one oocyte. Therefore, many natural cycles are needed to collect and freeze an adequate number of oocytes.

25.3.2 Controlled Ovarian Stimulation

Alternatively, a very small dose of gonadotropins can be administrated in order to establish ovarian stimulation with minimal doses of follicle stimulating hormone (FSH), in order to grow more than one follicle. This results in more than one oocyte per cycle, and therefore fewer sessions of pickup are needed [11–14].

Back in 2005, Oktay and colleagues used aromatase inhibitors and selective estrogen receptor modulators in the context of controlled ovarian stimulation, trying to reduce the blood concentrations of both these hormones [15, 16]. Thereafter a large number of trials has been conducted on this issue by using letrozole and tamoxifen. The aim is the achievement of the lowest levels of estradiol possible, without reducing the number of follicles and therefore the oocytes collected, which is particularly beneficial in patients with positive estrogen receptors [14, 17, 18].

Although existing data is not much, we have to assume that it is a theoretically beneficial option. A protocol for letrozole is similar to a short antagonist protocol where the additional medication is only letrozole, which commences on day-2 of stimulation, and gonadotropins are added on day-4 [18]. However, different trials have tested if onset of the procedure has to be on a specific day of the menstrual cycle. It is now well documented that these protocols can be started at any time and therefore can be initiated as soon as needed.

In November 2013 a Cochrane Database review was published to assess the efficacy of protocols containing tamoxifen or letrozole for ovarian stimulation compared to protocols without them in premenopausal women with estrogen receptor (ER) positive breast cancer who want to cryopreserve oocytes or embryos. No randomized prospective studies were found [14]. Thus the official opinion is that there is no evidence to suggest the advantage of the use of these drugs in ovarian stimulation compared to ovulation without them. However, the fact is that they drastically reduce estradiol levels and this might be proved beneficial in the future. In July 2018, ASCO supported the use of flexible ovarian stimulation protocols for oocyte collection that are feasible on a cycle day–independent schedule. Additionally, in estrogen positive breast cancer, these interventions will not increase the risk of cancer recurrence since aromatase inhibitor assisted protocols keep estrogen levels depressed. Current studies do not indicate an augmented risk of recurrence of cancer from regimens including ovarian stimulation supplemented by aromatase inhibitor [8].

25.3.3 In Vitro Maturation

Another recently available method is in vitro maturation (IVM). One of its main indications is in women who will undergo chemotherapy for breast cancer, where fertility has to be preserved by oocyte cryopreservation without any delay and without hormonal stimulation [19]. The concept is that if it is not allowed to give medication to the ovaries to grow and mature oocytes, then the oocytes can be extracted from the body and matured in the laboratory, avoiding any increase in the estradiol levels and any consequent harm to the body of the breast cancer patient [20, 21].

In IVM, immature oocytes are collected transvaginally under ultrasound guidance from antral follicles located in non-stimulated or poorly stimulated ovaries. These oocytes are matured in-vitro in the laboratory under controlled conditions for 24 to 48 hours, followed if necessary by the standard procedure of either fertilization with intracytoplasmic sperm injection (ICSI), or cryopreservation for future use [20, 22]. Despite the satisfactory pregnancy rates with this method in appropriately selected patients, overall pregnancy rates remain lower than those of conventional in vitro fertilization (IVF). Using this method, a few thousands of healthy infants have been born worldwide in infertile women, without any increase in abortion rates or other abnormalities [21–25]. These studies were carried out regardless of cancer survivor's issues.

25.3.4 Oocyte Cryopreservation

Cryopreservation of oocytes is an alternative to cryopreservation and storage of embryos, an ideal method for young women with breast cancer who do not have a partner and do not want to use donor sperm. The mature or matured oocytes are stored after being frozen via a procedure called vitrification. They will later be thawed, fertilized, and transferred in the uterine cavity if and when necessary. This can avoid the moral and religious dilemmas associated with embryo storage and disposal, and does not require IVF. Since the first birth report from a cryopreserved human oocyte in 1986 [26], more than a thousand of healthy babies have been born worldwide [27], while the delivery rates ranged from 14-34% after oocyte thawing and IVF. However, these were not about breast cancer survivors.

Oocyte cryopreservation is now included in the guidelines of ASCO 2018 "Fertility Preservation in Patients with Cancer" as an option. According to this recommendation, oocyte cryopreservation should be performed in centers with enough expertise. Since 2012, ASRM does not consider this procedure as an experimental method [8].

25.3.5 Embryo Cryopreservation

After the oocyte collection and if a partner exists, fertilization of the collected oocyte(s) may take place before freezing, creating the ability to freeze embryos and not oocytes [15, 28].

Embryo freezing-thawing and transfer have higher pregnancy rates [28], and is the most tested and well-established conventional method for maintaining fertility since it has been applied for more than 25 years now[28]. Based on 2016 Society for Assisted Reproductive Technology (SART) and Center for Disease Control (CDC) data, live birth rates are 46.6% for women younger than 35 years, 44% for women 35-37 and 38.3% for women 38-40 years of age, so millions of children have been born with this method. Success of this method in breast cancer survivors needs to be assessed in the future.

25.3.6 Cryopreservation and Transplantation of Ovarian Tissue

A recent option is the maintenance of fertility by cryopreservation of ovarian tissue and its transplantation, and one of its main indications is in patients undergoing treatment for any stage of breast cancer [29]. The idea is practically the following: since we do not want the ovaries to be exposed to the toxic and detrimental effects or consequences of the various therapies, a good solution would be to take them out of the body and store them somewhere safely; then to put them back in the body in order to restore normal ovarian function and fertility when the patient is cured [30–32].

In women who will be treated for breast cancer, fertility can be preserved without delay with this technique; so it can be performed immediately after diagnosis without any hormonal stimulation.

Briefly described, a laparoscopy is performed during which tissue pieces of the ovarian cortex are obtained. These pieces are then cut into smaller portions of 2 × 5 × 5 mm, which are the final grafts. The density of the follicles is recorded, and finally the grafts are frozen [30–32]. Thereafter chemotherapy and/or radiotherapy for breast cancer treatment follow as per planned. After years, if the woman has got infertile and wants to restore her fertility, the small grafts of ovarian tissue are thawed and re-transplanted with an orthotopic or heterotopic transplantation. In the case of orthotopic ovarian tissue transplantation the tissue is placed via laparoscopy on the ovaries where it naturally belongs. In the case of heterotopic ovarian tissue transplantation, the tissue could be transplanted in different places, such as under the skin. In the first case, natural fertility and comprehensive ovarian function might be restored, while in the second, pregnancy can occur only through in vitro fertilization [30–33]. Heterotopic transplantation is an attractive option since it avoids invasive procedures in the abdomen like laparoscopy and makes oocyte collection easier. It is practical and cost effective particularly when it is required to be repeated due to the reduced lifespan of the ovarian grafts [30–33].

Restoring fertility using cryopreserved and thawed ovarian tissue is a challenge because it involves many technical and scientific obstacles and difficulties that the gynaecologist has to overcome. Based on the existing data [29, 30], although successful, ovarian tissue transplantation after its cryopreservation has led so far to the delivery of more than 37 healthy newborns [30]. Therefore in humans, ovarian tissue transplantation is still considered to be an experimental process until its effectiveness is proven. However, this technique is already considered non-experimental in some countries, and its experimental status is undergoing evaluation in the United States.

25.3.7 Oocyte Donation

The last option is egg donation. This, of course, is an option for childbearing, not for fertility. It involves oocyte donation from a healthy donor, preferably less than 30 years of age. The donor will be given gonadotropins in order to achieve a satisfactory number of follicles developing in her ovaries, and will eventually undergo oocyte pick-up, while the recipient is having her endometrium prepared for transfer of the embryo.

25.4 Conclusion

Over the years, more and more young women are facing the problem of breast cancer early in their lives. The modern therapeutic strategies that oncologists apply nowadays have significantly improved survival rates and life expectancy. After the priority of curing the disease, fertility will be the patient's next request from her doctor. Therefore, there is a certain need for coordinated action protocols in all these cases.

ASCO has set out relevant guidelines in 2013, and has updated them recently in July 2018 [8]. The basic aspects can be summarized as in Table 25.1.

Regarding the available options for fertility preservation and restoration, current data are summarized in Table 25.2.

Table 25.1 Summary of ASCO guidelines about fertility preservation while treating cancer

Oncologists, radiotherapists and gynecologists should inform the patient about the possibility of fertility impairment or even loss, prior to performing chemotherapy.
It is their obligation, first of all, to inform the patient for the option of preserving fertility before chemotherapy.
If interested, the patient should be referred to assisted reproduction specialists. These will inform her about all the existing options and prospects in detail.
After the end of the treatment cycles, the patient's ovarian reserve should be re-estimated.
Early and prompt information given on these sensitive issues significantly reduces the patient's stress and improves her quality of life.

Table 25.2 Summary of available options for fertility preservation and restoration in breast cancer

Method	Points
Cryopreservation of embryos	The most widely tested, acceptable and successful method
Cryopreservation of oocytes	Not experimental since October 2012; appropriate for cases with no partner, not accepting sperm donation, or with moral or religious dilemmas for embryo freezing
Ovarian suppression	Insufficient data on the efficacy; should not be used as the only option; might be administrated with the aim of reducing chemotherapy-induced ovarian insufficiency
Controlled ovarian stimulation	Can be implemented via a variety of protocols; can be started at any time during the cycle
Ovarian stimulation protocols + AI or SERM	Does not seem to increase the likelihood of disease recurrence in hormone dependent estrogen receptor positive breast cancer
Cryopreservation of ovarian tissue and re-transplantation	is considered an experimental method; should be performed in specialized centers with the appropriate experience

AI aromatase inhibitor, *SERM* selective estrogen receptor modulator

The essential message is that eventually there are reassuring and effective solutions for fertility preservation in a woman with breast cancer. We have to focus on the correct approach, which always includes providing comprehensive and reliable information to the patient; followed by the state of the art therapeutic strategies that, without jeopardizing her life, covers the patient's uncertainty and anxiety about her future fertility. Thus, fertility preservation can be applied to young breast cancer patients with a variety of techniques, as well as their combinations. More powerful and well-designed studies are needed to document the effectiveness of the newer and most promising methods. It is likely that the list of various techniques will expand in the future, and there is still a need for research that will provide answers to the current technical and scientific questions, and overcome the obstacles.

References

1. Francis PA, Regan MM, Fleming GF, Láng I, Ciruelos E, Bellet M et al (2015) Adjuvant ovarian suppression in premenopausal breast cancer. N Engl J Med 372(5):436–446
2. Moore HC, Unger JM, Phillips KA, Boyle F, Hitre E, Porter D et al (2015) Goserelin for ovarian protection during breast-cancer adjuvant chemotherapy. N Engl J Med 372(10):923–932
3. Del Mastro L, Ceppi M, Poggio F, Bighin C, Peccatori F, Demeestere I et al (2014) Gonadotropin-releasing hormone analogues for the prevention of chemotherapy-induced premature ovarian failure in cancer women: systematic review and meta-analysis of randomized trials. Cancer Treat Rev 40(5):675–683
4. Vitek WS, Shayne M, Hoeger K, Han Y, Messing S, Fung C (2014) Gonadotropin-releasing hormone agonists for the preservation of ovarian function among women with breast cancer who did not use tamoxifen after chemotherapy: a systematic review and meta-analysis. Fertil Steril 102(3):808–815
5. Elgindy E, Sibai H, Abdelghani A, Mostafa M (2015) Protecting ovaries during chemotherapy through gonad suppression: a systematic review and meta-analysis. Obstet Gynecol 126(1):187–195
6. Munhoz RR, Pereira AA, Sasse AD, Hoff PM, Traina TA, Hudis CA et al (2016) Gonadotropin-releasing hormone agonists for ovarian function preservation in premenopausal women undergoing chemotherapy for early-stage breast cancer: a systematic review and meta-analysis. JAMA Oncol 2(1):65–73
7. Lambertini M, Ceppi M, Poggio F, Peccatori FA, Azim HA Jr, Ugolini D et al (2015) Ovarian suppression using luteinizing hormone-releasing hormone agonists during chemotherapy to preserve ovarian function and fertility of breast cancer patients: a meta-analysis of randomized studies. Ann Oncol 26(12):2408–2419
8. Oktay K, Harvey BE, Partridge AH, Quinn GP, Reinecke J, Taylor HS et al (2018) Fertility preservation in patients with cancer: ASCO clinical practice guideline update. J Clin Oncol 36(19):1994–2001
9. Loren AW, Mangu PB, Beck LN, Brennan L, Magdalinski AJ, Partridge AH et al (2013) Fertility preservation for patients with cancer: American Society of Clinical Oncology clinical practice guideline update. J Clin Oncol 31(19):2500–2510
10. Cardoso F, Loibl S, Pagani O, Graziottin A, Panizza P, Martincich L et al (2012) The European Society of Breast Cancer Specialists recommendations for the management of young women with breast cancer. Euro J Cancer 48(18):3355–3377
11. Muñoz E, González N, Muñoz L, Aguilar J, Velasco JA (2015) Ovarian stimulation in patients with breast cancer. Cancer Med Sci 9:504
12. Baynosa J, Westphal LM, Madrigrano A, Wapnir I (2009) Timing of breast cancer treatments with oocyte retrieval and embryo cryopreservation. J Am Coll Surg 209(5):603–607
13. Ben-Aharon I, Gafter-Gvili A, Leibovici L, Stemmer SM (2010) Pharmacological interventions for fertility preservation during chemotherapy: a systematic review and meta-analysis. Breast Cancer Res Treat 122(3):803–811
14. Dahhan T, Balkenende E, van Wely M, Linn S, Goddijn M (2013) Tamoxifen or letrozole versus standard methods for women with estrogen-receptor positive breast cancer undergoing oocyte or embryo cryopreservation in assisted reproduction. Cochrane Database Syst Rev 11:CD010240
15. Oktay K, Buyuk E, Libertella N, Akar M, Rosenwaks Z (2005) Fertility preservation in breast cancer patients: a prospective controlled comparison of ovarian stimulation with tamoxifen and letrozole for embryo cryopreservation. J Clin Oncol 23(19):4347–4353
16. Oktay K (2005) Further evidence on the safety and success of ovarian stimulation with letrozole and tamoxifen in breast cancer patients undergoing in vitro fertilization to cryopreserve their embryos for fertility preservation. J Clin Oncol 23(16):3858–3859
17. Meirow D, Raanani H, Maman E, Paluch-Shimon S, Shapira M, Cohen Y et al (2014) Tamoxifen co-administration during controlled ovarian hyperstimulation for in vitro fertilization in breast cancer patients increases the safety of fertility-preservation treatment strategies. Fertil Steril 102(2):488–495
18. Azim AA, Costantini-Ferrando M, Oktay K (2008) Safety of fertility preservation by ovarian stimulation with letrozole and gonadotropins in patients with breast cancer: a prospective controlled study. J Clin Oncol 26(16):2630–2635
19. Cruz MR, Prestes JC, Gimenes DL, Fanelli MF (2010) Fertility preservation in women with breast cancer undergoing adjuvant chemotherapy: a systematic review. Fertil Steril 94(1):138–143

20. Cha KY, Chian RC (1998) Maturation in vitro of immature human oocytes for clinical use. Hum Reprod Update 4(2):103–120
21. Rao GD, Tan SL (2005) In vitro maturation of oocytes. Sem Reprod Med 23(3):242–247
22. Lim KS, Chae SJ, Choo CW, Ku YH, Lee HJ, Hur CY et al (2013) In vitro maturation: clinical applications. Clin Exp Reprod Med 40(4):143–147
23. Huang JY, Tulandi T, Holzer H, Tan SL, Chian RC (2008) Combining ovarian tissue cryobanking with retrieval of immature oocytes followed by in vitro maturation and vitrification: an additional strategy of fertility preservation. Fertil Steril 89(3):567–572
24. Cao YX, Chian RC (2009) Fertility preservation with immature and in vitro matured oocytes. Semin Reprod Med 27(6):456–464
25. Chian RC, Huang JY, Gilbert L, Son WY, Holzer H, Cui SJ et al (2009) Obstetric outcomes following vitrification of in vitro and in vivo matured oocytes. Fertil Steril 91(6):2391–2398
26. Chen C (1986) Pregnancy after human oocyte cryopreservation. Lancet 327(8486):884–886
27. Noyes N, Porcu E, Borini A (2009) Over 900 oocyte cryopreservation babies born with no apparent increase in congenital anomalies. Reprod Biomed Online 18(6):769–776
28. Rienzi L, Gracia C, Maggiulli R, LaBarbera AR, Kaser DJ, Ubaldi FM et al (2017) Oocyte, embryo and blastocyst cryopreservation in ART: systematic review and meta-analysis comparing slow-freezing versus vitrification to produce evidence for the development of global guidance. Hum Reprod Update 23(2):139–155
29. Kim SS, Klemp J, Fabian C (2011) Breast cancer and fertility preservation. Fertil Steril 95(5):1535–1543
30. Stoop D, Cobo A, Silber S (2014) Fertility preservation for age-related fertility decline. Lancet 384(9950):1311–1319
31. Donnez J, Dolmans MM (2009) Cryopreservation of ovarian tissue: an overview. Minerva Med 100(5):401–413
32. Sonmezer M, Oktay K (2010) Orthotopic and heterotopic ovarian tissue transplantation. Best Pract Res Clin Obstet Gynaecol 24(1):113–126
33. Pacheco F, Oktay K (2017) Current success and efficiency of autologous ovarian transplantation: a meta-analysis. Reprod Sci 24(8):1111–1120

Pregnancy after Breast Reconstruction

26

Heba Alkhashnam, Francoise Rimareix, Chafika Mazouni, Nicolas Leymarie, Benjamin Sarfati, Jean-Francois Honart, and Frédéric Kolb

Abstract

Breast cancer is one of the most commonly diagnosed malignancies in women of childbearing age. One of the surgeries performed in the management of the disease is mastectomy, which might negatively affect body image and quality of life, and breast reconstruction is seen as a remedy to this problem. However, for the women who are interested in having children after treatment of breast cancer, the debate is whether they can have a successful pregnancy and delivery after breast reconstruction.

Keywords

Breast reconstruction · Breast cancer · Pregnancy · Breastfeeding · Implant · Latissimus dorsi flap · Transverse rectus abdominis myocutaneous flap · Deep inferior epigastric artery perforator flap · Fat grafting · Transverse myocutaneous gracilis flap

H. Alkhashnam · F. Rimareix (✉) · C. Mazouni
N. Leymarie · B. Sarfati · J.-F. Honart · F. Kolb
Department of Plastic and Breast Surgery, Gustave Roussy Cancer Campus, Grand Paris, France
e-mail: heba.alkhashnam@gustaveroussy.fr; Francoise.rimareix@gustaveroussy.fr; chafika.mazouni@gustaveroussy.fr; Nicolas.Leymarie@gustaveroussy.fr; Benjamin.Sarfati@gustaveroussy.fr; jeanfrancois@gustaveroussy.fr; Feredric.Kolb@gustaverussy.fr

26.1 Overview

Currently, breast cancer is detected earlier owing to the higher awareness of the general population and because of screening principles that have been established in many areas. However, owing to the increasing age of pregnancy, many women diagnosed with the disease have not yet given birth to a child and are planning for it in the future [1].

Breast conserving surgery (BCS) is not always an option in the management of breast cancer and mastectomy remains one of the main treatments. This operation not only affects the patient physically but also psychologically because of the disturbed self-image; hence, breast reconstruction is becoming one of the main parts of the treatment of breast cancer and helps patients have a better quality of life.

However, in young patients who do not opt for BCS or who have contraindications for it and undergo mastectomy, the question is whether they can become pregnant after breast reconstruction. This subject is discussed below considering different techniques of reconstruction.

S. Alipour, R. Omranipour (eds.), *Diseases of the Breast during Pregnancy and Lactation*, Advances in Experimental Medicine and Biology 1252,
https://doi.org/10.1007/978-3-030-41596-9_26

26.2 Pregnancy after Different Methods of Breast Reconstruction

Immediate breast reconstruction (IBR) or delayed breast reconstruction (DBR) does not prevent a successful pregnancy. For IBR, simple techniques could be used (implant or expander) to avoid major complications; whereas in DBR, all techniques are possible and will not affect a future pregnancy. Even with the transverse rectus abdominis myocutaneous (TRAM) flap technique where the rectus abdominis muscle is harvested, pregnancy is still possible [2].

Before carrying out mastectomy as treatment of breast cancer in a woman who plans for subsequent pregnancy, all pertinent techniques of breast reconstruction should be discussed with the patient, considering the quality and laxity of her thoracic wall skin and her comorbidities (smoking, high body mass index, and other diseases such as diabetes or thromboembolic blood diseases). In addition, breast asymmetry after pregnancy must be described for the patient so that she is aware that there might be a need to perform further surgeries after delivery.

26.3 Pregnancy after Prosthesis Reconstruction

Either an implant or an expander-based reconstruction is possible. In the latter instance, it is important to inform the patient that performing MRI is considered unsafe when tissue expanders are in place, owing to the metallic port. Possible complications include displacement of the port and implant, image artifact, local heating, and device malfunction [3]. A systematic review performed in 2019 has investigated the issue and conclude that while controversies still exist about the safety of MRI with tissue expanders, it can be carried out by following a detailed protocol [4]. The expander could be exchanged for the definitive implant before or after pregnancy, but it is preferable to finalize the reconstruction when the patient's weight is stable.

Pregnancy affects the whole body, with the breast being particularly targeted. Breast hypertrophy, ptosis, areolar enlargement, nipple hypertrophy, and increased pigmentation of the nipple-areola complex usually occur during pregnancy [5], resulting in asymmetry between the two breasts (see also Chaps. 1 and 2). Consequently, a second surgery, such as the implant change of the reconstructed breast or mammoplasty of the contralateral breast might be needed about 6 to 12 months after pregnancy.

26.3.1 Autologous Reconstruction

26.3.1.1 After Abdominal Flaps

The abdomen is the most common area from which tissue is harvested for autologous breast reconstruction [6–10].

The transverse rectus abdominis myocutaneous (TRAM) flap could be harvested as a pedicled or free flap. The whole rectus abdominis muscle or part of it (muscle sparing) together with the overlying adipose tissue and skin are harvested based on the superior epigastric vessels and transposed to reconstruct the breast. On the site of harvest of the rectus muscle, the posterior rectus fascia should be reinforced either with the contralateral anterior fascia or with a mesh to prevent abdominal wall weakness and hernias.

Earlier reports of pregnancy after a TRAM flap reconstruction described a hernia with an attenuated and weak abdominal wall during cesarean section [11]. However, even if a portion of the abdominal wall fascia is removed with a part of the muscle, the potential adverse effects on pregnancy and labor are not very significant, and successful pregnancies and labor have been reported after a TRAM flap reconstruction [2, 11–13] (Fig. 26.1).

Considering the morbidity of the donor site and the potential weakness of the abdominal wall, a one year delay is preferable between the TRAM flap reconstruction and the pregnancy, although pregnancy has also been achieved uneventfully within a shorter interval [11, 12].

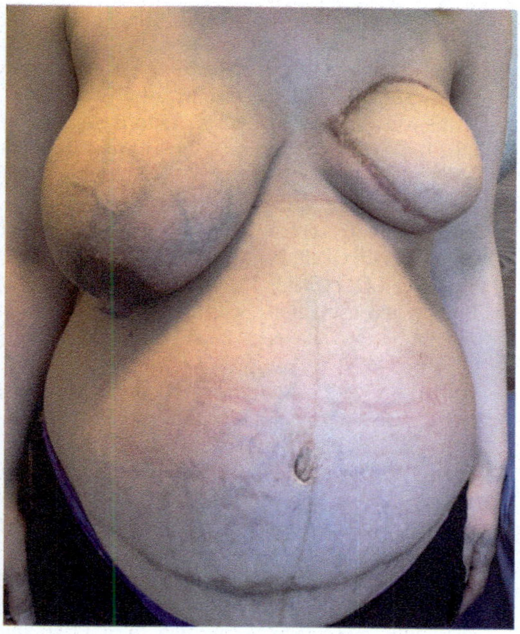

Fig. 26.1 Pregnancy occurring 8 years after left breast TRAM flap reconstruction in a 35 years-old patient. The patient is planning to undergo right breast symmetrization after breastfeeding. (Courtesy of Dr. Ramesh Omranipour)

With the advancement of microsurgery and perforator flaps, the deep inferior epigastric perforator (DIEP) flap has gradually become the superior choice for autologous breast reconstruction because it reduces donor site morbidity, abdominal wall complications, and postoperative recovery time [10]. Here the skin and underlying adipose tissue are harvested on a perforator of the inferior epigastric vessels, leaving the rectus abdominis muscle intact. The flap is transferred to the chest area as a free flap where vascular anastomosis is performed and is then reshaped to reconstruct the breast.

Pregnancy after DIEP flap reconstruction is possible, and there would not be any consequences on the delivery as the muscles are intact; vaginal or cesarean births are possible in these circumstances [14, 15].

26.3.1.2 After Dorsal Flaps

The latissimus dorsi (LD) musculocutaneous flap provides a readily available local source of well-vascularized muscle and fat that can be used in conjunction with tissue expanders and implants to reconstruct the breast after mastectomy in both an immediate and a delayed fashion [16]. It involves harvesting the LD muscle together with the overlying adipose tissue and skin and then transposing it anteriorly to reconstruct the breast.

On the contrary, the thoracodorsal artery perforator flap spares the LD muscle and only the skin and adipose tissue are raised on a perforator. This flap too has been reported in breast reconstruction [17]. Another free flap that has been used is the lumbar artery perforator flap [18, 19].

There is no damage to the abdominal wall with these techniques, and the result is stable during pregnancy as the increase of flap volume is proportional to weight gain. Consequently, there is absolutely no effect on future pregnancies or labor. In cases where an implant is used with these flaps, the implant might need to be changed after the pregnancy owing to asymmetry between the two breasts.

26.3.1.3 After Lower Limb Flaps

The transverse myocutaneous gracilis flap, the inferior gluteal artery perforator or superior gluteal artery perforator flaps, the profunda artery perforator flap, the anterolateral thigh flap, and the fascia lata perforator flaps are all different options for breast reconstruction [20, 21]. These flaps also do not affect the abdominal wall and do not have adverse effects on pregnancy.

26.3.1.4 After Fat Grafting

Reconstruction with fat grafting is possible before pregnancy, but the patient must know that the reconstructed breast will increase in volume during pregnancy and get smaller after delivery.

Contraindications for fat grafting, which consist of insufficient excision and a mammographic density classification of ACR 3 or 4, must be respected; as there is a debate about increased risk of recurrence of estrogen-dependent cancer during pregnancy, particularly if the pregnancy is precocious to the diagnosis of cancer [22, 23]. We have also raised the question of recurrence owing to growth factors of fat grafting during pregnancy but there is no literature yet on this issue.

26.4 Immediate Breast Reconstruction during Pregnancy

Owing to the complexities associated with surgical decision making in the treatment of breast cancer in pregnancy, a multidisciplinary team-based approach should occur early in treatment planning [24–26].

As radiotherapy is contraindicated during pregnancy (see also Chap. 16), mastectomy is frequently unavoidable (see also Chap. 12). In this situation, IBR can be performed using an implant or an expander. This allows a simple reconstructive option with fewer complications and less operation time. The volume of the reconstructed breast will change after delivery, so it is preferable to use an expander and change it to the definitive implant around 6–12 months after delivery, when the patient's weight is stable. Here again, it is important to inform the patient about the concerns of MRI studies while the expander is in place, because of its metallic valve.

IBR improves the quality of life of the patient. It is a possibility we can offer to our patients even during pregnancy. However, the reconstruction technique must be simple, without risks of failure, infection, or necrosis; therefore, it is imperative to discuss and explain all the risks to the patient. Free flaps, on the contrary, are avoided during pregnancy owing to high thromboembolic risk and lengthy procedures.

References

1. Partridge AH, Gelber S, Peppercorn J, Ginsburg E, Sampson E, Rosenberg R et al (2008) Fertility and menopausal outcomes in young breast cancer survivors. Clin Breast Cancer 8(1):65–69
2. Alipour S, Eskandari A (2015) Systematic review of effects of pregnancy on breast and abdominal contour after TRAM/DIEP breast reconstruction in breast cancer survivors. Breast Cancer Res Treat 152(1):9–15
3. Nava MB, Bertoldi S, Forti M, Catanuto G, Vergnaghi D, Altomare L et al (2012) Effects of the magnetic resonance field on breast tissue expanders. Aesthetic Plast Surg 36(4):901–907
4. Dibbs R, Culo B, Tandon R, Hilaire HS, Shellock FG, Lau FH (2019) Reconsidering the "MR Unsafe" breast tissue expander with magnetic infusion port: A case report and literature review. Arch Plast Surg 46(4):375–380
5. Gümüş N (2008) Severe influence of early pregnancy on newly reconstructed breast. Breast 17(4):429–431
6. Kwok AC, Simpson AM, Ye X, Tatro E, Agarwal JP (2019) Immediate unilateral breast reconstruction using abdominally based flaps: analysis of 3,310 cases. J Reconstr Microsurg 35(01):74–82
7. Yu P (2016) Breast reconstruction at the MD Anderson cancer center. Gland Surg. 5(4):416–421
8. Vasconez LO, Psillakis J, Johnson-Giebeik R (1983) Breast reconstruction with contralateral rectus abdominis myocutaneous flap. Plast Reconstr Surg 71(5):668–677
9. Drever JM (1983) The lower abdominal transverse rectus abdominis myocutaneous flap for breast reconstruction. Ann Plast Surg 10(3):179–185
10. Zhang A, Dayicioglu D (2018) Outcomes of 270 consecutive deep inferior epigastric perforator flaps for breast reconstruction. Ann Plast Surg 80(6 Suppl):S388–S394
11. Lawrence WT, McDonald HD (1986) Pregnancy after breast reconstruction with a transverse rectus abdominis musculocutaneous flap. Ann Plast Surg 16(4):354–355
12. Chen L, Hartrampf CR Jr, Bennet GK (1993) Successful pregnancies following TRAM flap surgery. Plast Reconstr Surg 91(1):69–71
13. Parodi PC, Osti M, Longhi P, Rampino E, Anania G, Riberti C (2001) Pregnancy and tram-flap breast reconstruction after mastectomy: a case report. Scand J Plast Reconstr Surg Hand Surg 35(2):211–215
14. Patel KM, Basci D, Nahabedian MY (2013) Multiple pregnancies following deep inferior epigastric perforator (DIEP) flap breast reconstruction. J Plast Reconstr Aesthet Surg 66(3):434–436
15. Moshrefi S, Kanchwala S, Momeni A (2018) Should planned/desired pregnancy be considered an absolute contraindication to breast reconstruction with free abdominal Flaps? A retrospective case series and systematic review. J Plast Reconstr Aesthet Surg. 71(9):1295–1300
16. Hammond DC (2007) Latissimus dorsi flap breast reconstruction. Clin Plast Surg 34(1):75–82
17. Hamdi M, Van Landuyt K, Monstrey S, Blondeel P (2004) Pedicled perforator flaps in breast reconstruction: a new concept. Br J Plast Surg 57(6):531–539
18. de Weerd L, Elvenes OP, Strandenes E, Weum S (2003) Autologous breast reconstruction with a free lumbar artery perforator flap. Br J Plast Surg 56(2):180–183
19. Honart JF, Leymarie N, Sarfati B, Alknashnam H, Rem K, Rimareix F et al (2018) Lumbar artery perforator flap for breast reconstruction. Ann Chir Plast Esthet 63(1):25–30
20. Patel NG, Ramakrishnan V (2017) Microsurgical tissue transfer in breast reconstruction. Clin Plast Surg 44(2):345–359
21. Lefèvre M, Sarfati B, Honart JF, Alkashnam H, Rimareix F, Leymarie N et al (2017) Fasciae latae

perforator flap for breast reconstruction: An attractive alternative in case of DIEP contraindication. Ann Chir Plast Esthet 62(1):97–103
22. Kranick JA, Schaefer C, Rowell S, Desai M, Petrek JA, Hiatt RA et al (2010) Is pregnancy after breast cancer safe? Breast J 16(4):404–411
23. Lambertini M, Kroman N, Ameye L, Cordoba O, Pinto A, Benedetti G et al (2017) Long-term safety of pregnancy following breast cancer according to estrogen receptor status. J Natl Cancer Inst 110(4):426–429
24. Chirappapha P, Thaweepworadej P, Ngamphaiboon N, Sukprasert M, Sukarayothin T, Leesombatpaiboon M (2017) Breast reconstruction in pregnancy: a case report of multidisciplinary team approach in immediate autologous flap reconstruction for pregnancy-associated breast cancer. Clin Case Rep 5(9):1450
25. Caragacianu DL, Mayer EL, Chun YS, Caterson S, Bellon JR, Wong JS et al (2016) Immediate breast reconstruction following mastectomy in pregnant women with breast cancer. J Surg Oncol 114(2):140–143
26. Lohsiriwat V, Peccatori FA, Martella S, Azim HA Jr, Sarno MA, Galimberti V et al (2013) Immediate breast reconstruction with expander in pregnant breast cancer patients. Breast 22(5):657–660

27 Pregnancy and Lactation: Risk or Protective Factors for Breast Cancer?

Bruna Migliavacca Zucchetti, Fedro A. Peccatori, and Giovanni Codacci-Pisanelli

Abstract

Pregnancy and lactation represent the most effective protective elements against breast cancer; counter-intuitively breast cancer incidence shows a small but noticeable increase up to 5 years after delivery. The cumulative effect is however favourable and women show a reduction in breast cancer risk which is proportional to the total duration of lactation and to the number of full-term pregnancies.

Keywords

Breast cancer · Breastfeeding · Cancer risk · Pregnancy · Protective factors

B. Migliavacca Zucchetti
European Institute of Oncology IRCCS, Milan, Italy

Medical Oncology Department, Hospital Sirio-Libanes, Sao Paulo, Brazil
e-mail: brunazucchetti@icloud.com

F. A. Peccatori (✉)
Gynecologic Oncology Program, European Institute of Oncology IRCCS, Milan, Italy
e-mail: fedro.peccatori@ieo.it

G. Codacci-Pisanelli
Department of Medical and Surgical Sciences and Biotechnology, Sapienza University of Rome, Rome, Italy
e-mail: Givonni.codacci-pisanelli@uniroma1.it

27.1 Overview

The increasing incidence of breast cancer has been partially attributed to changing patterns of lifestyle including reproductive factors. Later marriage, fewer pregnancies, and shorter duration of breastfeeding are known to influence the risk of breast cancer.

27.2 Pregnancy and Breastfeeding as Protective Factors for Breast Cancer

27.2.1 Pregnancy

The protective role of pregnancy on breast cancer development was established 50 years ago, with the publication of a paper discussing the protective role of parity on breast cancer and the varying effect of age at first pregnancy [1]. This observation; however, was not novel: the effect of pregnancy (and breastfeeding) on breast cancer was suggested by Ramazzini in the 18th century. He was the first to report that breast cancer was particularly frequent in convents. This information is still relevant because even today nuns have a higher mortality risk from breast cancer [2]. Later studies established that protection is increased by an increase in the number of pregnancies and time of breastfeeding. The

S. Alipour, R. Omranipour (eds.), *Diseases of the Breast during Pregnancy and Lactation*, Advances in Experimental Medicine and Biology 1252,
https://doi.org/10.1007/978-3-030-41596-9_27

mechanism behind this protective effect is not clear, and a role for mammary stem cells has been suggested [3].

Parous women have a reduced mortality for all cancers and for breast cancer specifically [4], and this effect is particularly evident for full-term pregnancy at an early age. Recently, reproductive behaviors have been studied in relation to different subtypes of breast cancer, and it has been shown that parity reduces the risk of luminal breast cancer but not that of HER2-positive or triple-negative breast cancer (TNBC). On the contrary, age at first pregnancy has a different effect in that old age at first pregnancy increases the risk of luminal breast cancer but not that of HER2-positive breast cancer or TNBC [5]. The effect of parity in *BRCA*-mutated women appears to be less consistent, with differences according to the gene involved [6].

27.2.2 Breastfeeding

Several studies have shown the protective effect of breastfeeding on breast cancer, and it has been suggested that incidence of breast cancer in developed countries could be impressively reduced if childbearing behavior (parity, duration of breastfeeding) of women in developed countries was similar to that of women in developing countries [7]. A detailed analysis of molecular subtypes of breast cancer showed that not all types are reduced to the same extent and that breastfeeding reduces the risk of luminal breast cancer and TNBC [5, 8, 9] but not that of HER2-positive cancer [5]. Breastfeeding also plays a protective role for women carrying a *BRCA* mutation, but this has been shown in *BRCA-1*–mutated cases, and not yet in BRCA-2-positive patients [10].

27.3 Pregnancy as a Risk Factor for Breast Cancer

Most breast cancers are sporadic, and the etiology of the disease is not well understood; although it is now clear that some external factors such as the pattern of reproductive behavior can modulate the risk. Many epidemiological studies have indicated that the long-term protective effect of pregnancy on the risk of breast cancer is preceded by a short-term adverse effect, with an increase in breast cancer risk for the first 5 years after delivery as compared with other periods afterward [11] (see also Chap. 23). As reported by Bruzzi et al. [12], the relative risk for breast cancer in women who had given birth to a child during the previous 3 years was 2.66 compared with women whose last childbirth had occurred 10 years ago, or more. The risk slowly decreased thereafter.

Along with these data, a study on Swedish women published by Lambe et al. [13] comprising approximately 75,000 patients reported that primiparous women were at a higher risk of breast cancer than nulliparous women for up to 15 years after childbirth and at a lower risk thereafter. The excess of risk was most pronounced among those who were older at the time of first delivery (odds ratio, 1.26; 95% CI, 1.10–1.44; 5 years after delivery among women who were 35 years old at the first delivery). Some years later, the same group reported a case-control study [14] comparing primiparous with nulliparous women and showed that a transient increase in maternal breast cancer risk peaked 5 years following delivery (odds ratio, 1.49; 95% CI, 1.01–2.20) and leveled off 15 years postpartum. Women who had given birth to two children had a transient increase in risk that was lower at its peak than that of primiparous women, occurring about 3 years following the second delivery. This time window of 5 years postpartum may define the latent period required for pregnancy hormones to promote the progression of normal breast cells toward early stages of malignant transformation.

Despite this slight increase in breast cancer risk after delivery, an extensive body of epidemiological studies has proved a strong and lifelong protective effect of early full-term pregnancy [1, 11, 15, 16]. This protective effect is at least 50% for a pregnancy occurring before the age of 20 years, meaning that women that had an early pregnancy develop 50% fewer cancers compared with nulliparous women. On the other hand, there is an overall increase in the risk of breast tumors for first pregnancies after the age of 35 years [17].

Regarding breast cancer subtype, a meta-analysis [5] of 15 studies including 21,941 cancer patients and 864,177 controls showed that parity was associated with a 25% risk reduction of developing a luminal subtype (OR, 0.75; 95% CI, 0.70–0.81; $P < .001$), but advanced age at first birth was associated with an increased risk of developing a luminal subtype (OR, 1.15; 95% CI, 1.00–1.32; $P = .05$). The data above show that there is a nonlinear relationship between breast cancer incidence and time interval since delivery.

References

1. MacMahon B, Cole P, Lin TM, Lowe CR, Mirra AP, Ravnihar B et al (1970) Age at first birth and breast cancer risk. Bull World Health Organ 43(2):209–221
2. Britt K, Short R (2012) The plight of nuns: hazards of nulliparity. Lancet 379(9834):2322–2323
3. Dall G, Risbridger G, Britt K (2017) Mammary stem cells and parity-induced breast cancer protection-new insights. J Steroid Biochem Mol Biol 170:54–60
4. Merritt MA, Riboli E, Murphy N, Kadi M, Tjønneland A, Olsen A et al (2015) Reproductive factors and risk of mortality in the European Prospective Investigation into Cancer and Nutrition; a cohort study. BMC Med 13(1):252
5. Lambertini M, Santoro L, Del Mastro L, Nguyen B, Livraghi L, Ugolini D et al (2016) Reproductive behaviors and risk of developing breast cancer according to tumor subtype: a systematic review and meta-analysis of epidemiological studies. Can Treat Rev 49:65–76
6. Terry MB, Liao Y, Kast K, Antoniou AC, McDonald JA, Mooij TM et al (2019) The influence of number and timing of pregnancies on breast cancer risk for women with BRCA1 or BRCA2 mutations. JNCI Cancer Spectr 2(4):pky078
7. Collaborative Group on Hormonal Factors in Breast Cancer Breast cancer and breastfeeding: collaborative reanalysis of individual data from 47 epidemiological studies in 30 countries, including 50302 women with breast cancer and 96973 women without the disease. Lancet 360(9328):187–195
8. Giudici F, Scaggiante B, Scomersi S, Bortul M, Tonutti M, Zanconati F (2017) Breastfeeding: a reproductive factor able to reduce the risk of luminal B breast cancer in premenopausal white women. Eur J Cancer Prev 26(3):217–224
9. Sisti JS, Collins LC, Beck AH, Tamimi RM, Rosner BA, Eliassen AH (2016) Reproductive risk factors in relation to molecular subtypes of breast cancer: results from the nurses' health studies. Int J Cancer 138(10):2346–2356
10. Kotsopoulos J, Lubinski J, Salmena L, Lynch HT, Kim-Sing C, Foulkes WD et al (2012) Breastfeeding and the risk of breast cancer in BRCA1 and BRCA2 mutation carriers. Breast Cancer Res 14(2):R42
11. Subramani R, Lakshmanaswamy R (2017) Pregnancy and breast cancer. Prog Mol Biol Transl Sci 151:81–111
12. Bruzzi P, Negri E, La Vecchia C, Decarli A, Palli D, Parazzini F et al (1988) Short term increase in risk of breast cancer after full term pregnancy. BM. 297(6656):1096–1098
13. Lambe M, Hsieh CC, Trichopoulos D, Ekbom A, Pavia M, Adami HO (1994) Transient increase in the risk of breast cancer after giving birth. N Engl J Med 331(1):5–9
14. Liu Q, Wuu J, Lambe M, Hsieh SF, Ekbom A, Hsieh CC (2002) Transient increase in breast cancer risk after giving birth: postpartum period with the highest risk (Sweden). Cancer Causes Control 13(4):299–305
15. Albrektsen G, Heuch I, Hansen S, Kvåle G (2005) Breast cancer risk by age at birth, time since birth and time intervals between births: exploring interaction effects. Br J Cancer 92(1):167–175
16. Rojas K, Stuckey A (2016) Breast cancer epidemiology and risk factors. Clin Obstet Gynecol 59(4):651–672
17. Meier-Abt F, Bentires-Alj M (2014) How pregnancy at early age protects against breast cancer. Trends Mol Med 20(3):143–153

Psychological Aspects of Pregnancy and Lactation in Patients with Breast Cancer

28

Ali-Akbar Nejatisafa, Flavia Faccio, and Ronak Nalini

Abstract

Young breast cancer patients face numerous challenges during the cancer trajectory. As in the last decade, women tend to delay pregnancies to a later time in life, and clinicians are often faced with young breast cancer patients who want to start a family or complete it. Becoming a mother is a delicate developmental process in which the woman redefines and restructures her identity as she gets prepared for her new role and responsibilities. When there is a history of cancer or cancer diagnosis is communicated during the pregnancy, fears, worries, and concerns emerge and specific support may be necessary. Follow-ups during the post-partum period are also recommended as lactation issues should not be overlooked. In this chapter, we analyze the psychological aspects of cancer survivors and women with pregnancy-associated breast cancer, and the management of these issues.

Keywords

Breast cancer · Breastfeeding · Mental Health · Pregnancy · Psychological Aspects

28.1 Overview

The first known reference to the definition of breast cancer is excerpted from an Egyptian papyrus written 3500 years ago. It seems the significance of breasts for women magnifies their fear towards this disease. For example, it is referenced that 2500 years ago Atossa, the wife of Darius I, the king of Persia, hid her breast mass for some time. She was perceivably worried about dying and dysmorphic looks, about the loss of sexual attraction and her enticement. These fears and worries resemble the current concerns of our patients, 2500 years later [1].

Breast cancer is the most common cancer among women of reproductive age and the first, or second most common pregnancy-associated malignancy (after cervical cancer). Pregnancy-associated breast cancer (PABC) is described as breast cancer diagnosed during pregnancy or

A.-A. Nejatisafa (✉)
Department of Psychiatry, Division of Psychosomatic Medicine, Psychosomatic Research Center, Tehran University of Medical Sciences, Tehran, Iran
e-mail: nejatisafa@tums.ac.ir

F. Faccio
Applied Research Division for Cognitive and Psychological Science, European Institute of Oncology IRCCS, Milan, Italy

R. Nalini
Department of Internal Medicine, Division of Hematology-Oncology, Kermanshah University of Medical Sciences, Kermanshah, Iran

S. Alipour, R. Omranipour (eds.), *Diseases of the Breast during Pregnancy and Lactation*, Advances in Experimental Medicine and Biology 1252,
https://doi.org/10.1007/978-3-030-41596-9_28

within one year postpartum and it is estimated to account for up to 3% of all breast cancers [2, 3]. Following the trend of the last decade, women tend to postpone motherhood to a later age, therefore the incidence of PABC may increase over time [4, 5] (see also Chap. 9). The natural changes in the breast during pregnancy limit usefulness of mammography and breast examination (see also Chaps. 2 and 3). This may play a role in delaying the diagnosis for 1–2 months [6, 7]. Current evidence does not support the termination of pregnancy in patients with PABC [8, 9].

Young breast cancer patients face numerous challenges during the cancer trajectory. As in the last decade, women tend to delay pregnancies to a later time in life, and clinicians are often faced with young breast cancer patients who want to start a family or complete it. Becoming a mother is a delicate developmental process in which the woman redefines and restructures her identity as she gets prepared for her new role and responsibilities. When there is a history of cancer or cancer diagnosis is communicated during the pregnancy, fears, worries, and concerns emerge and specific support may be necessary. Follow-ups during the postpartum period are also recommended as lactation issues should not be overlooked.

28.2 The Psychological Impact of Breast Cancer on Pregnant Mothers

Breast cancer is the most widely scrutinized cancer in women with respect to its psychological impact. This is partly due to the fact that it is the most common form of cancer in women in the world. On the other hand, we live in a "body culture", where the emphasis is laid on physical appearance, and this disease involves a part of the body that is associated with attractiveness, femininity, sexuality, maternity, and consequently with body image, self-image, and self-esteem. Psychosocial impact of breast cancer is divided into three broad areas: psychological discomfort (anxiety, depression, and anger), changes in life patterns (related to physical discomfort and disfigurement, fertility, marital disruption, and altered activity level), and fears and concerns (related to body mutilation, recurrence of the disease, uncertainty about the future, possibility of death, and existential issues) [10].

Cancer during pregnancy places the mother in a challenging and delicate position. A new life is developing inside her, it feels like fulfillment and a gift; but simultaneously her own life is in danger. Moreover, it is a difficult situation for health professionals, as two people are involved: the mother and the fetus. Cancer treatment has to consider the needs of the mother while always being mindful of the child.

When a breast cancer diagnosis is communicated to a pregnant woman, a period of crisis begins, at least in the short term. In the first step of the crisis, the patient will listen to the doctor describing the diagnosis and treatments and in the second step, she is asked to adapt to her new life with cancer.

28.2.1 Breaking Bad News

An operational definition of the bad news has been presented by Robert Buckman: "any news that adversely and seriously affects an individual's view of his or her future" [11]. The way in which bad news is disclosed to the patients could affect their perception of the disease, their psychological coping with illness, satisfaction with care and extent of hope [12–18]. Currently, it is vastly approved that being informed about the disease is the patients' legal and ethical right and hiding the information about the disease may lead to distrust towards the physician [11, 12, 19, 20].

Breaking bad news is a difficult task for clinicians, patients, and families because it is accompanied by distress on both sides. To overcome this complex task, clinicians should have enough mastery in advanced communication skills. One framework that clinicians may find helpful is that developed by Baile and Buckman [12, 21]. This guideline includes the major points to be consid-

Table 28.1 SPIKES- the six-step protocol for delivering bad news

Steps		Component	Measures
1	S	*SETTING UP* the Interview	Maximize privacy, avoid interruption, respect confidentiality and provide support
2	P	Assessing the Patient's *PERCEPTION*	Demonstrate how much the patient knows, how serious she/he thinks the illness is, and how much it will affect the future?
3	I	Obtaining the Patient's *INVITATION*	Declare how much the patient wants to know.
4	K	Giving KNOWLEDGE and Information to the Patient	Keep in mind the objectives for the consultation: diagnosis, treatment plan, prognosis, and support. Listen to the patient's agenda.
5	E	Addressing the Patient's *EMOTIONS* with empathic responses	Respond to the Patient's Feelings. Responses can vary from silence to distress, denial or anger. Observe the patient and give her time. Empathetic reflection.
6	S	*STRATEGY* and Summary	Make a Plan or Strategy and Explain it. Identify coping strategies of the patient and reinforce them. Tell them what happens next.

Adapted from Baile and Buckman (2000, pp. 305–8) [21]

ered when giving bad news to patients and/or their relatives (Table 28.1).

In some advanced cases women with PABC are presented with two bad news: the cancer diagnosis, and that they have to decide whether to continue or not their pregnancy. Both termination and continuation of pregnancy are risky decisions that should be shared between the patient, family members, and the clinician. The clinician should consider the patient's and her family personal values like religion, the meaning of family and motherhood and their personal relationship. The following topics should be discussed in the consultation sessions for the process of breaking bad news and shared decision making:

- Potential treatment options and their possible complications during cancer therapy (see also Chaps. 12–16).
- The impact of pregnancy on the natural course of breast cancer (see also Chap. 11).
- The impact of breast cancer and its treatment on pregnancy, birth and the child's development (see also Chap. 21).
- Termination of pregnancy has no further benefit to cancer therapy and does not increase survival per se (see also Chap. 21).
- The situation of giving birth where the father may become a single parent and the children may lose their mother after an expected time (see also Chap. 11).

28.2.2 Adaptation

Adjustment to cancer is facing up to the problems that arise from the disease, which include changes in family and work situations, pain, and disability due to cancer and its treatments as well as living as cancer survivors. Among other things, patients with PABC are also asked to be ready for motherhood, which by itself brings personal and developmental changes that can manifest as psychological challenges during the process of developing a relationship with their child.

Adaptation to cancer is a dynamic process. The initial response usually is denial or despair, followed by dysphoria that may be associated with anxiety, depression, insomnia, poor concentration; and would usually terminate to acceptance and resuming usual activities over months. Patients vary widely in how they cope with cancer over time. Thus, it is very important to recognize early on factors that predict poor/good adaptation, in order to facilitate recognition of vulnerable patients [12, 22].

Adaptation to cancer depends on three groups of factors: (1) Community factors, which include society attitude and perception of cancer and its treatments; (2) Patient's factors that include the individual (psychological profile and abilities), interpersonal (relational issues and support from others), and socioeconomic (availability of resources for support) factors; (3) Cancer factors, which include the stage of cancer at diagnosis, type, and severity of symptoms including pain, the extent of treatments and adverse effects, the effect of cancer on body image and the possibility of rehabilitation [12, 22].

After breast surgery, some mothers may avoid physical contact with their children to prevent them from noticing their body transformations. Other mothers may be advised against physical efforts, such as picking up or holding their children. These situations can cause negative thoughts and feelings in breast cancer patients [12, 23]. Predictors of poor adaptation are social isolation, low socioeconomic status, drug abuse, prior psychiatric history, prior experience with cancer, recent losses/bereavement, low flexibility in the use of coping strategies, pessimistic outlook and no value system [24].

The findings of a small group of qualitative studies on PABC show that the majority of patients report that the situation has both a positive and negative impact on their life, if we consider the trajectory of patients that experience disease and pregnancy from the stage of diagnosis through survivorship, both as individuals and as families [12, 25–27]. Emotional resilience plays an important role in coping with this difficult situation. Patients feel that motherhood increases their fighting spirit, providing a reason to battle the illness and live, and through this belief, the motivation for vitality and survival increases. Many women believe that being a cancer survivor makes them meaningful and powerful individuals and better parents [28].

28.2.3 Ethical Issues

The literature about ethical issues in patients with PABC usually uses the term "fetal-maternal conflict" which pertains to the conflicting situation for a mother to choose between her own versus the fetus' benefits. As the majority of mothers prefer the beneficence of their children, indeed, this ethical dilemma or conflict belongs to the clinician who has to balance between two sides of ethical obligation: mother versus fetus [29].

The most important question in this context is whether we should consider the fetus as a patient? Various religions have different approaches to this issue. Some have proposed that the fetus becomes a person at a specific time during the pregnancy. However, this approach does not solve the problem because agreement on personhood on the basis of age of the embryo is elusive. Another approach to the personhood of the fetus is to consider the chance of its viability outside the uterus. However viability is the function of medical care and technological capacity. As a result, there is no universal worldwide gestational age that guarantees viability [29, 30].

The ethical models and principles help clinicians answer these issues in our pluralistic society. These models often use a diversity of approaches that are applicable to clinical situations. For example, according to Beauchamp and Childress' (2001) approach, two specific principles should be considered in this ethical situation:

Respect for Autonomy The mother should have the right to choose different options freely, based on her values and beliefs. On the other hand, there is no autonomy right for the fetus.

Beneficence This requires the clinician to evaluate correctly the different treatment options and to implement those which offer the patient the greatest balance of benefits over risks. There is also a beneficence obligation of the clinician towards the fetus.

A framework for medical ethics, incorporating these two ethical principles and the concept of the fetus as a patient, would allow developing a comprehensive guideline for clinical judgment and the process of shared decision-making by the clinician, the patient, and her family regarding the management of PABC [30].

28.2.4 Psycho-Oncological Care

Caring for a woman suffering from PABC is a challenge for the following reasons: dealing with a patient with contradictory emotions, on the one hand, fear of death and hopelessness and on the other the joy of motherhood; considerable concern of the patient about the future of her own and her child. A multidisciplinary approach to the patient is the sine qua non for medical care in this clinical situation. All of the team members should pay special attention to the sense of security of the patient. There is a regular dialogue between the patient and the experts for exchanging opinions and information in this context. All of the team members should have general counseling and communication skills to deal with this complex situation. More specific psycho-oncologic care that is provided by psychiatrist or clinical psychologist should apply for more complex problems of the patients and her family in different stages of the cancer journey. Various psychotherapeutic interventions are available for helping these patients including education, solution-focused therapy, hypnosis, cognitive-behavior therapy, mindfulness-based cognitive therapy, existential psychotherapy, group therapy and finally pharmacotherapy. Each method should be tailored to patients with PABC and their family.

28.3 Pregnancy in Breast Cancer Survivors: Mental Health Issues

More than 25% of breast cancer cases occur in premenopausal women [31]. Many young breast cancer survivors have major concerns when faced with this diagnosis, especially those who have not yet completed their families. On the one hand, they may have a set of concerns about the safety of pregnancy with regard to cancer recurrence and mortality [32]. Most of the studies have failed to find an impact of pregnancy on the outcome and mortality of breast cancer survivors and in some studies, it is considered a protective factor [33–36]. Patients are typically advised to wait for 2–3 years after treatment ends before conceiving, especially those with estrogen receptor negative cancers. Nonetheless, the evidence for this waiting time is weak [34, 37]. In spite of these optimistic results, several investigators questioned the quality and power of the available research in terms of sample size, limited data source and selection and recall bias [38, 39] (see also Chap. 23).

One of the pivotal biases in these studies is the selection bias termed the 'healthy mother bias,' which is a kind of selection bias and presumes that breast cancer survivors who decide to become pregnant are a self-selected, healthier group based on their prognosis [40]. However, a nested case-control study conducted on the database of participants in the Women's Healthy Eating and Living (WHEL) study did not find evidence of a healthy mother bias. However, mental health, although not traditionally discussed in the context of this bias, was marginally significantly better among women who had children, suggesting that mental health is an important component of overall health and that its relationship to post-cancer pregnancy should be evaluated in future research [41].

This statement is in accordance with previous studies showing that the presence of medical risk is generally associated with psychological suffering in pregnant women and to a poorer representation of the child-to-be, as well as themselves as mothers [42]. A recent meta-analysis of 43 studies compared the prevalence of negative feelings and mood disorders in two populations, cancer survivors and healthy controls. The prevalence of depression in cancer survivors was 11.6%, compared to 10.2% in healthy control patients. The prevalence of anxiety was higher: 17.9% in cancer survivors, compared to 13.9% in healthy controls [43]. As negative mood states such as depression and anxiety and higher psychological burden are frequently encountered in young breast cancer survivors, it is vital to see whether these have any impact on pregnant patients' mental well-being when they become mothers.

Another fundamental aspect of this population is partner support as no support system can increase the risk for poor pregnancy outcomes and mental health issues. Previous studies have shown that low levels of partner support were associated with higher levels of antenatal anxiety and depression during pregnancy [44] and that low levels of shared communication, of relationship satisfaction, and of emotional and instrumental support increased the risk of developing depression [45].

A systematic review of studies about childbearing attitude and decisions of young breast cancer survivors showed that childbearing after breast cancer is an important issue for survivors. While on one hand, some women welcomed the idea of becoming a mother and bearing (more) children, some women were against motherhood. The third group of women were hesitant and felt that they could not make the decision yet [46]. Ambivalence may be heightened by fears of the woman regarding her own prognosis in terms of recurrence or survival, or may be related to fear that their child might have a birth defect due to previous oncological treatment or that the child will be born with a greater susceptibility to cancer. Ambivalence can also be partially explained by the high levels of unmet informational needs for young breast cancer survivors regarding fertility and menopause [47]. In addition, a kind of disharmony may exist between young survivor's worry about childbearing ability and how this concern is approached by the clinicians [47, 48].

To sum up it seems that, according to the current evidence, the younger is the cancer survivor the more are the psychosocial needs; especially those concerning physical changes due to treatment, and associated reproductive complications [49]. Motherhood seems to be a relevant matter for breast cancer survivors; therefore, it is essential to develop and improve educational tools and incorporate psychosocial support services for young breast cancer survivors.

28.4 Psychological Aspects of Lactation in Breast Cancer Survivors

The majority of research shows that breastfeeding is accompanied with important health benefits for women, including decreased risk for metabolic syndromes and certain reproductive cancers, and it is considered a protective factor for breast cancer incidence [31, 50] (see also Chap. 27). However, there is also evidence of substantial costs associated with breastfeeding, such as spending several hours per day to breastfeed or not feeling at ease with breastfeeding in public, which inevitably forces mothers to stay at home. Even though in recent decades breastfeeding promotion has become a key element in health programs, it is important to be aware that if changes in the mothers' personal and social roles occur, breastfeeding might not be the best choice [51]. Specifically, mothers who survive breast cancer encounter a unique situation with physical and emotional aspects that might affect their decision and/or ability to breastfeed.

Although breastfeeding after breast cancer treatment is possible, lactation may be limited [25, 52] (see also Chap. 27). Only a few studies have investigated patient's worries, concerns and fears related to breastfeeding and they suggest that breast cancer survivors experience two kinds of barriers with regards to breastfeeding: practical barriers and emotional barriers. Practical barriers include the impact of the anatomical and physiological changes of the breast on the ability to breastfeed and the cost of breastfeeding on patients' social and occupational life [52, 53]. A qualitative study conducted by Ives et al. sustains that some women with PABC experienced breastfeeding as an additional challenge to the management of cancer-related issues [25]. They also felt that they did not belong to the maternity ward as they perceived more difficulties adjusting to feeding schedules compared to women sharing the ward space. For women with pregnancy-associated breast cancer, the transition to bottle-feeding can be perceived as forced as occurring before time [25, 54]. The delicate decision women make between delaying chemotherapy treatment for some weeks in order to continue breastfeeding or stopping it to start the treatment plan is another aspect that should also be considered during counselling.

Emotional barriers include worrying about the possible difficulty of recognizing breast cancer recurrence due to physiologically swollen breasts during pregnancy; worrying about transfer of the disease to the baby through their milk; conflicting thoughts about the failure to comply with the social expectation to breastfeed; and concern about showing a scarred breast to the infant or to others in case of public breastfeeding. Women who had undergone surgery in the past also reported feelings of isolation, as their "different" breasts were noticed by staff, and feared that they might not be able to provide enough milk for their newborn from one side, thus harbouring feelings of inadequacy, of not being "good enough" mothers and of being betrayed by their own body [6, 55]. These worries might also be manifested by women with PABC as a perception of decreased milk production is reported by women who had undergone chemotherapy treatment [54]. Figure 28.1 summarizes these barriers to breastfeeding.

In addition, studies point out that breastfeeding not only protects the mother against psychosocial stressors during the postpartum period but also increases the quality and intensity of the mother-infant relationship [51]. For this reason, healthcare professionals often encourage breastfeeding in breast cancer survivors [6, 55].

In Western cultures, breastfeeding is widely envisioned as one of the main means through which women start to develop a bond with their child [56]. However, this is only one of the aspects of the development of the mother-child relationship, as this complex and delicate process is actually constructed during different developmental phases (i.e. from the third trimester of pregnancy up to the first 3 years of the child) and through various emotional aspects (i.e. from attunement to empathy). This said, the inability to breastfeed is perceived by some cancer survivors as a barrier towards developing a positive relationship with their child [55, 56].

Following the evidence presented above, it seems that during the delicate transition of breast cancer patients into motherhood, the breast can be considered "good", due to its nurturing role, or "evil", as it is taking something away from the experience of being a mother. Therefore, concerns about safety of breastfeeding, uncertainty about nursing from one side only and misinformation are only some of the issues that can increase worries and anxiety in women with past or current breast cancer. These issues and the related psychological aspects can be managed by providing appropriate counselling to avoid nega-

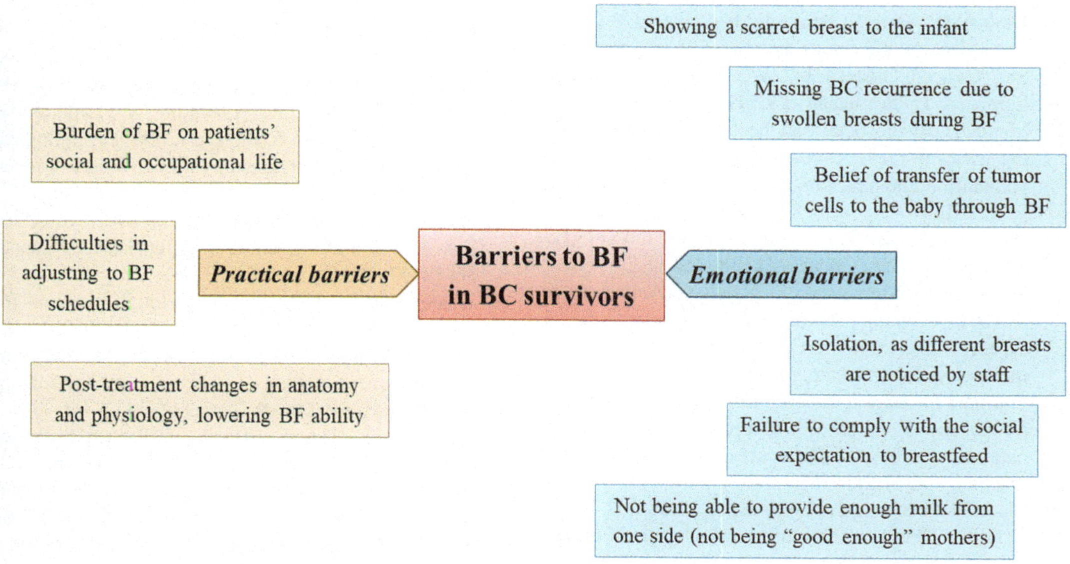

Fig. 28.1 Practical and emotional barriers to breastfeeding in breast cancer survivors. *BC* breast cancer, *BF* breastfeeding

tive consequences on the mother's and infant's well-being.

References

1. Galgut C (2010) The psychological impact of breast cancer: a psychologist insight as a patient, 1st edn. Radcliffe, Oxford
2. Liberman L, Giess CA, Dershaw DD, Deutch BM, Petreck JA (1994) Imaging of pregnancy-associated breast cancer. Radiology 191:245–248
3. Ishida T, Yokoe T, Kasumi F, Sakamoto G, Makita M, Tominaga T et al (1992) Clinicopathologic characteristics and prognosis of breast cancer patients associated with pregnancy and lactation: analysis of case-control study in Japan. Jpn J Cancer Res 83(11):1143–1149
4. Smith LH, Dalrymple JL, Leiserowitz GS, Danielsen B, Gilbert WM (2001) Obstetrical deliveries associated with maternal malignancy in California, 1992 through 1997. Am J Obstet Gynecol 184:1504–1512
5. Goldhirsch A, Gelber RD (2004) Life with consequences of breast cancer: pregnancy during and after endocrine therapies. Breast 13:443–445
6. Helewa M, Levesque P, Provencher D, Lea RH, Rosolowich V, Shapiro HM (2002) Breast cancer, pregnancy,and breastfeeding. J Obstet Gynaecol Cancer 24:164–180
7. Woo JC, Yu T, Hurd TC (2003) Breast cancer in pregnancy: a literature review. Arch Surg 138:91–98
8. Berry DL, Theriault RL, Holmes FA, Parisi VM, Booser DJ, Singletary SE et al (1999) Management of breast cancer during pregnancy using a standardized protocol. J Clin Oncol 17:855–861
9. King RM, Welch JS, Martin JK Jr, Coulam CB (1985) Carcinoma of the breast associated with pregnancy. Surg Gynecol Obstet 160:228–232
10. Lemieux J, Bordeleau LJ, Goodwin PJ (2007) Medical, psychosocial and health-related quality of life issues in breast cancer survivors. In: Cancer survivorship, today and tomorrow. Springer, New York, pp 122–144
11. Buckman RA (2005) Breaking bad news: the SPIKES strategy. Commun Oncol 2(2):138–142
12. Yaghmaie M, Ahmadvand M, Nejatisafa A, Pashaiefar H (2017) Genetic, hematologic and psychological aspects of leukemia. In: Cancer genetics and psychotherapy. Springer, Cham, pp 667–755
13. Fujimori M, Akechi T, Morita T, Inagaki M, Akizuki N, Sakano Y et al (2007) Preferences of cancer patients regarding the disclosure of bad news. Psycho-Oncology 16(6):573–581
14. Meredith C, Symonds P, Webster L, Lamont D, Pyper E, Gillis CR et al (1996) Information needs of cancer patients in west Scotland: cross sectional survey of patients' views. BMJ 313(7059):724–726
15. Valizadeh L, Zamanzadeh V, Rahmani A, Howard F, Nikanfar AR, Ferguson C (2012) Cancer disclosure: experiences of Iranian cancer patients. Nurs Health Sci 14(2):250–256
16. Yun YH, Lee CG, S-y K, Lee S-w, Heo DS, Kim JS et al (2004) The attitudes of cancer patients and their families toward the disclosure of terminal illness. J Clin Oncol 22(2):307–314
17. Parker PA, Baile WF, de Moor C, Lenzi R, Kudelka AP, Cohen L (2001) Breaking bad news about cancer: patients' preferences for communication. J Clin Oncol 19(7):2049–2056
18. Butow PN, Brown RF, Cogar S, Tattersall MH, Dunn SM (2002) Oncologists' reactions to cancer patients' verbal cues. Psychooncology 11(1):47–58
19. Mueller PS (2002) Breaking bad news to patients: the SPIKES approach can make this difficult task easier. Postgrad Med 112(3):15–18
20. Arbabi M, Roozdar A, Taher M, Shirzad S, Arjmand M, Mohammadi MR et al (2010) How to break bad news: physicians' and nurses' attitudes. Iran J Psychiatry 5(4):128
21. Baile WF, Buckman R, Lenzi R, Glober G, Beale EA, Kudelka AP (2000) SPIKES – a six-step protocol for delivering bad news: application to the patient with cancer. Oncologist 5(4):302–311
22. Holland JC, Rowland JH (1990) Handbook of psychooncology: psychological care of the patient with cancer. Oxford University Press, Oxford
23. Tavares R, Brandão T, Matos PM (2018) Mothers with breast cancer: a mixed-method systematic review on the impact on the parent-child relationship. Psychooncology 27(2):367–375
24. Spencer S, Carver CS, Price A (1998) Psychological and social factors in adaptation. Psychooncology 2:211–222
25. Ives AD, Musiello T, Saunders C (2012) The experience of pregnancy and early motherhood in women diagnosed with gestational breast cancer. Psychooncology 21(7):754–761
26. Schover LR (2000) Psychosocial issues associated with cancer in pregnancy. Semin Oncol 27(6):699–703
27. Zanetti-Dällenbach R, Tschudin S, Lapaire O, Holzgreve W, Wight E, Bitzer J (2006) Psychological management of pregnancy-related breast cancer. Breast 15:S53–S59
28. Schover LR, Rybicki LA, Martin BA, Bringelsen KA (1999) Having children after cancer. Cancer 86(4):697–709
29. Waalen J (1991) Pregnancy poses tough questions for cancer treatment. J Natl Cancer Inst 83:900–901
30. Chervenak FA, McCullough LB (1985) Perinatal ethics: a practical method of analysis of obligations to mother and fetus. Obstet Gynecol 66:442–446
31. Collaborative Group on Hormonal Factors in Breast Cancer (2002) Breast cancer and breastfeeding: collaborative reanalysis of individual data from 47 epidemiological studies in 30 countries, including 50302 women with breast cancer and 96973 women without the disease. Lancet 360(9328):187–195
32. Partridge AH, Gelber S, Peppercorn J, Sampson E, Knudsen K, Laufer M et al (2004) Web-based survey

of fertility issues in young women with breast cancer. J Clin Oncol 22(20):4174–4183
33. Gelber S, Coates AS, Goldhirsch A, Castiglione-Gertsch M, Marini G, Lindtner J et al (2001) Effect of pregnancy on overall survival after the diagnosis of early-stage breast cancer. J Clin Oncol 19(6):1671–1675
34. Ives A, Saunders C, Bulsara M, Semmens J (2007) Pregnancy after breast cancer: population based study. Br Med J 334(7586):194
35. Velentgas P, Daling JR, Malone KE, Weiss NS, Williams MA, Self SG et al (1999) Pregnancy after breast carcinoma: outcomes and influence on mortality. Cancer 85(11):2424–2432
36. Kroman N, Jensen MB, Wohlfahrt J, Ejlertsen B (2008) Pregnancy after treatment of breast cancer- a population- based study on behalf of Danish Breast Cancer Cooperative Group. Acta Oncol 47(4):545–549
37. Clark RM, Chua T (1989) Breast cancer and pregnancy: the ultimate challenge. Clin Oncol (R Coll Radiol) 1(1):11–18
38. Upponi SS, Ahmad F, Whitaker IS, Purushotham AD (2003) Pregnancy after breast cancer. Eur J Cancer 39(6):736–741
39. Barthelmes L, Davidson LA, Gaffney C, Gateley CA (2005) Pregnancy and breast cancer. Br Med J 330(7504):1375–1378
40. Sankila R, Heinavaara S, Hakulinen T (1994) Survival of breast cancer patients after subsequent term pregnancy: healthy mother effect. Am J Obstet Gynecol 170(3):818–823
41. Gorman JR, Roesch SC, Parker BA, Madlensky L, Saquib N, Newman VA et al (2010) Physical and mental health correlates of pregnancy following breast cancer. Psychooncology 19:517–524
42. Brandon AR (2007) Maternal and fetal representations, dimensions of personality, and prenatal attachment in women hospitalized with high-risk pregnancy. J Am Psychoanal Assoc 55(1):253–259
43. Mitchell AJ, Ferguson DW, Gill J, Paul J, Symonds P (2013) Depression and anxiety in long-term cancer survivors compared with spouses and healthy controls: a systematic review and meta-analysis. Lancet Oncol 14:721–732
44. Cheng ER, Rifas-Shiman SL, Perkins ME, Rich-Edwards JW, Gillman MW, Wright R et al (2016) the influence of antenatal partner support on pregnancy outcomes. J Womens Health 25(7):672–679
45. Pilkington PD, Milne LC, Cairns KE, Lewis J, Whelan TA (2015) Modifiable partner factors associated with perinatal depression and anxiety: a systematic review and meta-analysis. J Affect Disord 178:165–180
46. Gonçalves V, Sehovic I, Quinn G (2014) Childbearing attitudes and decisions of young breast cancer survivors: a systematic review. Hum Reprod Update 20(2):279–292
47. Thewes B, Meiser B, Taylor A, Phillips KA, Pendlebury S, Capp A et al (2005) Fertility- and menopause-related information needs of younger women with a diagnosis of early breast cancer. J Clin Oncol 23(22):5155–5165
48. Braun M, Hasson-Ohayon I, Perry S, Kaufman B, Uziely B (2005) Motivation for giving birth after breast cancer. Psychooncology 14(4):282–296
49. Thewes B, Butow P, Girgis A, Pendlebury S (2004) The psychosocial needs of breast cancer survivors; a qualitative study of the shared and unique needs of younger versus older survivors. Psychooncology 13(3):177–189
50. Shantakumar S, Terry MB, Teitelbaum SL, Britton JA, Millikan RC, Moorman PG et al (2007) Reproductive factors and breast cancer risk among older women. Breast Cancer Res Treat 102(3):365–374
51. Hahn-Holbrook J, Schetter CD, Haselton M (2013b) Breastfeeding and maternal mental and physical health. In: Women's health psychology. Wiley, Hoboken, pp 414–439
52. Schover LR (1994) Sexuality and body image in younger women with breast cancer. J Natl Cancer Inst Monogr 16:177–182
53. Moran MS, Colasanto JM, Haffty BG, Wilson LD, Lund MW, Higgins SA (2005) Effects of breast-conserving therapy on lactation after pregnancy. Cancer J 11(5):399–403
54. Stopenski S, Aslam A, Zhang X, Cardonick E (2017) After chemotherapy treatment for maternal cancer during pregnancy, Is breastfeeding possible? Breastfeed Med 12(2):91–97
55. Camune B, Gabzdyl E (2007) Breast-feeding after breast cancer in childbearing women. J Perinat Neonatal Nurs 21(3):225–233
56. Ferrari F, Faccio F, Peccatori FA, Pravettoni G (2018) Psychological issues and construction of the mother-child relationship in women with cancer during pregnancy: a perspective on current and future directions. BMC Psychol 6:10

Index

S. Alipour, R. Omranipour (eds.), *Diseases of the Breast during Pregnancy and Lactation*, Advances in Experimental Medicine and Biology 1252,
https://doi.org/10.1007/978-3-030-41596-9

The manufacturer's authorised representative in the EU is Springer Nature Customer Service Centre GmbH, Europaplatz 3, 69115 Heidelberg, Germany. If you have any concerns regarding our products, please contact ProductSafety@springernature.com

Printed and bound by CPI Group (UK) Ltd, Croydon, CR0 4YY
30/09/2026
02225491-0019